Caterina Faggio is an associate professor of physiology in the Department of Chemical, Biological, Pharmaceutical and Environmental Sciences, University of Messina, Italy. She is the departmental Erasmus coordinator and the Vice President of the Italian Society of Experimental Biology (SIBS). Some of her research interests include the mechanisms of osmoregulation and acid secretion in fish gastrointestinal tracts by electrophysiology, effects of environmental contaminants and food additives on salt transports in the teleost intestine, cell volume regulation in the fish intestine and hepatocytes and digestive cells of mussels by videometric methods.

Pharmaceuticals in Aquatic Environments

Toxicity, Monitoring, and Remediation Technologies

Edited by
Vinod Kumar Garg, Ashok Pandey, Navish Kataria,
and Caterina Faggio

CRC Press
Taylor & Francis Group
Boca Raton London New York

CRC Press is an imprint of the
Taylor & Francis Group, an **informa** business

Designed cover image: © Shutterstock Images

First edition published 2024
by CRC Press
6000 Broken Sound Parkway NW, Suite 300, Boca Raton, FL 33487–2742

and by CRC Press
4 Park Square, Milton Park, Abingdon, Oxon, OX14 4RN

CRC Press is an imprint of Taylor & Francis Group, LLC

ISBN: 9781032413815 (hbk)
ISBN: 9781032420783 (pbk)
ISBN: 9781003361091 (ebk)

DOI: 10.1201/9781003361091

Typeset in Times
by Apex CoVantage, LLC

Contents

Contributors

Ghazanfar Abbas
University of Veterinary & Animal Sciences, Jhang, Pakistan

Haitham G. Abo-Al-Ela
Suez University, Suez, Egypt

Muhammad Adil
University of Veterinary & Animal Sciences, Jhang, Pakistan

Shreanshi Agrahari
Banaras Hindu University, Varanasi, India

Rabia Ashraf
University of Agriculture, Faisalabad, Pakistan

Jana Blahova
University of Veterinary Sciences Brno, Brno, Czech Republic

Mario Alberto Burgos-Aceves
Universidad Autónoma de San Luis Potosí, Mexico

Giulia Caioni
University of L'Aquila, L'Aquila, Italy

Rishabh Chalotra
Central University of Punjab, Bathinda, India

Rishabh Chaudhary
Central University of Punjab, Bathinda, India

Renu Daulta
Guru Jambheshwar University of Science and Technology, Hisar, India

Divya Dhillayan
Guru Jambheshwar University of Science and Technology, Hisar, India

Caterina Faggio
University of Messina, Messina, Italy

Francesca Falco
Institute of Biological Resource and Marine Biotechnology (IRBIM), National Research Council (CNR), Italy

Farwa
Donghua University, Shanghai, China, and University of Agriculture Faisalabad, Pakistan

Vinod Kumar Garg
Central University of Punjab, Bathinda, India

Ravindra Kumar Gautam
Banaras Hindu University, Varanasi, India

Israr Masood ul Hasan
Donghua University, Sanghai, China, and University of Agriculture Faisalabad, Pakistan

César Arturo Ilizaliturri-Hernández
Universidad Autónoma de San Luis Potosí, Mexico

Mavara Iqbal
University of Veterinary & Animal Sciences, Jhang, Pakistan

Shamsa Kanwal
University of Veterinary & Animal Sciences, Jhang, Pakistan

Navish Kataria
J.C. Bose University of Science and Technology YMCA, Faridabad, India

Raj Kumari
I.T.S. College of Pharmacy, Ghaziabad, India

Pavla Lakdawala
University of Veterinary Sciences Brno, Brno, Czech Republic

Carmine Merola
University of Teramo, Teramo, Italy

Monia Perugini
University of Teramo, Teramo, Italy

Himani Sabherwal
J. C. Bose University of Science and Technology YMCA, Faridabad, India

Meenakshi Sharma
I.T.S. College of Pharmacy, Ghaziabad, India

Wen-Jun Shi
South China Normal University, Guangzhou, China

Ankit Kumar Singh
Banaras Hindu University, Varanasi, India

Neetu Singh
Maharshi Dayanand University, Rohtak, India

Randhir Singh
Central University of Punjab, Bathinda, India

Muhammad Sufhan Tahir
University of Agriculture, Faisalabad, Pakistan

Ida Tiwari
Banaras Hindu University, Varanasi, India

Surender Singh Yadav
Maharshi Dayanand University, Rohtak, India

Li Yao
China National Analytical Center, Guangzhou, China

Jian-Liang Zhao
South China Normal University, Guangzhou, China

1 Pharmaceuticals in the Aquatic Environment
Introduction

Divya Dhillayan, Himani Sabherwal, Navish Kataria, and Vinod Kumar Garg

CONTENTS

1.1 INTRODUCTION

Pharmaceuticals are used to diagnose, treat, prevent, and immunise both humans and animals against diseases. These have unquestionably played a significant role in improving human life. Hundreds to thousands of pharmaceuticals (effective organic and inorganic chemicals) have been utilised for the treatment and cure of different diseases (Bilal et al., 2019). The commonly used pharmaceuticals are encapsulated in Figure 1.1 and include psycho-stimulants, antibiotics, oestrogens, anti-epileptics, cardiovascular pharmaceuticals (b-blockers/diuretics), lipid drugs, analgesics/anti-inflammatories, and hormonal compounds (Fent, 2008). Population growth, rising investments in the health and wellness sector, research and development, the global market, and communities in developed nations have contributed to rising pharmaceutical use (Boeckel et al., 2014). Hence, it is unavoidable that the surging usage of pharmaceuticals has a risk of their spillage into the water and soil through various routes.

Pharmaceuticals enter in environment matrices via improper disposal, metabolic excretion, human application, and veterinary uses. It is important to understand that pharmaceuticals are present in municipal sewage regardless of location, and in each geographic area their types, quantity, and relative abundance may vary (Daughton, 2008). The pharmaceutical residues reach surface water

DOI: 10.1201/9781003361091-1

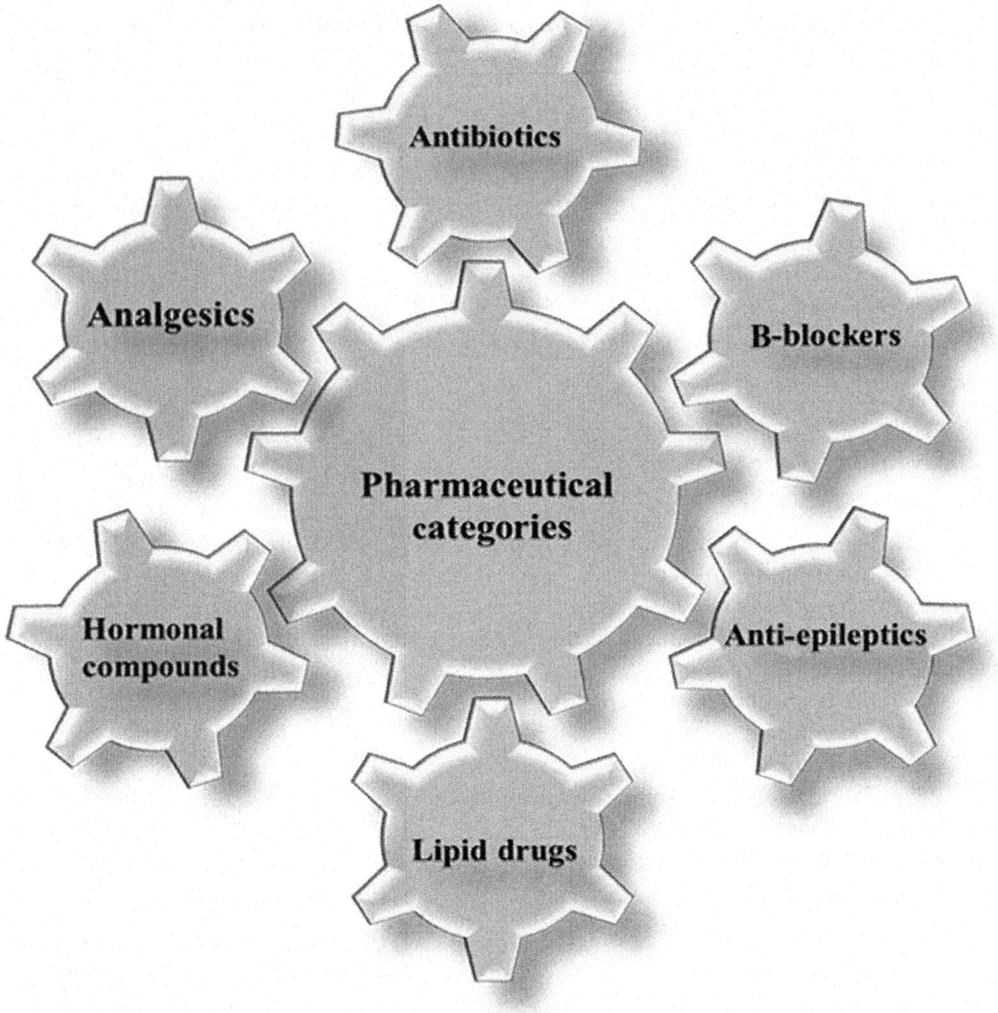

FIGURE 1.1 Different categories of pharmaceuticals.

and ground water, thereby affecting the water quality. Several pharmaceuticals used to treat disorders associated with urbanisation, such as cholesterol management, stress treatment, mental illness, asthma, thyroid, ulcers, etc. are found in sewage of urban areas. There are some regularly utilised pharmaceutical classes in animal pharmacology, such as azoles, antibiotics, Synthroid, Crestor, Naxium, etc. India is the country where one in three pills consumed worldwide are produced. It has been reported that 80% of the bulk formulations production in India are consumed locally (Kallummal et al., 2012). Figure 1.2 displays that Asia consumes the most antibiotics overall, while Eastern Europe consumes the least. Similarly, the analgesic drug class is primarily used in Eastern Europe followed by Latin America. The lipid drug class has been used minimally worldwide, but both Western Europe and Latin America have experienced it equally. In contrast, the oestrogenic drug class is primarily used in Africa (Der et al., 2016; aus der Beek et al., 2016).

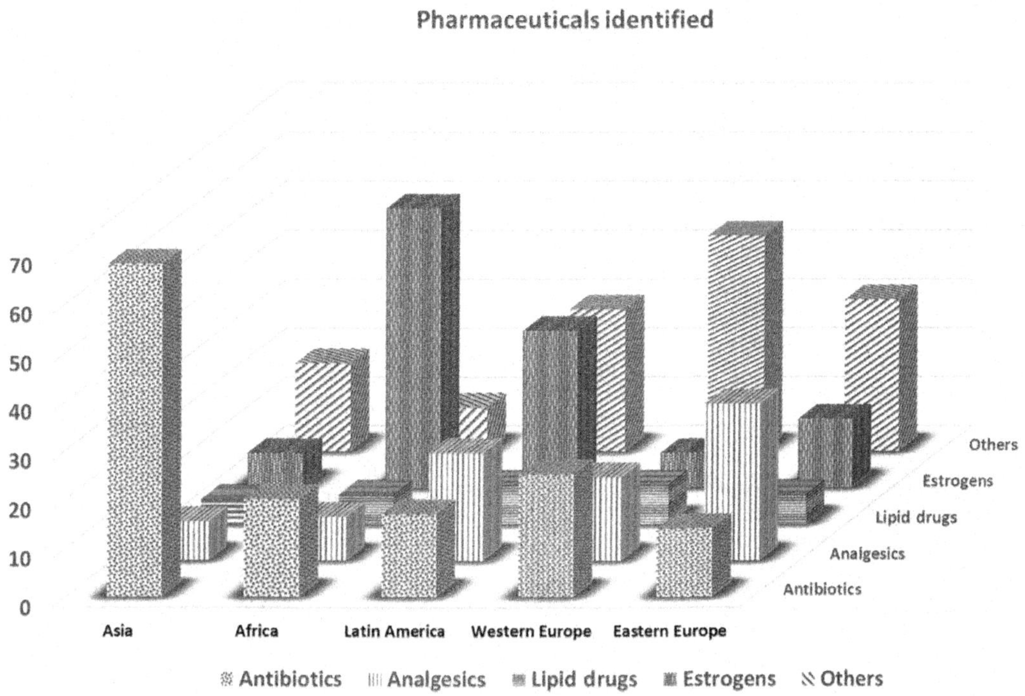

FIGURE 1.2 Pharmaceutical consumption pattern at global level.

Note: MEC = measured environmental concentration.

1.2 SOURCES AND OCCURRENCE OF PHARMACEUTICALS

Since the 1980s, when the first pharmaceutically active contaminants were found in aquatic systems, numerous pharmaceutically active substances have been detected in water systems. The pharmaceutical contamination sources are majorly categorised as point-source and non-point-source. Point-source pollution, which is defined as a single, recognisable source with multiple origins, can be estimated using mathematical modelling. On the other hand, non-point-source based contamination is challenging to locate because it occurs at various geographic scales (Kumar et al., 2023). Waste from sources like landfills, biomedical facilities, and industries comes under point-sources. Non-point-sources include agricultural runoff in aquatic water bodies, a uncontrolled domestic wastewater, and atmospheric deposition. Various sources of pharmaceutical contamination are shown in Figure 1.3.

Specifically localised where production takes place, pharmaceutical production industries are considered as a point-source of contamination. Due to their incredibly high efflux concentrations, these units are particularly problematic. Because of the inadequate industrial effluent treatment in developing countries, this is more serious (Rehman et al., 2015).

Many drain and sewage streams from the pharmaceutical and residential sectors contain a variety of dangerous compounds known as "emerging contaminants" (ECs), which impair entire living ecosystems and decrease biodiversity. Table 1.1 lists the drugs that are frequently detected in wastewater, surface water, groundwater, and other liquids.

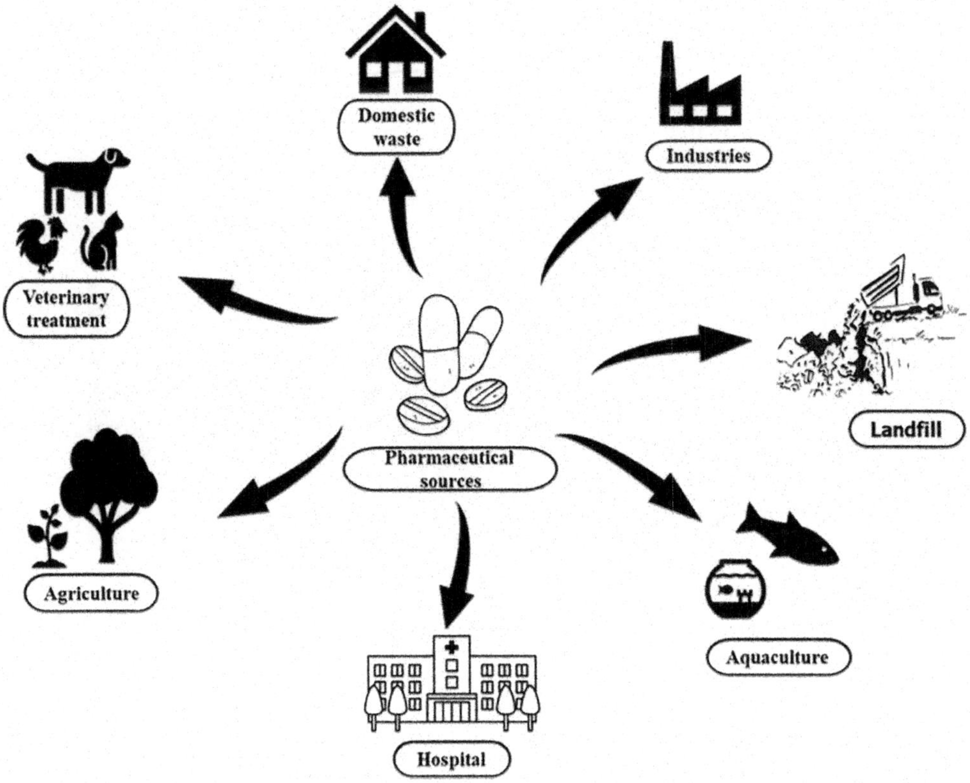

FIGURE 1.3 Potential sources of pharmaceutical contamination.

TABLE 1.1
Pharmaceutical Contamination Detected in Water Sources

Pharmaceutical	Water system	Concentration range	Reference
Ibuprofen (Analgesic)	Surface water	70 ng/L–2.7 μg/L	Flippin et al. (2007)
Naproxen (NSAID)	Surface water	70 ng/L–0.39 μg/L	Thibaut et al. (2006)
Diclofenac (NSAID)	Surface water	6.2 ng/L–1.8 μg/L	Hoeger et al. (2005)
Ketoprofen (NSAID)	Surface water	120 ng/L	Thibaut et al. (2006)
Propranolol (Beta-blocker)	Surface water	0.012–0.59 μg/L	Huggett et al. (2002)
Tetracyclines (Antibiotic)	Surface water	0.11–0.34 μg/L	Grondel et al. (1985)
Paracetamol (Analgesic)	Sewage effluent	6000 ng l^{-1}	Ternes (1998)
Erythromycin (Antibiotic)	Sewage effluent	6000 ng l^{-1}	Hirsch et al. (1999)
Carbamazepine (Anti-epileptic)	Surface water	1100 ng l^{-1}	Ternes (1998)
Bisoprolol (B-blocker)	Surface water	2900 ng l^{-1}	Ternes (1998)
Bezafibrate (Lipid regulator)	Surface water	3100 ng l^{-1}	Ternes (1998)
Fenofibrate (Lipid regulator)	Ground water	45 ng l^{-1}	Heberer (2002)
Sulfamethoxazole (Antibiotic)	Ground water	470 ng l^{-1}	Hirsch et al. (1999)
Diclofenac (Analgesic)	Drinking water	6 ng l^{-1}	Stumpf et al. (1999)
Bleomycin (Anti-neoplastic)	Drinking water	13 ng l^{-1}	Aherne et al. (1990)
Diazapam (Psychiatric drug)	Drinking water	10 ng l^{-1}	Christensen (1998)

Location and seasonal variations affect the relative frequency of finding various pharmaceuticals. Six factors which affect a drug's ecological footprint include:

1. Size and distribution of the human population along with age distribution (demography);
2. Healthcare infrastructure;
3. Existence and size of the manufacturing sector;
4. Connectivity of sewage treatment plants;
5. Environment matrices in which effluents discharge;
6. Current importance of policies and regulation.

Pharmaceutical concentrations in the environment can change on an hourly, daily, seasonal, geographic, temporal, and socio-economic basis. Variations depend on usage habits, geographic regions, hospital and industrial facility contributions, and sewer degradation.

1.3 PATHWAY AND TRANSPORTATION OF PHARMACEUTICAL WASTE

Pharmaceuticals are discharged in the environment from a diverse range of activities. The pharmaceuticals are initially transported in each environmental component through the aquatic system as well as food chain due to low volatility of pharmaceuticals (Fent et al., 2006). Domestic, industrial, agricultural, and hospital effluents all contain pharmaceutical traces. The effluent from here enters the aquatic ecosystem after passing through water treatment plants. Discarded pharmaceuticals end up in landfills, where leachates enter the groundwater and pollute it. Through the soil, untreated discharged water from agricultural fields enters the natural aqueous environment (surface water).

Animals and people are the primary consumers of pharmaceuticals, making them the main contributors to pharmaceutical pollution. The parent pharmaceuticals compounds are directly released from the body through the urinary tract or faeces also (Wang et al., 2021). The unused or expired medications are finally disposed into the environment as household waste, flushed down drains or toilets (Daughton & Ternes, 1999). Pharmaceuticals that are disposed of with household waste end up in landfills as shown in Figure 1.4. After that, when landfill sites are

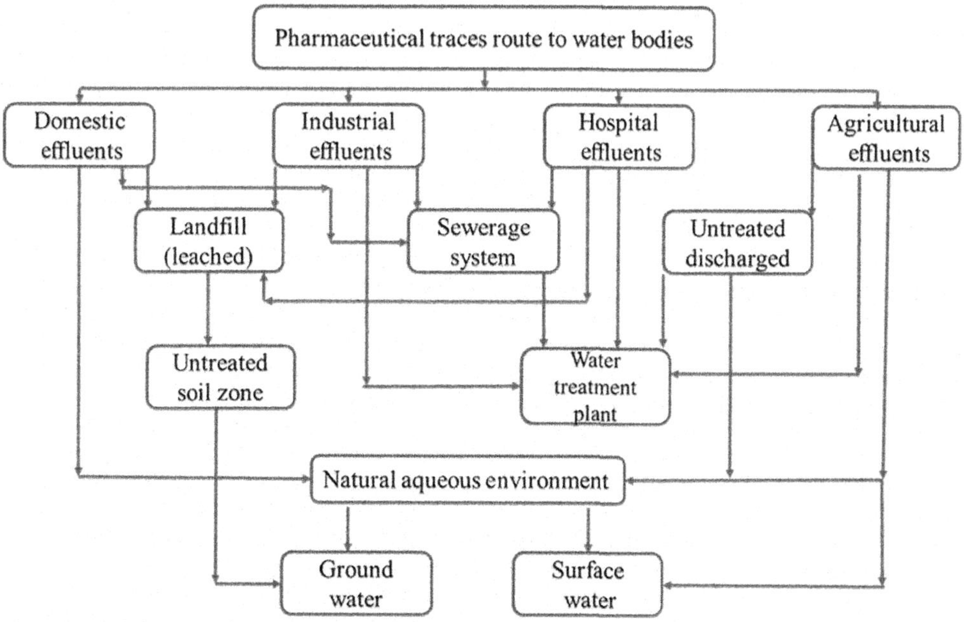

FIGURE 1.4 Schematic diagram showing pathways, and transportation of pharmaceutical pollution.

not properly sealed, they turn into groundwater leachates (Schwarzbauer et al., 2002). In addition to pharmaceuticals, a wide range of other contaminants can be found in landfills and their leachates. This transport is verified by the fact that they are present in groundwater beneath or downstream from landfills (Buszka et al., 2009). In order to be treated, the wastewater treatment plants (WWTPs) receive pharmaceutical waste from hospitals and domestic sources. Pharmaceuticals that are non-polar and less polar in nature adsorb heavily to the sediments and solids in WWTPs and are then removed with the sludge. Sludges that are used as manure on agricultural fields can also introduce contaminants into soil, surface, and groundwater systems (Jones et al., 2005). Pharmaceuticals and personal care products (PPCPs) are also transferred to crops when treated wastewater or biosolids are applied to agricultural soils. Then, through the food chain, these pollutants from crops reach various trophic levels. In the organisms that consume them, pharmaceuticals have a propensity to bio-concentrate, bio-accumulate, and bio-magnify. The water enters the aquatic ecosystem through agricultural runoff and WWTPs, where the pharmaceuticals it contains interact with the aquatic life.

Applications for pharmaceutical aquaculture offer direct access to aquatic environments. Pharmaceuticals are used in aquaculture by being incorporated into the fish food. To produce beneficial effects, large doses of antibiotics are used. Pharmaceuticals move through the environment by a variety of chemical and physical transport mechanisms after being released. Aquatic life is directly or indirectly interconnected to the food chain or food web. Terrestrial animals are connected to aquatic life through the food chain, where they consume fish that have already consumed pharmaceuticals from planktons and zooplanktons. The cycle goes on to higher trophic levels, contributing to detrimental effects to both plants and animals who are exposed to pharmaceutical waste (Bojarski et al., 2020; Ebele et al., 2017).

Anaesthesia, anti-inflammatory, antibiotic, antiparasitic, and antifungal drugs, as well as hormones and sedatives, are all used by veterinarians. These are released into the environment as a result of livestock feeding, ineffective drug disposal, or treatment of animals. The primary entry point into the terrestrial environment is through veterinary drug applications in reared livestock. Animal waste lagoons that collect animal waste or areas where this waste is spread on agricultural fields often contain veterinary antibiotics and hormones (Bártíková et al., 2016). These accumulate and move around in groundwater. Pharmaceuticals go through physicochemical and biotic transformations even though they are intended to be chemically stable. High pharmaceutical stabilities indicate a comparatively high persistence in environmental conditions (Khetan & Collins, 2007). Other crucial pharmaceutical transport processes include leaching, sorption, desorption, and degradation. The mobility of pharmaceuticals in soil-water matrices is influenced by a number of factors including shapes and sizes, compound solubilities, hydrophobicity, speciations, partitioning coefficients of pharmaceuticals at various pH values, sorption/desorption rates, and binding stabilities of drugs towards soil (Sarmah et al., 2006).

1.4 TOXIC EFFECTS OF PHARMACEUTICALS

Pharmaceuticals go through physio-chemical and biotic alterations despite being chemically stable by design (Khetan & Collins, 2007). This means that understanding of a drug's deconjugation, biodegradability, metabolic pathways, conjugation, sorption, and persistence is necessary in order to identify its environmental fate. The high persistency of pharmaceuticals in the environment are directly connected with the chemical stabilities of pharmaceuticals (Heberer, 2002). Continuously discharging partially treated wastewater into the environment runs the risk of causing unpredictable ecological effects because it may act strangely when mixed with other environmental contaminants. Aquatic organisms are more susceptible to the effects of most pharmaceuticals because they are highly soluble in water (Pal et al., 2010). A comprehensive analysis of pharmaceutical pollution in the aquatic environment is made more difficult by the fact that, in some instances, the concentration of transformation products may be higher than that of the parent

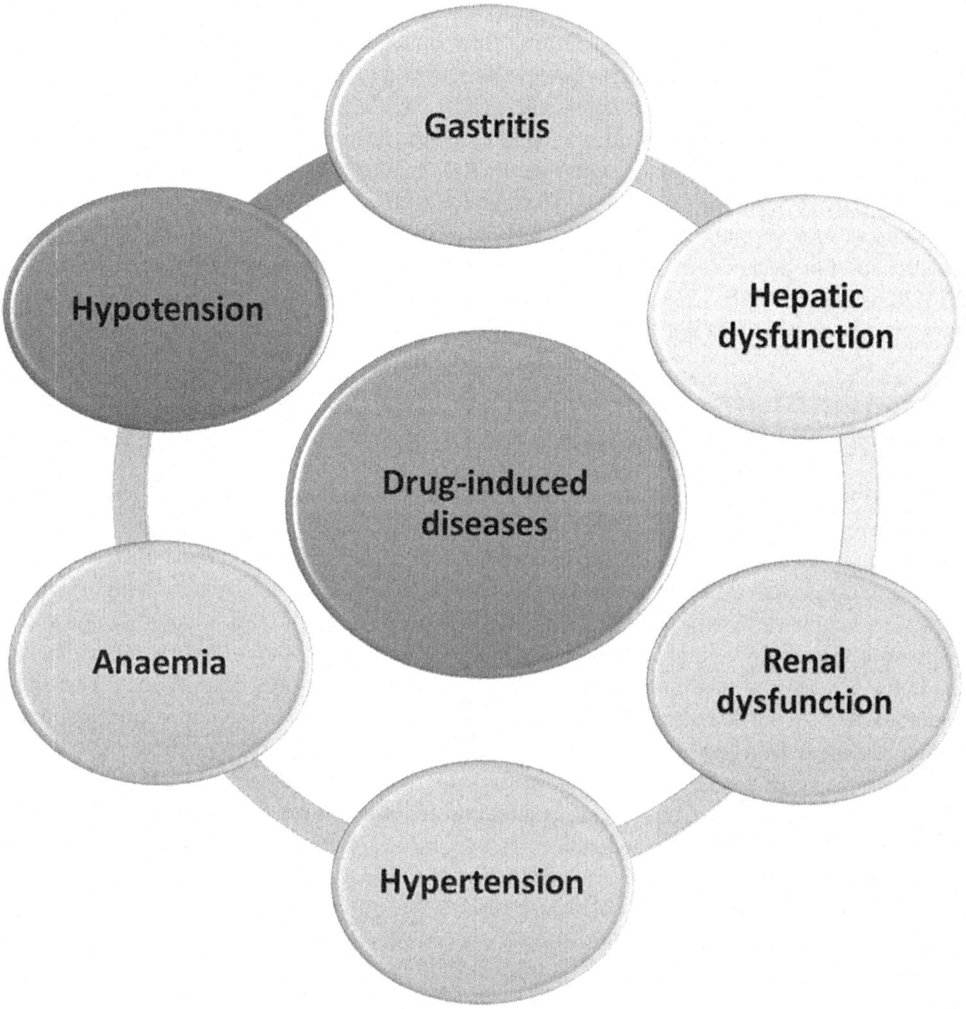

FIGURE 1.5 Various diseases caused by pharmaceuticals.

compound. Examples of such a phenomenon include atorvastatin, carbamazepine, and diclofenac (Langford & Thomas, 2011).

Because most pharmaceuticals are highly soluble, aquatic organisms are more vulnerable to their effects (Pal et al., 2010). As a result, numerous studies have been undertaken to examine the effects on aquatic organisms like fish, algae, and mussels. One of the organisms most susceptible to the high levels of pharmaceuticals is fish. According to some studies, substances like diclofenac and 17a-ethinylestradiol may cause structural disruption in the kidney and intestine as well as change the expression of genes that are connected to the major metabolic processes (Mehinto et al., 2010). Sulphonamides, tetracyclines, and macrolides are just a few of the antibiotics that have been shown to negatively impact the development and growth of algae (Bilal et al., 2019). In general, aquatic organisms exposed to pharmaceuticals may experience a range of negative effects, such as oxidative stress, early mortality, growth inhibition, reduction of energy level and change in feeding behaviour, metabolic disorder, decline of immunity inflammatory responses, and neurotransmission dysfunction (Girona, 2021). Various diseases caused by pharmaceuticals contamination in water bodies are shown in Figure 1.5. The oestrogens' existence in aquatic system can enhance the risk of testicular cancer, breast cancer (gynecomastia), and male infertility (Rai et al., 2016; Sankhla & Kumar, 2019). Fluorouracil, an anticancer drug that can cross the blood-placenta barrier and have teratogenic and embryotoxic effects, is known to be harmful to pregnant women due to the presence in drinking water (Chan et al., 2017; Zarei et al., 2020). Similarly, in the South East Asia region, the well-known incident of decrease in vulture populations has been observed due to the high amount of diclofenac existing in the food chain (Oaks et al., 2004). Ethinylestradiol causes feminising effects on male fish (Gross-Sorokin et al., 2006). Through the web of the food chain, pharmaceuticals also have caused several secondary negative effects on living organisms due to their transportation in the food chain. For instance, diclofenac could increase the mortality rate of oriental white-backed vulture adults and subadults (5–86%) in the Indian subcontinent by causing severe visceral gout and renal failure (Swan et al., 2006).

1.5 ANALYTICAL TECHNIQUES FOR QUANTIFICATION OF PHARMACEUTICALS IN WASTEWATER

Nowadays, the concentration of diverse pharmaceuticals and their derivatives has constantly increased in different water systems such as drinking water, wastewater, groundwater, and surface water (Nabeel et al., 2020). Due to the complexity of the molecules and drug formulations that pharmaceutical research is creating, there is great potential interest in any novel and highly selective analytical technique (Pluym et al., 1992). It is now a critical scientific task to identify and evaluate these pharmaceuticals in various waters, which necessitates the use of extremely complex analytical methods for detection at nanograms per litre (ng/L) scales (Thomaidis et al., 2012). Therefore, it is crucial to develop quick and responsive detection technology for effective monitoring and measurements of diverse ECs (Agüera et al., 2013). Recently, various analytical methods have been created and put to use for the detection and quantification of various pharmaceuticals in water bodies. Analysis processes and methods show a significant role throughout the drug development process, from marketing to post-marketing, whether it be to comprehend the physical and chemical stability of the drug. In recent years, titrimetry, spectrometry, chromatography, and capillary electrophoresis have all been used as assay methods in monographs (Figure 1.6). The literature has also shown the use of electroanalytical methods. The selected analytical techniques—such as ultraviolet/visible

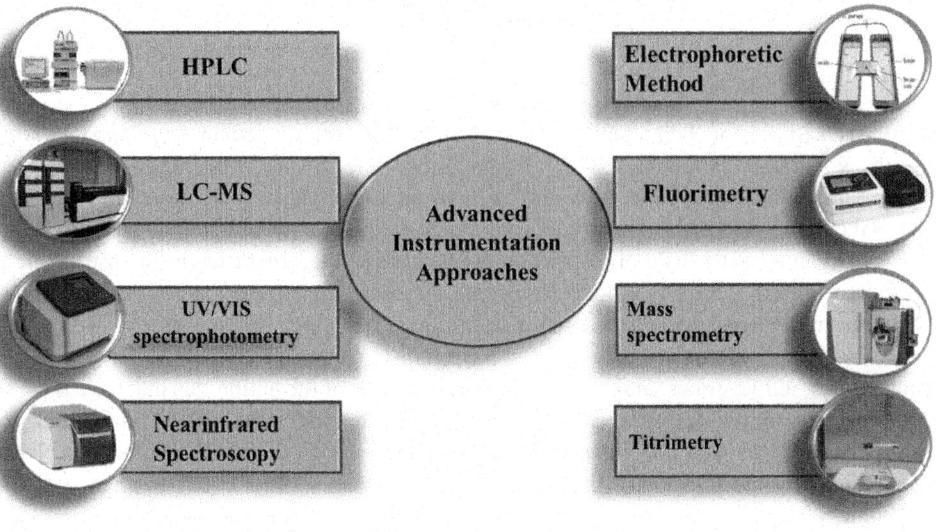

FIGURE 1.6 Various analysis methods used for monitoring of pharmaceuticals in water.

(UV/VIS) spectrophotometry, electrochemical methods, liquid chromatography-mass spectroscopy (LC-MS), titrimetric methods, near-infrared (NIF) spectroscopy, high-resolution liquid chromatography (HPLC), electrophoretic methods, fluorimetry are used for quantitative research on pharmaceuticals in liquid and solid samples.

1.5.1 Titrimetric Techniques

Although titrimetry was widely used in the past, a literature search reveals that there are not many current methods that use it for pharmaceutical analysis. Titrimetry has several benefits, including quick analysis times and inexpensive equipment needs. However, when compared to contemporary separation methods like HPLC or capillary electrophoresis, titrimetric methods exhibit a lack of selectivity, which is likely what caused them to become less popular today.

1.5.2 UV/VIS Spectrophotometric Methods

Numerous UV/VIS spectrophotometric tests have been developed to measure the active ingredients in pharmaceuticals. Since most pharmaceuticals have chromophore groups, their ultraviolet determination can be done without the use of a derivatisation reaction. Spectrophotometric methods are still appealing due to the widespread availability of the equipment; the ease of the procedures; the economy, speed, precision, and accuracy of the method (Bonfilio et al., 2010).

1.5.3 Near Infrared Spectroscopy

A quick and non-destructive method that provides multi-component analysis of virtually any matrix is near infrared spectroscopy (NIRS). The growing pharmaceutical interest in NIRS is likely due to its key advantages over other analytical techniques, including the ease of sample preparation without the need for pre-treatments, the likelihood of separating the sample measurement position by using fibre optic probes, and the expectation of chemical and physical sample parameters from a single spectrum.

1.5.4 Electrochemical Methods

Over the past few years, there has been a significant increase in the use of electrochemical techniques in the analysis of drugs and pharmaceuticals. More advanced instrumentation and a greater understanding of electrochemical techniques themselves can be partly attributed to the resurgence of interest in them.

1.5.5 High Performance Liquid Chromatography

High Performance Liquid Chromatography (HPLC) is widely used to monitor the chemical compounds in order to understand behaviour of different molecules in complex pharmaceutical media found in the environmental systems. It must be noted that better specificity, precision, and accuracy for complex mediums can only be attained by appropriate pre-treatment or tests conducted before the HPLC analysis. Because of this, paying for high levels of specificity, precision, and accuracy is also expensive.

1.5.6 Gas Chromatography

Gas chromatography–mass spectrometry is a sensitive, exact, repeatable, quantitative, and adaptable technique that is highly preferable for the detection of pharmaceuticals in complex compounds. This method is crucial in the analysis of pharmaceuticals and drugs. However, the use of CG is restricted to molecules that can undergo derivatisation reactions to produce thermally stable products or to volatile and thermally stable compounds.

1.5.7 Phosphorimetry and Fluorimetry

The pharmaceutical industries are constantly searching for delicate analytical methods that use micro samples. One method that achieves high sensitivity without sacrificing specificity or precision is fluorescence spectrometry. A gradual rise in the number of articles discussing the use of fluorimetry (Nafisur Rahman, 2012) and phosphorimetry (De Souza et al., 2013) in quantitative analysis of various drugs in dosage forms and biological fluids has been noticed in the recent past.

1.6 SUMMARY

Pharmaceuticals play an important role in the modern society. They help in combatting various diseases, nutrient deficiency and in aiding growth. The pharmaceutical companies have been growing swiftly in recent years. There are several reasons behind this, including population growth that ultimately increases the demand for pharmaceuticals, rapid industrialisation constantly producing effluents in the environment, lack of effective regulations and guidelines, and the usage of pharmaceuticals by different populations living in the world. The unregulated dumping of pharmaceuticals in the receiving bodies of water and soil from numerous sources is causing a number of toxic effects to the humans, animals, and plants who are directly or indirectly dependent on the water and soil. This book summarises the various sources and pathways of pharmaceutical waste, their occurrence and toxicity on humans, plants, and marine organisms. Various analytical techniques and advance instrumentation methods for the analysis of pharmaceuticals in water is discussed in the book.

REFERENCES

Agüera, A., Martínez Bueno, M. J., & Fernández-Alba, A. R. (2013). New trends in the analytical determination of emerging contaminants and their transformation products in environmental waters. *Environmental Science and Pollution Research*, *20*(6), 3496–3515. https://doi.org/10.1007/s11356-013-1586-0

Aherne, G. W., Hardcastle, A., & Nield, A. H. (1990). Cytotoxic drugs and the aquatic environment: Estimation of bleomycin in river and water samples. *Journal of Pharmacy and Pharmacology*, *42*(10), 741–742.

Aus der Beek, T., Weber, F. A., Bergmann, A., Hickmann, S., Ebert, I., Hein, A., & Küster, A. (2016). Pharmaceuticals in the environment-global occurrences and perspectives. *Environmental Toxicology and Chemistry*, *35*(4), 823–835. https://doi.org/10.1002/etc.3339

Bártíková, H., Podlipná, R., & Skálová, L. (2016). Veterinary drugs in the environment and their toxicity to plants. *Chemosphere*, *144*, 2290–2301. https://doi.org/10.1016/j.chemosphere.2015.10.137

Beek, T., Weber, F.-A., Bergmann, A., Hickmann, S., Ebert, I., Hein, A., & Küster, A. (2016). Pharmaceuticals in the environment—Global occurrences and perspectives. *Environmental Toxicology and Chemistry*, *35*(4), 823–835. https://doi.org/10.1002/etc.3339

Bilal, M., Mehmood, S., Rasheed, T., & Iqbal, H. M. N. (2019). Antibiotics traces in the aquatic environment: Persistence and adverse environmental impact. *Current Opinion in Environmental Science & Health*, *13*, 68–74. https://doi.org/10.1016/j.coesh.2019.11.005

Boeckel, T. P. Van, Gandra, S., Ashok, A., Caudron, Q., Grenfell, B. T., Levin, S. A., & Laxminarayan, R. (2014). Global antibiotic consumption 2000 to 2010: An analysis of national pharmaceutical sales data. *The Lancet Infectious Diseases*, *3099*(14), 1–9. https://doi.org/10.1016/S1473-3099(14)70780-7

Bojarski, B., Kot, B., & Witeska, M. (2020). Antibacterials in aquatic environment and their toxicity to fish. *Pharmaceuticals*, *13*(8), 1–24. https://doi.org/10.3390/ph13080189

Bonfilio, R., de Araújo, M. B., & Salgado, H. R. N. (2010). Recent applications of analytical techniques for quantitative pharmaceutical analysis: A review. *WSEAS Transactions on Biology and Biomedicine*, *7*(4), 316–338.

Buszka, P. M., Yeskis, D. J., Kolpin, D. W., Furlong, E. T., Zaugg, S. D., & Meyer, M. T. (2009). Waste-indicator and pharmaceutical compounds in landfill-leachate-affected ground water near Elkhart, Indiana, 2000–2002. *Bulletin of Environmental Contamination and Toxicology*, *82*(6), 653–659. https://doi.org/10.1007/s00128-009-9702-z

Chan, G. G., Koch, C. M., & Connors, L. H. (2017). Blood proteomic profiling in inherited (ATTRm) and acquired (ATTRwt) forms of transthyretin-associated cardiac amyloidosis. *Journal of Proteome Research*, 16(4), 1659–1668.

Christensen, F. M. (1998). Pharmaceuticals in the environment—A human risk? *Regulatory Toxicology and Pharmacology, 28*(3), 212–221.

Daughton, C. G. (2008). Pharmaceuticals as environmental pollutants: The ramifications for human exposure. *International Encyclopedia of Public Health, 5*. https://doi.org/10.1016/B978-012373960-5.00403-2

Daughton, C. G., & Ternes, T. A. (1999). Pharmaceuticals and personal care products in the environment: Agents of subtle change? *Environmental Health Perspectives, 107*(Suppl. 6), 907–938. https://doi.org/10.1289/ehp.99107s6907

De Souza, C. F., Martins, R. K. S., Da Silva, A. R., Da Cunha, A. L. M. C., & Aucélio, R. Q. (2013). Determination of enrofloxacin by room-temperature phosphorimetry after solid phase extraction on an acrylic polymer sorbent. *Spectrochimica Acta—Part A: Molecular and Biomolecular Spectroscopy, 100*, 51–58. https://doi.org/10.1016/j.saa.2012.01.043

Ebele, A. J., Abou-Elwafa Abdallah, M., & Harrad, S. (2017). Pharmaceuticals and personal care products (PPCPs) in the freshwater aquatic environment. *Emerging Contaminants, 3*(1), 1–16. https://doi.org/10.1016/j.emcon.2016.12.004

Fent, K. (2008). Effects of pharmaceuticals on aquatic organisms. *Pharmaceuticals in the Environment: Sources, Fate, Effects and Risks*, 175–203.

Fent, K., Weston, A. A., & Caminada, D. (2006). Ecotoxicology of human pharmaceuticals. *Aquatic Toxicology, 76*(2), 122–159. https://doi.org/10.1016/j.aquatox.2005.09.009

Flippin, J. L., Huggett, D., & Foran, C. M. (2007). Changes in the timing of reproduction following chronic exposure to ibuprofen in Japanese medaka, Oryzias latipes. *Aquatic Toxicology, 81*(1), 73–78.

Girona, U. D. (2021). Microplastics as vectors of pharmaceuticals in aquatic organisms—An overview of their environmental implications. *Case Studies in Chemical and Environmental Engineering, 3*. https://doi.org/10.1016/j.cscee.2021.100079

Grondel, J. L., Gloudemans, A. G. M., & Van Muiswinkel, W. B. (1985). The influence of antibiotics on the immune system. II. Modulation of fish leukocyte responses in culture. *Veterinary Immunology and Immunopathology, 9*(3), 251–260.

Gross-Sorokin, M. Y., Roast, S. D., & Brighty, G. C. (2006). Assessment of feminization of male fish in english rivers by the environment agency of England and Wales. *Environmental Health Perspectives, 114*(Suppl. 1), 147–151. https://doi.org/10.1289/ehp.8068

Heberer, T. (2002). Tracking persistent pharmaceutical residues from municipal sewage to drinking water. *Journal of Hydrology, 266*, 175–189.

Hirsch, R., Ternes, T., Haberer, K., & Kratz, K. L. (1999). Occurrence of antibiotics in the aquatic environment. *Science of the Total Environment, 225*(1–2), 109–118.

Hoeger, B., Köllner, B., Dietrich, D. R., & Hitzfeld, B. (2005). Water-borne diclofenac affects kidney and gill integrity and selected immune parameters in brown trout (Salmo trutta f. fario). *Aquatic Toxicology, 75*(1), 53–64.

Huggett, D. B., Brooks, B. W., Peterson, B., Foran, C. M., & Schlenk, D. (2002). Toxicity of select beta adrenergic receptor-blocking pharmaceuticals (B-blockers) on aquatic organisms. *Archives of Environmental Contamination and Toxicology, 43*, 229–235.

Jones, O. A., Lester, J. N., & Voulvoulis, N. (2005). Pharmaceuticals: A threat to drinking water? *Trends in Biotechnology, 23*(4), 163–167. https://doi.org/10.1016/j.tibtech.2005.02.001

Kallummal, M., & Bugalya, K. (2012). Trends in India's trade in pharmaceutical sector: Some insights. *Center for WTO Studies Working Paper, 200*(2).

Khetan, S. K., & Collins, T. J. (2007). Human pharmaceuticals in the aquatic environment: A challenge to green chemistry. *Chemical Reviews, 107*(6), 2319–2364

Kumar, S., Yadav, S., Kataria, N., Chauhan, A. K., Joshi, S., Gupta, R, & Show, P. L. (2023). Recent advancement in nanotechnology for the treatment of pharmaceutical wastewater: Sources, toxicity, and remediation technology. *Current Pollution Reports*, 1–33. https://doi.org/10.1007/s40726-023-00251-0

Langford, K., & Thomas, K. V. (2011). Input of selected human pharmaceutical metabolites into the Norwegian aquatic environment. *Journal of Environmental Monitoring*, 416–421. https://doi.org/10.1039/c0em00342e

Mehinto, A. C., Hill, E. M., & Tyler, C. R. (2010). Uptake and biological effects of environmentally relevant concentrations of the nonsteroidal anti-inflammatory pharmaceutical diclofenac in rainbow trout (oncorhynchus mykiss). *Environmental Science and Technology, 44*(6), 2176–2182. https://doi.org/10.1021/es903702m

Nabeel, F., Rasheed, T., Bilal, M., Li, C., Yu, C., & Iqbal, H. M. N. (2020). Bio-inspired supramolecular membranes: A pathway to separation and purification of emerging pollutants. *Separation and Purification Reviews, 49*(1), 20–36. https://doi.org/10.1080/15422119.2018.1500919

Oaks, J. L., Gilbert, M., Virani, M. Z., Watson, R. T., Meteyer, C. U., Rideout, B. A., Shivaprasad, H. L., Ahmed, S., Chaudhry, M. J. I., Arshad, M., Mahmood, S., Ali, A., & Khan, A. A. (2004). Diclofenac residues as the cause of vulture population decline in Pakistan. *Nature, 427*(6975), 630–633. https://doi.org/10.1038/nature02317

Pal, A., Gin, K. Y. H., Lin, A. Y. C., & Reinhard, M. (2010). Impacts of emerging organic contaminants on freshwater resources: Review of recent occurrences, sources, fate and effects. *Science of the Total Environment, 408*(24), 6062–6069. https://doi.org/10.1016/j.scitotenv.2010.09.026

Pluym, A., Van Ael, W., & De Smet, M. (1992). Capillary electrophoresis in chemical/pharmaceutical quality control. *Trends in Analytical Chemistry, 11*(1), 27–32. https://doi.org/10.1016/0165-9936(92)80116-N

Rahman, N., Khatoon, A., & Rahman, H. (2012). Studies on the development of spectrophotometric method for the determination of haloperidol in pharmaceutical preparations. *Quim Nova, 35*(2), 392–397.

Rai, M. K., Shahi, G., Meena, V., Meena, R., Chakraborty, S., Singh, R. S., & Rai, B. N. (2016). Removal of hexavalent chromium Cr (VI) using activated carbon prepared from mango kernel activated with H3PO4. *Resource-Efficient Technologies, 2*, S63–S70. https://doi.org/10.1016/j.reffit.2016.11.011

Rehman, M. S. U., Rashid, N., Ashfaq, M., Saif, A., Ahmad, N., & Han, J. I. (2015). Global risk of pharmaceutical contamination from highly populated developing countries. *Chemosphere, 138*, 1045–1055. https://doi.org/10.1016/j.chemosphere.2013.02.036

Sankhla, M. S., & Kumar, R. (2019). Contaminant of heavy metals in groundwater & its toxic effects on human health & environment. *International Journal of Environmental Sciences & Natural Resources, 18*(5). https://doi.org/10.19080/IJESNR.2019.18.555996

Sarmah, A. K., Meyer, M. T., & Boxall, A. B. A. (2006). A global perspective on the use, sales, exposure pathways, occurrence, fate and effects of veterinary antibiotics (VAs) in the environment. *Chemosphere, 65*(5), 725–759. https://doi.org/10.1016/j.chemosphere.2006.03.026

Schwarzbauer, J., Heim, S., Brinker, S., & Littke, R. (2002). Occurrence and alteration of organic contaminants in seepage and leakage water from a waste deposit landfill. *Water Research, 36*(9), 2275–2287. https://doi.org/10.1016/S0043-1354(01)00452-3

Stumpf, M., Ternes, T. A., Wilken, R. D., Rodrigues, S. V., & Baumann, W. (1999). Polar drug residues in sewage and natural waters in the state of Rio de Janeiro, Brazil. *Science of the Total Environment, 225*(1–2), 135–141.

Swan, G. E., Cuthbert, R., Quevedo, M., Green, R. E., Pain, D. J., Bartels, P., Cunningham, A. A., Duncan, N., Meharg, A. A., Oaks, J. L., Parry-Jones, J., Shultz, S., Taggart, M. A., Verdoorn, G., & Wolter, K. (2006). Toxicity of diclofenac to Gyps vultures. *Biology Letters, 2*(2), 279–282. https://doi.org/10.1098/rsbl.2005.0425

Ternes, T. A. (1998). Occurrence of drugs in German sewage treatment plants and rivers. *Water research, 32*(11), 3245–3260.

Thibaut, R., Schnell, S., & Porte, C. (2006). The interference of pharmaceuticals with endogenous and xenobiotic metabolizing enzymes in carp liver: An in-vitro study. *Environmental Science & Technology, 40*(16), 5154–5160.

Thomaidis, N. S., Asimakopoulos, A. G., & Bletsou, A. A. (2012). Emerging contaminants: A tutorial minireview. *Global Nest Journal, 14*(1), 72–79. https://doi.org/10.30955/gnj.000823

Wang, H., Xi, H., Xu, L., Jin, M., Zhao, W., & Liu, H. (2021). Ecotoxicological effects, environmental fate and risks of pharmaceutical and personal care products in the water environment: A review. *Science of the Total Environment, 788*, 147819. https://doi.org/10.1016/j.scitotenv.2021.147819

Zarei, S., Salimi, Y., Repo, E., Daglioglu, N., Safaei, Z., Güzel, E., & Asadi, A. (2020). A global systematic review and meta-analysis on illicit drug consumption rate through wastewater-based epidemiology. *Environmental Science and Pollution Research International, 27*, 36037–36051.

2 Sources and Occurrence of Pharmaceuticals Residue in the Aquatic Environment

Rishabh Chaudhary, Rishabh Chalotra, and Randhir Singh

CONTENTS

DOI: 10.1201/9781003361091-2

2.1 INTRODUCTION

Large quantities of pharmaceutical products intended for human and animal use are manufactured globally per year. Moreover, approximately 4000–5000 active pharmaceutical compounds are manufactured every year around the globe, but their production, usage, as well as their impact on the environment are not systematically observed. Utilization of pharmaceutical products varies significantly from one nation to the next. It is estimated that around 50% of all produced medications are never used by humans or animals. However, among the remaining percentage of medicines which are being used, a significant amount is eliminated from the body either as an active metabolite or in a state that has not been metabolized.

Pharmaceutical residues are another term used for excreted metabolites or active metabolites, depending on which category they fall under. These residues are excreted via bodily functions and get mixed into the environment via different mechanisms (Dasenaki and Thomaidis, 2015).

> Pharmaceutical residues are the substances or products or part of products/drugs which are released from the human and animal body in the form of urine or fecal residue and enter in the environmental via air, water, and soil route. These residues further affect the health of other animals as well as human life by causing severe unfavorable effects and toxicity.
>
> (Rosman et al., 2018)

Since the 1990s, pharmaceutical residues have been given consideration. These residues include a wide variety of compounds used for either personal health or aesthetic objectives, as well as those employed by the agricultural industry to promote the growth or health of livestock (Togola and Budzinski, 2008). According to the European Union, over 20 million tonnes of pharmaceutical residues are created every year, and these residues play a significant role in the pollution of water and land. Moreover, environmental pollution by pharmaceutical residues is present globally also. In this chapter, various sources of pharmaceutical residues in the environment, various categories of pharmaceutical waste, their routes to entry into aquatic environments, nuclear medicines, antibiotics, disposal of pharmaceutical waste, biodegradation, etc., are discussed.

2.2 SOURCES OF PHARMACEUTICAL RESIDUES IN ENVIRONMENT

Besides human and animal, there are a number of sources which contribute to the generation of pharmaceutical residues. Hospitals, manufacturing units, residential areas, aquaculture, agriculture, landfills, sewage treatment plants, biosolids are some examples of these sources which contribute significantly to aquatic environmental pollution (Fatta-Kassinos et al., 2011; Cacace et al., 2019) (Figure 2.1). Various other sources of pharmaceutical waste are listed in Table 2.1.

There are several means via which pharmaceutical chemicals enter the environment:

- The inappropriate and direct disposal of pharmaceutical products by individuals or patients, including the discharge of urine or faeces into the open environment.
- Disposal of undesired, expired pharmaceuticals and waste or junk by drugstores and pharmacies, including syringes, injectables, oral prescription dosages, and waste from surgical procedures directly into open spaces or water sources (Caban and Stepnowski, 2021).
- Disposal of waste from dairy products, including pharmaceutical residues from the veterinary medication of animals as well as from contaminants from household water, sewage, molecular farming, and pest control drugs. The seepage of contaminants from landfills with defects.
- Another source includes dead animals which had been medicated with medicines or animals which were killed due to consumption of hazardous chemicals.

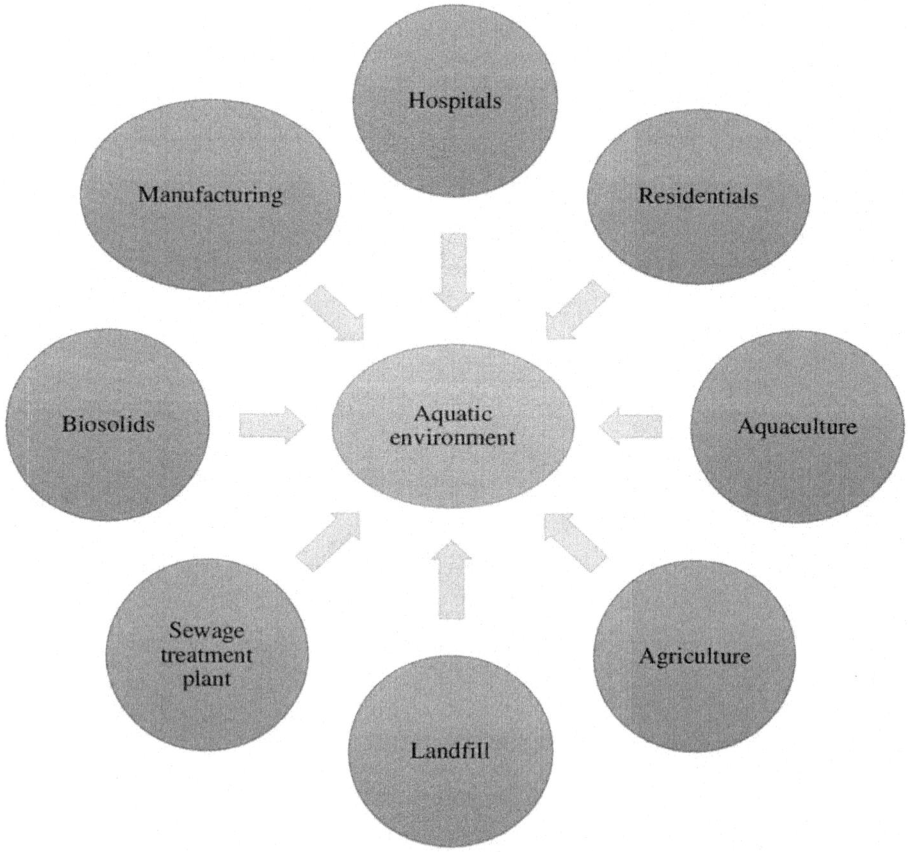

FIGURE 2.1 Sources of pharmaceutical residues affecting the aquatic environment.

TABLE 2.1
Various Sources of Pharmaceutical Waste

S. No.	Pharmaceutical wastes
1	Expired drugs
2	Patients' discarded personal medication
3	Waste materials containing excess drugs in syringes
4	Waste materials containing excess drugs in IV bags
5	Waste materials containing excess drugs in tubing
6	Waste materials containing excess drugs in vials
7	Chemotherapy drug residues
8	Open container drugs that cannot be used
9	Containers that held acutely hazardous waste drugs
10	Drugs that are discarded
11	Contaminated garments
12	Contaminated absorbents
13	Contaminated spill clean-up material

Source: Jaseem et al. (2017).

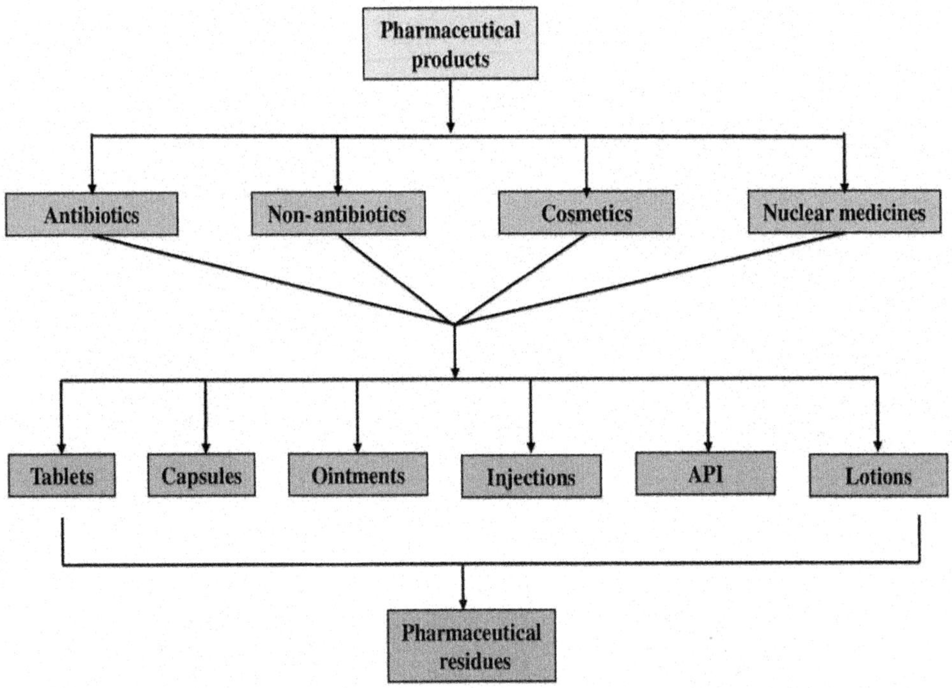

FIGURE 2.2 The different categories of pharmaceutical residues.

2.3 CATEGORIZATION OF PHARMACEUTICAL RESIDUES

Pharmaceuticals residues are categorized in different ways depending on the environmental hazards and sources of pharmaceutical waste. Generally, pharmaceutical residues are classified on the basis of different dosage forms such as non-antibiotics compounds, antibiotics compounds, cosmetics compounds, nuclear medicines. All these products are available in different formulations such as tablets, capsules, syrups, emulsions, ointments, injections, active pharmaceutical ingredients, and lotions, and lead to the generation of pharmaceutical residues (Figure 2.2).

2.4 HOW PHARMACEUTICAL RESIDUES REACH THE AQUATIC ENVIRONMENT/ROUTES OF ENTRY OF PHARMACEUTICALS INTO THE ENVIRONMENT

The term "aquatic ecosystem" or "environment" refers to any environment that is water-based and in which living organisms, such as flora and fauna, interact with the surrounding water. These aquatic organisms are dependent on the water for their sustenance, shelter, reproduction, and several other functions which are necessary for their survival. However, once these pharmaceutical residues are released into the water, they are freely dispersed throughout both upper and lower layers of water. As a result, these pharmaceutical residues gain entry into the aquatic organisms. These animals, such as fish, are consumed by humans, and the residues from the fish then transfer into the humans, causing a variety of physical and chemically related deformities in the human system (Matongo et al., 2015).

Pharmaceutical residues from manufacturing units, and waste from incineration generally also enters the aquatic environment. These pathways also allow pharmaceuticals to reach landfills by way of ground water and eventually enter the aquatic environment (Chalotra et al., 2022). The

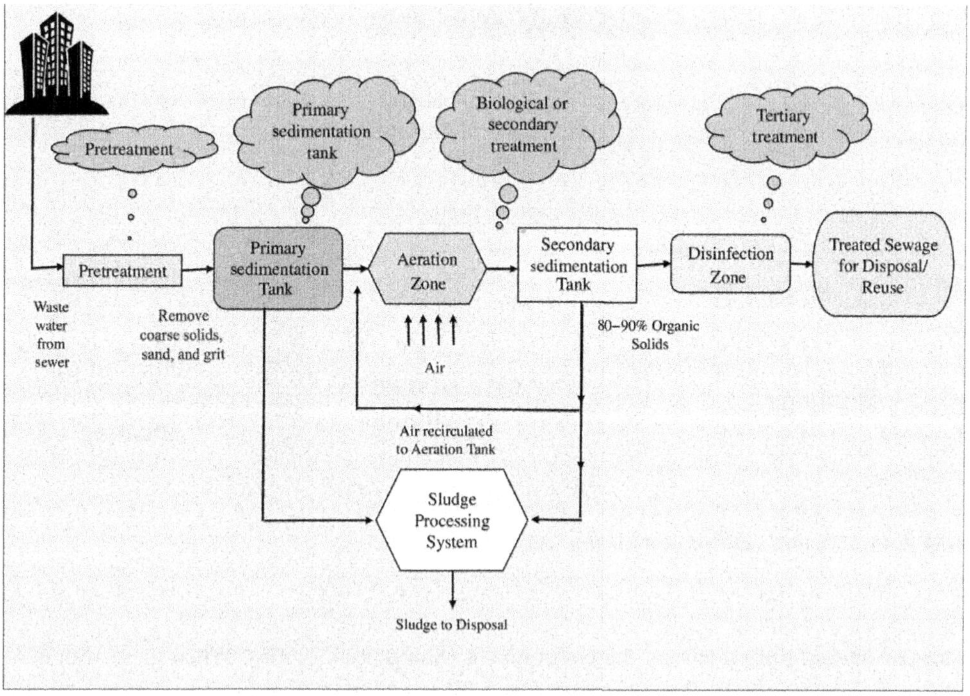

FIGURE 2.3 Basic functioning of a sewage treatment plant.

pharmaceutical products which are used by humans and animals ultimately end up in the sewage system via excretion as urine and faeces. These pharmaceutical residues also make their way into the surface water and ground water by though leakage and STPs (Sewage Treatment Plants) (Puckowski et al., 2016). The most common sewage plant can be seen in Figure 2.3.

These residues are swallowed or pierced by the flora and fauna, which leads to the development of certain physically and chemically associated alterations or biochemical changes in the biological system. These changes may be caused by either direct or indirect exposure to the residues. In addition, the consumption of these aquatic creatures by humans and other animals leads to further malformations in the species concerned. Moreover, consumption of waste materials by aquatic species sometimes leads to the death of an organism. The decomposition of the dead corpse in the water further pollutes both the top and the lower layers of the water and this is one of the major factors responsible for water pollution. The death of various aquatic species each year all over the globe is directly attributable to the contamination of water, which in turn leads to the development of economic and health crises (Ort et al., 2010). Therefore, this cycle continues and results in permanent changes in aquatic life and surrounding environment.

2.5 NUCLEAR MEDICINES

The field of medicine known as nuclear medicine is general, and helps with the diagnosis and treatment of a wide range of illnesses, including cancer (Ziessman et al., 2013). Isotopes, radioactive compounds, or nuclear substances are often used in the practice of nuclear medicine. Within the context of this study, the dosage of radiation-based therapy that was given intravenously or orally to the patient is considered. These radioactive compounds often include some of the residues that, during the process of metabolism, are changed into other active or inactive forms and then excreted out of the body through urine and faeces. These excreted pharmaceutical residues are responsible

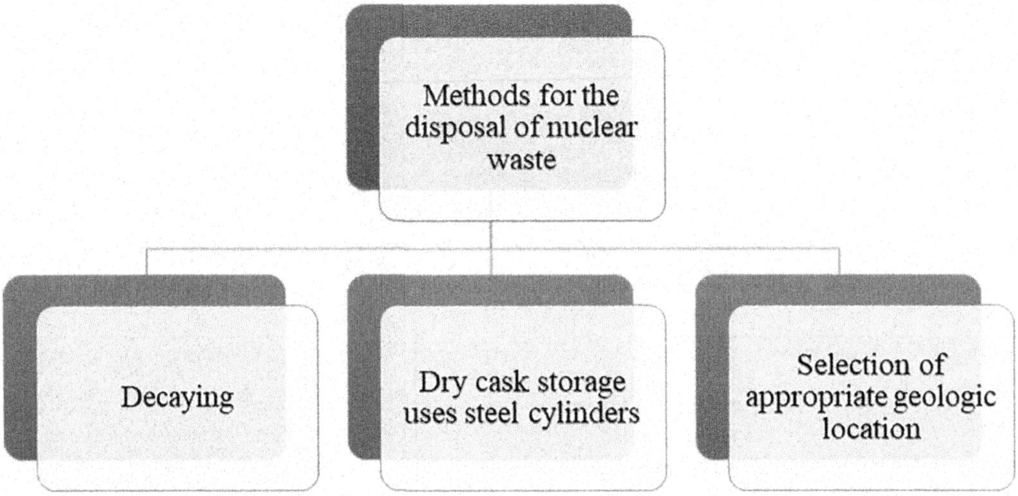

FIGURE 2.4 Different methods of nuclear waste disposal.

for subsequent harmful effects on the environment (Sharp et al., 2005). When these residues are excreted out from the body, these residues travel over a number of different intermediary routes before entering the aquatic environment. These residues become active once again when they are ingested or pierced by aquatic creatures; they are the cause of the development of cancer and other disorders. These aquatic animals are then consumed by humans as well as other animals that are not aquatic, including birds, which results in the further development of diseases in humans and other animals (Hennig and Wollner, 1974).

2.5.1 Disposal of Nuclear Medicine Waste

Worldwide, the disposal of nuclear waste is still a major challenge because nuclear medicine residues are the most hazardous residues in nature and cannot be easily disposed. Even naturally they cannot be disposed and there is need of standardized protocols for disposal of these challenging wastes (Ravichandran et al., 2011).

Radioactive waste or nuclear medicine waste is generally found in hospitals and clinical areas in different forms such as liquids, capsules, and gases. All the radioactive waste materials are generally kept in the "hot lab" (Ojovan and Steinmetz, 2022). The most important procedure for the disposal of nuclear medicine waste is "decaying" (Darda et al., 2021).

In major hospitals, conventional methods include simple storage in the form of dry casks such as steel cylinders and an appropriate selection of geological location (Ojovan et al., 2019) (Figure 2.4).

2.5.1.1 Decaying

The typical process that results in the loss of energy from an atomic nucleus by the emission of radioactive particles is referred to as decay. When the nucleus is unstable, it has radioactive properties. However, once it loses its charge or energy from the unstable nucleus and the nucleus becomes stable, the particles are called radio-inactive.

This technique is often used at high-level hospitals or research centres as well as nuclear power plants. It is a very sophisticated method for getting rid of radioactive pharmaceutical waste. The process may then be further subcategorized into three different types: gamma decay, beta decay, and alpha decay (Brill et al., 1985) (Figure 2.5).

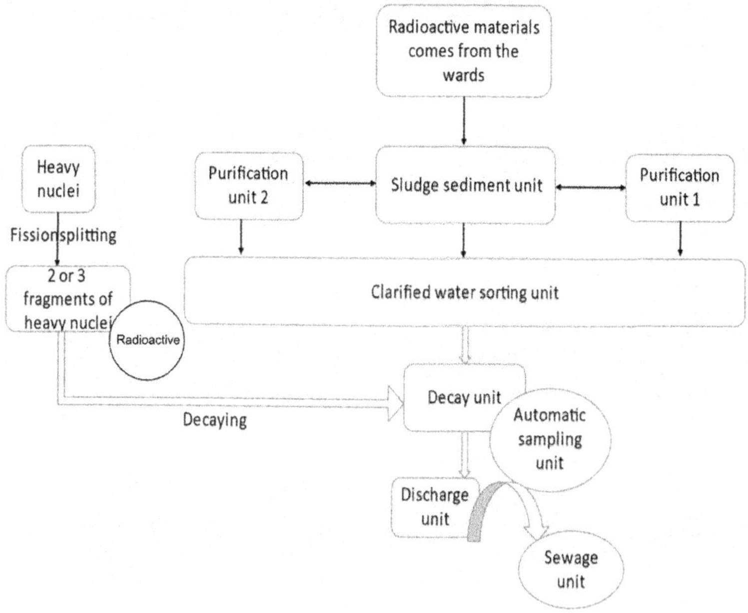

FIGURE 2.5 The decaying procedure for the disposal of nuclear medicine waste.

2.5.1.1.1 Alpha Decay

Alpha decay is a type of radioactive decay in which the parent nucleus is split into two daughter nuclei with the assistance of the emission of the nucleus of the helium atoms at position 21. This type of decay can only occur under certain circumstances (Rao et al., 2022).

2.5.1.1.2 Beta Decay

Because of the breakdown of the nucleus, which results in the emission of beta particles, this kind of decay results in the transformation of parent cells into daughter cells (Corkhill and Hyatt, 2018).

2.5.1.1.3 Gamma Decay

Gamma decay is a form of radioactive decay that involves the disintegration of the parent nucleus into the daughter cells with the aid of the emission of gamma rays (United States Nuclear Regulatory Commission, 1996).

2.5.1.2 Simple Storage

Keeping things in storage is the traditional approach to the problem of waste management and disposal. Steel cylinders are utilized for the storage of the casks in this approach. These cylinders are used not only for the storage of radioactive waste, but also for the storage of inert gases or water. In order to proceed with the processing, the identical cylinder that is filled with waste from nuclear medicine is inserted in the cylinder made of concrete. This sort of method is not only economical, but also it does not need any kind of specialized local or transportation infrastructure. Because of this, radioactive waste may be conveniently kept away from the location where it was generated. These cylinders can be utilized in further processing steps since they are suitable for reprocessing (Shrader-Frechette, 1993).

2.5.1.3 Selection of Appropriate Geologic Location

A significant quantity of radioactive waste materials are disposed after utilization. With the use of traditional mining technologies, this method involves selecting deep geological zones that are either stable or inactive for the purpose of storing waste items for an extended period of time.

These radioactive waste leftovers are being deposited in tunnels at this time. This method has not yet been perfected and more research has to be done on it. As a result, a great number of nations all over the globe continue to be divided about using the deep geologic method (Siegel and Sparks, 2002).

2.6 ANTIBIOTICS

These are the pharmaceutical products which either prevent bacteria from multiplying or kill them. Through the process of fermentation, these chemicals are excreted from the microorganisms and used against the microorganism in clinical settings. The development of antimicrobial resistance is now the most common public health concern on a global scale. Virtually all bacteria and germs have the ability to defend themselves against certain medications. When these antibiotics residues are discharged from pharmaceutical manufacturing units, these antibiotic residues gain entry into human and animal bodies though ground water or soil sources. This is again an important factor in the development of antibiotic resistance among human and animals (Kümmerer, 2003), along with the overuse of antibiotics. This antibiotic resistance has become a serious issue because the sensitivity of microorganisms to antibiotics is decreasing due to their continuous exposure to antibiotics. Moreover, antibiotics are the only option to treat or prevent microbial infections.

Antibiotics from all these sources eventually gain entry into the water and organisms such as fish, reptiles, and people who consume this water and use it for other routine needs as well. These antibiotics residues enter the body of the organism via the mouth or by penetration through the skin and lead to the development of several ailments in humans as well as affecting the aquatic environment (Hutchings et al., 2019).

When antibiotics residues make their way into aquatic environment, a wide variety of aquatic life, including fungi and plants which are already present in the aquatic zone, are adversely impacted and ultimately perish as a result of the excessive pollution caused by these residues. These antibiotics alter or destroy the growth of fungi and other microorganisms present in the aquatic environment. Moreover, aquatic organisms like fish which depend on the fungi and other plants as a source of food, are also severely affected. As a result, the equilibrium in the aquatic environment is affected. Overall, these pharmaceutical residues have a very significant effect on aquatic life and the environment (Larsson, 2014).

The residues left behind by pharmaceutical products are responsible for the deaths of thousands of aquatic species each year. Each year, China and India are among the countries that manufacture the most antibiotics. In India, Puducherry and Chennai are the states that are primarily responsible for generating the majority (90%) of the antibiotics of the country, and manufacturing is carried out at a few of the noted manufacturing units like Medzone Formulations, Microlabs, Phoenix Biologicals, Aeon Formulations, Appasamy Ocular Devices, and Caplin Point Laboratories. The waste product of antibiotics is often discharged into the rivers located in the adjoining area, including the Gingee river, Pambai river, Suthukey canal, Ouster lake, and Thiruvandarkoil lake (Hamad, 2010). All of these rivers are present in close proximity to the production facilities.

After the completion of one batch of antibiotic production, the tanks are cleaned before the production of A second batch. Consequently, the tanks are washed and cleaned completely to remove all the content and traces of previous batches. The water used for the cleaning is usually discharged into the adjoining water bodies, which after mixing with river water affect the flora and fauna of the aquatic environment. These residues from the antibiotics production either kill the microorganisms in the aquatic system or lead to the development of resistance to the antibiotics. These antibiotic-resistant organisms become more dreadful when they come into contact with animals or people. Moreover, these resistant microorganisms are not affected by the antibiotics anymore and increase the severity of the diseases (Spellberg et al., 2013).

2.6.1 Effects of Antibiotics on Microorganisms in Aquatic Environments

Antibiotics were purposely developed to be effective against pathogenic microorganisms, and the advantages of antibiotics for the management and treatment of infectious diseases are indisputable. On the other hand, concern has been raised all over the globe due to the bioactive properties of antimicrobials, their increasing disruption in the environment, and their toxicity to organisms that are not the intended targets (Baquero et al., 2008).

Microorganisms are regarded to be the category of organisms that are most susceptible to and affected by the action of antimicrobial drugs. The bacteriostatic and bactericidal actions carried out by microorganisms not only change the composition of microbial populations but also the ecological functions which microorganisms play in the surrounding environment. The maintenance of biological processes, such as those involved in biogeochemical cycles, in both aquatic and terrestrial ecosystems relies majorly on the variety of microorganisms present in the ecosystem (Costanzo et al., 2005).

Antibiotics, even when administered in low doses, alter vital microbial activities for ecosystems. These activities include the nutrient cycle, the conversion of nitrogen, production of methane, reduction of sulfate, and decomposition of organic matter.

The contribution to the formation of antibiotic-resistant bacteria (ARB) and antibiotic resistance genes (GRA) is another environmental impact of this class of medications that has received a lot of attention from researchers (Kümmerer, 2009). Antibiotic resistance may develop naturally in the environment; however, it is speculated to be on the rise as a result of the improper use of antibiotics as well as their disposal in unsanitary conditions. According to the World Health Organization (WHO), the presence of ARB and GRA is one of the most concerning issues pertaining to public health in this century, with GRA being categorized as a developing environmental toxin.

It is generally agreed that, among the numerous ecological niches, the aquatic environment is the one in which the most resistant bacterial populations can be found. Because this habitat is prone to frequent changes, which might apply selection pressure, it supports the development of these bacteria in terms of their ability to adapt to antimicrobial agents.

Antibiotic-resistant opportunistic pathogens such as *Escherichia coli*, *Klebsiella pneumoniae*, *Acinetobacter spp.*, *Pseudomonas spp.*, and *Shigella spp.* have been discovered in a number of urban rivers, lakes, and streams that receive untreated effluent discharge from domestic, hospital, and industrial water treatment plants, as well as inefficient water treatment plants (Shao et al., 2018).

ARBs are most often spread across these ecosystems by the pollution caused by enteric bacteria that have been subjected to large doses of antibiotics inside the digestive tracts of people and animals (Rodriguez-Mozaz et al., 2020). Because of their high microbial density and the presence of contaminants including antibiotic residues, pharmaceuticals, and heavy metals, urban water treatment stations stand out as a source of these bacteria. This is because of the high concentration of microbes in these facilities. These bacteria are subjected to selection pressure as a result of the combination of these medications, which speeds up the process of developing and spreading resistance.

These bacteria, when found in their natural environments, operate as a source of resistance and have the potential to spread GRAs. According to the research, the spread of GRAs across bacterial populations via vertical gene transfer (VGT) and horizontal gene transfer (HGT) contributes to an increase in the number of bacteria in water that are resistant to the drug. During the process of VGT, also known as the transfer of genetic material from one parent to another during reproduction, bacterial DNA may become mutated. On the other hand, HGT refers to the intra- or inter-specific transmission of genes encoding antimicrobial resistance across bacteria via the use of mobile genetic elements such as plasmids and transposons (Hassoun-Kheir et al., 2020).

As was mentioned earlier, microorganisms that live in aquatic environments that have been damaged may develop a variety of resistance mechanisms in order to adapt to the stress that is caused by different types of pollutants, such as heavy metals. According to the findings of many studies, metal contamination in the natural environment may play a significant part in the development and dissemination of antimicrobial resistance. Because antibiotic resistance genes and metal resistance genes are frequently found in the same moving elements, the selective pressure exerted by metals in these conditions can select antibiotic-resistant isolates in a manner that is analogous to that exerted by antibiotics (Grenni et al., 2018).

Confirming that antibiotic resistance may be produced by the selection pressure of heavy metals in the environment, Martin-Laurent et al. (2019) discovered a conjugative plasmid with co-resistance to tetracycline and copper in a *P. aeruginosa* strain that came from a polluted river in southern Brazil. The plasmid was found in a *P. aeruginosa* strain that was isolated from the river. In polluted areas, Rasmussen and Sorensen found an increased presence of conjugative plasmids. They made the startling discovery that the genes for resistance to tetracycline and mercury were located on the same plasmid. Cullom and others support evidence of genes resistant to mercury and tetracycline in *K. pneumoniae* isolates from a polluted urban stream in the Brazilian state of Pernambuco (2020).

Because resistance genes can be passed horizontally from environmental microbes to human commensals, it is possible that not only the indiscriminate use of antibiotics, but also the environmental pollution by other chemicals, might pose dangers and harm to the environment and human health. This is because antibiotics are used without discrimination. In light of this problem, there is an obvious need to undertake research and create systems for monitoring and eradicating dangerous chemicals from the aquatic environment in order to restrict the impact that these substances have on the health of humans and the environment (Ribeiro et al., 2018).

2.7 DISPOSAL OF PHARMACEUTICAL WASTE

In the current climate, concerns over pharmaceutical residues or waste are among the most pressing, and almost every scientist is working to find a solution to this problem. However, there are many other options accessible for the disposal of trash. Although there are many different drugs on the market, each of which has its own unique formulation, and own unique method for disposal, the issue of pharmaceutical leftovers requires a solution that is both straightforward and economically effective. However, the disposal of some pharmaceutical residues that were acquired from nuclear medications and antibiotics requires further care and attention. In this section, some of the approaches that outline the way to get rid of waste from the pharmaceutical industry are laid forth.

2.7.1 Return to Manufacturer or Donor

This is the standard procedure for the elimination of trash. It is important that unused drugs be sent back to the company that manufactured them so that they may be disposed of safely. Because of the potential for harm, drugs that are close to reaching their expiration date should be sent back to the company that made them. If the wrappers of tablets are collected and sent back to the manufacturer via pharmacies, then these can be utilized again in the packing of other tablets. For example, in the case of tablets, the wrappers of tablets are generally split out in the dustbin of households. However, if these wrappers are collected and sent back to the manufacturer via pharmacies, then these can be utilized again in the packing of other tablets. The same is true for bottles made of plastic or glass; these containers may be returned to pharmacies, where they will be reused by the manufacturer in the process of filling new liquid dosage forms. Bottles made of other materials, such as aluminium, cannot be recycled. These are some of the major considerations that should be taken into account while finding a suitable manner to dispose of waste from medicines.

2.7.2 Landfills

When garbage is disposed of in landfills, the material is just deposited there without previously being subjected to any kind of treatment. When it comes to getting rid of solid trash, this is by far the most common and time-honoured approach. There are three distinct types of landfill (León and Parise, 2009).

2.7.2.1 Non-Engineered Open Uncontrolled Dump

This is by far the most common approach to getting rid of garbage in countries that are still developing. Using this method, the waste from the pharmaceutical industry is disposed of by dumping it straight into an open area that is not under any kind of supervision. It is possible that the surrounding region, including any water sources, may become polluted as a result.

2.7.2.2 Designed/Engineered Landfill

In this method, pharmaceutical waste that has not been processed is disposed of in a landfill that has been specifically built to reduce the likelihood of leaking into the surrounding community (Zuccato et al., 2006).

2.7.2.3 Landfill with Advanced Engineering

Landfill sites that are thoughtfully conceived and constructed provide protection to the aquifer as well as provide disposal of waste that is only moderately hazardous.

2.7.3 Waste Encapsulation and Immobilization

For the process of encapsulation, the unused drug must be compressed into a solid block and then placed within a drum made of plastic or metal. When these barrels have been filled to 75% of their capacity, the remaining 25% is filled with cement, and then they are discarded into the freshly formed municipal solid garbage. Antineoplastic waste should be appropriately enclosed, totally destroyed, and then burnt at a temperature of at least 1200 degrees Celsius before being disposed of (Azad et al., 2012).

2.7.4 Waste Inertization: Immobilization

Inertization is a kind of encapsulation that requires the removal of the paper, cardboard, and plastic packaging that was originally on the medications. It is necessary to remove pills from their blister or strip packaging before taking them. After that, the medications are mixed with water, cement, and lime to produce a homogeneous paste. This paste is the final product. After that, the paste is scattered throughout the municipal solid trash in the form of solid mass dispersion. The proportion of pharmaceutical waste is present in different compositions such as water, lime, and cement having ratio 65:15:15:5 respectively (Daughton, 2003).

2.7.5 Sewer

Some liquid medications, such syrups and intravenous fluids, may be diluted and flushed in little volumes over the course of time without having a substantial detrimental influence on the environment. This allows the drugs to be more easily disposed of.

2.7.6 Lighting a Fire in an Open Container

Burning pharmaceuticals at low temperatures in open containers, with paper or cardboard packaging that cannot be recycled, or if they contain harmful chemicals that might be discharged into the air, is not an acceptable method of disposal. The disposal of pharmaceutical waste in this way should be limited to just very tiny quantities since it is not used very often (Singh et al., 2021).

2.7.7 Low-Temperature Combustion

Incinerators with a high temperature and two chambers that are able to burn materials with more than 1% halogenated compounds are difficult to come by in many countries. These combustion centres are in compliance with the stringent emission control criteria established by the European

Union. In the event of an emergency, it is acceptable in the eyes of the responsible authorities to use incarcerators with two chambers that operate at a temperature of at least 850 degrees Celsius and have a combustion retention time of at least two seconds to treat pharmaceuticals in solid form that have expired. It is recommended to combine waste from pharmaceutical sources with a sizeable percentage of garbage from municipal sources (approximately 1:1000) (Stuart et al., 2012).

2.7.8 Using Currently Operating Industrial Plants for High-Temperature Incineration

Furnaces used in coal-fired thermal power plants and cement kilns, for instance, often run at temperatures that are higher than 850 degrees Celsius, have extended combustion retention periods, and discharge exhaust gases via towering chimneys. Many countries that are unable to finance facilities that safely dispose of chemical waste at a high cost have turned to industrial plants as a realistic and more inexpensive option. Because cement kilns normally generate between 1500 and 8000 metric tonnes of cement per day, considerable volumes of pharmaceutical waste may be disposed of in a short length of time (Yu et al., 2020).

2.7.9 Chemical Breakdown

In the event that a suitable incinerator is not available, chemical decomposition carried out in line with the instructions provided by the manufacturer may be utilized as an alternative to dumping the waste in a landfill. It is not recommended to use this procedure unless there is fast access to specialists with knowledge of chemicals. It is only possible to breakdown a tiny quantity of garbage since the procedure is both hard and time consuming (Lapworth et al., 2012).

2.8 DISPOSAL OF VETERINARY MEDICINES

Pharmaceutical waste includes any veterinary drugs or medical items with various routes of administration for such products, as well as any instruments for their administration. It may be found at the veterinarian, pharmacy, medicated feed maker, and animal keeper levels, including farms, breeders, kennels, and pets, as well as other animal owners. According to research by the US Geological Survey's hazardous substances hydrology programme (USGS) (Kolpin et al., 2002), a wide range of human and veterinary medications, as well as other home, industrial, and agricultural pollutants, were found in a water sample taken from 139 streams in 30 states between 1999 and 2000 and analyzed. At least one of the 95 targeted compounds was found in 80% of the streams that were analyzed, and more frequently than not, multiple substances were discovered. Insect repellents, hormones, caffeine, steroids, and a number of other substances were among the 95 compounds that were the focus of the study. Although there have been few studies on the effects of pharmaceutical waste on aquatic life, some studies have hypothesized that these wastes may affect the fertility of fish and other aquatic creatures as well as land animals who drink this water. The bioaccumulation of chemicals in the blood and tissues of aquatic animals is another effect of these chemicals or pharmaceutical waste. The main danger here is the rise of antibiotic resistance. The pharmaceutical waste was also discovered in very high concentrations in treated drinking water, in addition to rivers and streams. The consequences of exposure to these low-concentration pharmaceutical compounds on human health have not been studied. They still have some impacts, though.

2.9 BIODEGRADATION

It is defined as the breakdown of complex organic compounds in simpler molecules like carbon dioxide and water in presence of microorganisms like bacteria and fungi. All the pharmaceuticals and their metabolites are excreted from human body via faecal matter or urine after the administration of the drug. These waste materials go into waste water treatment plants and later get into the environment (Daughton, 2001) (Figure 2.6, Figure 2.7).

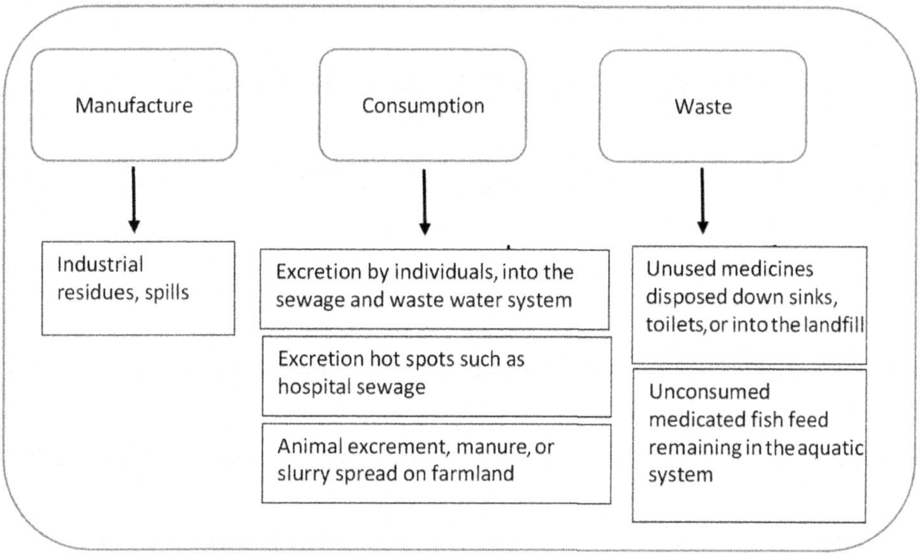

FIGURE 2.6 Waste disposal from manufacturer to aquatic life.

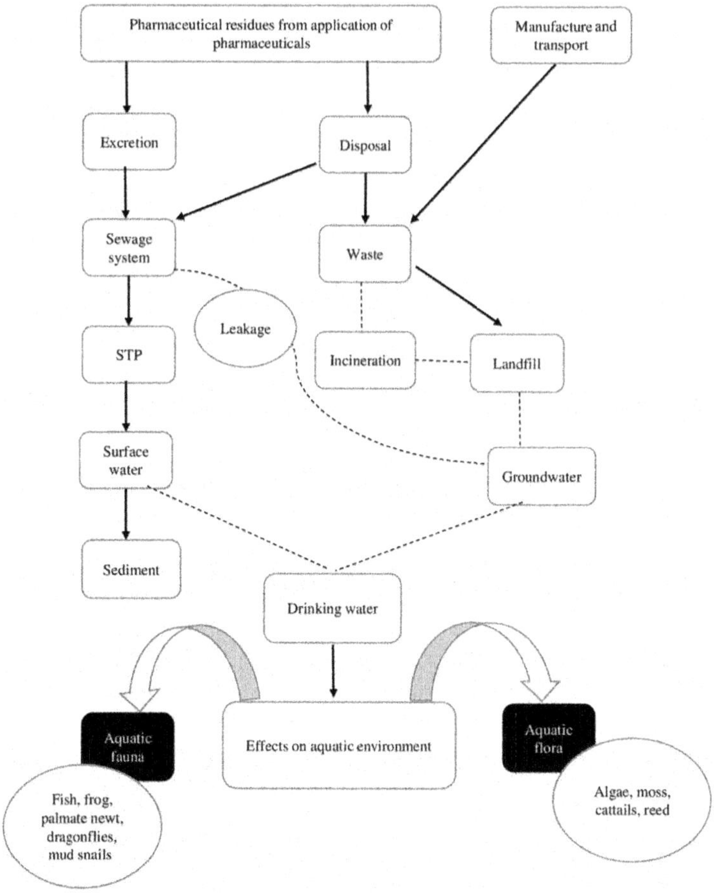

FIGURE 2.7 Pharmaceutical residue reaching the aquatic environment.

TABLE 2.2
Various Bio-Degradable Pharmaceutical Wastes

S. No.	Bio-degradable pharmaceutical waste
1	Chitosan polymerized drugs
2	Alginate polymerized drugs
3	Hydropropyl-methylcellulose
4	Poly lactic acid, co-glycolic acid
5	Dextrans
6	Hyaluronic acid
7	Polyester and polyether-based polymers
8	Poly(ε-caprolactone) (PCL)
9	Paper packaging

TABLE 2.3
Various Non-Bio-Degradable Pharmaceutical Waste

S. No.	Non-bio-degradable pharmaceutical waste
1	PVC foils
2	PVDC foils
3	Aluminium foils
4	Glass
5	Plastic
6	Antibiotics
7	Metal
8	Non-bio-degradable implants, etc.

2.9.1 Bio-Degradable Pharmaceutical Waste

These are defined as those waste materials which get degraded into the environment with due time interval. Common examples are included in Table 2.2 (Sharma and Kaushik, 2021).

2.9.2 Non-Bio-Degradable Pharmaceutical Waste

These are defined as those waste materials which do not get decomposed or degraded in the environment with due interval; common examples include those listed in Table 2.3 (Petrovic et al., 2009).

2.10 PHARMACEUTICAL WASTE CYCLE

In this cycle pharmaceutical waste management is discussed along with how it affects aquatic life, returns back to humans, and makes a complete cycle (Lalander et al., 2016), as mentioned in Figure 2.8.

2.11 CONCLUSION

Pharmaceutical waste is a very critical issue, as it ultimately goes into the environment and reaches aquatic life and affects the aquatic environment and the flora and fauna. These waste materials

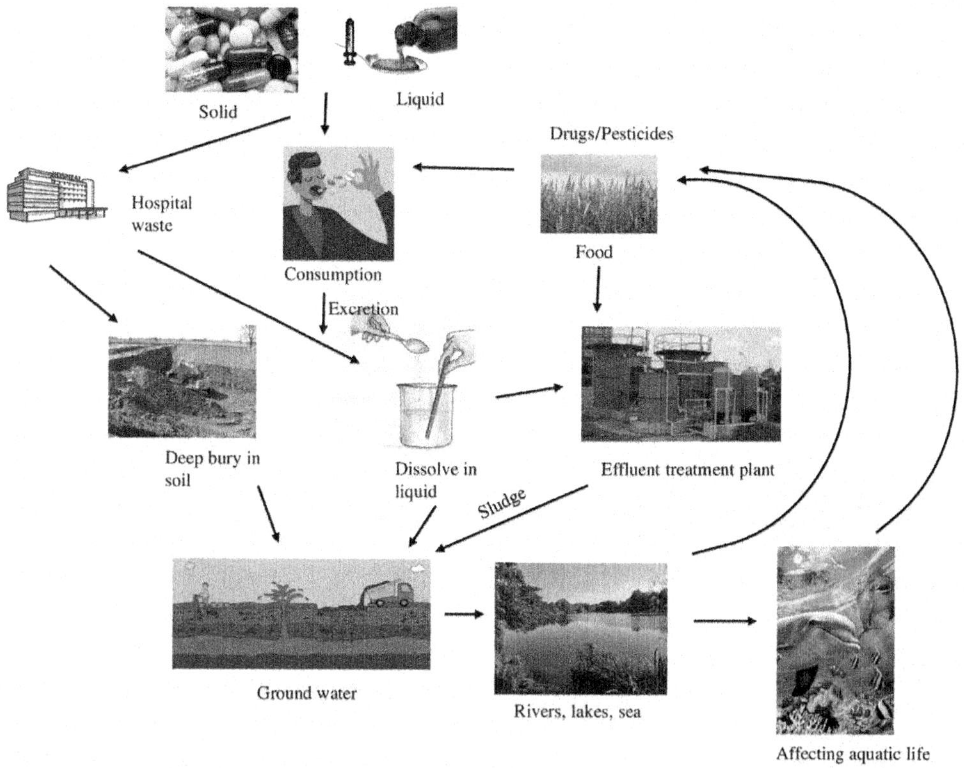

FIGURE 2.8 Pharmaceutical waste cycle.

have affected the growth and development of aquatic life. Pharmaceutical waste not only affects aquatic life alone, but directly or indirectly affects humans. A recent study has revealed that the microscopic plants known as diatoms absorb 10–20 billion tonnes of CO_2 every year as they float on the surface of the ocean. That is equal to the annual CO_2 absorption by all the world's evergreen rainforests. So, pharmaceutical waste might directly harm these diatoms. Preference should be given to this issue, so that better laws are made for the proper utilization and management of pharmaceutical waste.

REFERENCES

Azad, M. A. K., Akhtaruzzaman, M. R. H. A. M., Al-Mamun, S. M., Uddin, M. & Rahman, M. 2012. Disposal practice for unused medications among the students of the International Islamic University Malaysia. *Journal of Applied Pharmaceutical Science*, 2, 101–106.

Baquero, F., Martínez, J.-L., & Cantón, R. 2008. Antibiotics and antibiotic resistance in water environments. *Current Opinion in Biotechnology*, 19, 260–265.

Brill, D. R., Allen, E. W., Lutzker, L. G., Mckusick, K. A., Petersen, R. J., Powell, O. M. & Weir, G. J. 1985. Disposal of low-level radioactive waste: Impact on the medical profession. *JAMA*, 254, 2449–2451.

Caban, M. & Stepnowski, P. 2021. How to decrease pharmaceuticals in the environment? A review. *Environmental Chemistry Letters*, 19, 3115–3138.

Cacace, D., Fatta-Kassinos, D., Manaia, C. M., Cytryn, E., Kreuzinger, N., Rizzo, L., Karaolia, P., Schwartz, T., Alexander, J. & Merlin, C. 2019. Antibiotic resistance genes in treated wastewater and in the receiving water bodies: A pan-European survey of urban settings. *Water Research*, 162, 320–330.

Chalotra, R., Dhanawat, M., Maqbool, M., Lamba, N., Bibi, A. & Gupta, S. 2022. Phytochemistry and pharmacology of Iris kashmiriana. *Pharmacognosy Research*, 14.

Corkhill, C. & Hyatt, N. 2018. *Nuclear waste management.* IOP Publishing.
Costanzo, S. D., Murby, J. & Bates, J. 2005. Ecosystem response to antibiotics entering the aquatic environment. *Marine Pollution Bulletin*, 51, 218–223.
Cullom, A. C., Martin, R. L., Song, Y., Williams, K., Williams, A., Pruden, A. & Edwards, M. A. 2020. Critical review: Propensity of premise plumbing pipe materials to enhance or diminish growth of Legionella and other opportunistic pathogens. *Pathogens*, 9, 957.
Darda, S. A., Gabbar, H. A., Damideh, V., Aboughaly, M. & Hassen, I. 2021. A comprehensive review on radioactive waste cycle from generation to disposal. *Journal of Radioanalytical and Nuclear Chemistry*, 329, 15–31.
Dasenaki, M. E. & Thomaidis, N. S. 2015. Multi-residue determination of 115 veterinary drugs and pharmaceutical residues in milk powder, butter, fish tissue and eggs using liquid chromatography—tandem mass spectrometry. *Analytica Chimica Acta*, 880, 103–121.
Daughton, C. G. 2001. *Pharmaceuticals and personal care products in the environment: Overarching issues and overview.* ACS Publications.
Daughton, C. G. 2003. Cradle-to-cradle stewardship of drugs for minimizing their environmental disposition while promoting human health. II. Drug disposal, waste reduction, and future directions. *Environmental Health Perspectives*, 111, 775–785.
Fatta-Kassinos, D., Meric, S. & Nikolaou, A. 2011. Pharmaceutical residues in environmental waters and wastewater: Current state of knowledge and future research. *Analytical and Bioanalytical Chemistry*, 399, 251–275.
Grenni, P., Ancona, V. & Caracciolo, A. B. 2018. Ecological effects of antibiotics on natural ecosystems: A review. *Microchemical Journal*, 136, 25–39.
Hamad, B. 2010. The antibiotics market. *Nature Reviews Drug Discovery*, 9, 675.
Hassoun-Kheir, N., Stabholz, Y., Kreft, J.-U., De La Cruz, R., Romalde, J. L., Nesme, J., Sørensen, S. J., Smets, B. F., Graham, D. & Paul, M. 2020. Comparison of antibiotic-resistant bacteria and antibiotic resistance genes abundance in hospital and community wastewater: A systematic review. *Science of the Total Environment*, 743, 140804.
Hennig, K. & Wollner, P. 1974. Nuclear medicine.
Hutchings, M. I., Truman, A. W. & Wilkinson, B. 2019. Antibiotics: past, present and future. *Current Opinion in Microbiology*, 51, 72–80.
Jaseem, M., Kumar, P. & John, R. M. 2017. An overview of waste management in pharmaceutical industry. *The Pharma Innovation*, 6, 158.
Kolpin, D. W., Furlong, E. T., Meyer, M. T., Thurman, E. M., Zaugg, S. D., Barber, L. B. & Buxton, H. T. 2002. Pharmaceuticals, hormones, and other organic wastewater contaminants in US streams, 1999–2000: A national reconnaissance. *Environmental Science & Technology*, 36, 1202–1211.
Kümmerer, K. 2003. Significance of antibiotics in the environment. *Journal of Antimicrobial Chemotherapy*, 52, 5–7.
Kümmerer, K. 2009. Antibiotics in the aquatic environment—a review—part II. *Chemosphere*, 75, 435–441.
Lalander, C., Senecal, J., Calvo, M. G., Ahrens, L., Josefsson, S., Wiberg, K. & Vinnerås, B. 2016. Fate of pharmaceuticals and pesticides in fly larvae composting. *Science of the Total Environment*, 565, 279–286.
Lapworth, D., Baran, N., Stuart, M. & Ward, R. 2012. Emerging organic contaminants in groundwater: A review of sources, fate and occurrence. *Environmental Pollution*, 163, 287–303.
Larsson, D. J. 2014. Antibiotics in the environment. *Upsala Journal of Medical Sciences*, 119, 108–112.
León, L. M. & Parise, M. 2009. Managing environmental problems in Cuban karstic aquifers. *Environmental Geology*, 58, 275–283.
Martin-Laurent, F., Topp, E., Billet, L., Batisson, I., Malandain, C., Besse-Hoggan, P., Morin, S., Artigas, J., Bonnineau, C. & Kergoat, L. 2019. Environmental risk assessment of antibiotics in agroecosystems: Ecotoxicological effects on aquatic microbial communities and dissemination of antimicrobial resistances and antibiotic biodegradation potential along the soil-water continuum. *Environmental Science and Pollution Research*, 26, 18930–18937.
Matongo, S., Birungi, G., Moodley, B. & Ndungu, P. 2015. Pharmaceutical residues in water and sediment of Msunduzi River, Kwazulu-Natal, South Africa. *Chemosphere*, 134, 133–140.
Ojovan, M. I., Lee, W. E. & Kalmykov, S. N. 2019. *An introduction to nuclear waste immobilisation*. Elsevier.
Ojovan, M. I. & Steinmetz, H. J. 2022. Approaches to disposal of nuclear waste. *Energies*, 15, 7804.
Ort, C., Lawrence, M. G., Reungoat, J., Eaglesham, G., Carter, S. & Keller, J. 2010. Determining the fraction of pharmaceutical residues in wastewater originating from a hospital. *Water Research*, 44, 605–615.

Petrovic, M., De Alda, M. J. L., Diaz-Cruz, S., Postigo, C., Radjenovic, J., Gros, M. & Barcelo, D. 2009. Fate and removal of pharmaceuticals and illicit drugs in conventional and membrane bioreactor wastewater treatment plants and by riverbank filtration. *Philosophical Transactions of the Royal Society A: Mathematical, Physical and Engineering Sciences*, 367, 3979–4003.

Puckowski, A., Mioduszewska, K., Łukaszewicz, P., Borecka, M., Caban, M., Maszkowska, J. & Stepnowski, P. 2016. Bioaccumulation and analytics of pharmaceutical residues in the environment: A review. *Journal of Pharmaceutical and Biomedical Analysis*, 127, 232–255.

Rao, T. S., Panigrahi, S. & Velraj, P. 2022. Transport and disposal of radioactive wastes in nuclear industry. In *Microbial Biodegradation and Bioremediation*. Elsevier.

Ravichandran, R., Binukumar, J., Sreeram, R. & Arunkumar, L. 2011. An overview of radioactive waste disposal procedures of a nuclear medicine department. *Journal of Medical Physics/Association of Medical Physicists of India*, 36, 95.

Ribeiro, A. R., Sures, B. & Schmidt, T. C. 2018. Cephalosporin antibiotics in the aquatic environment: A critical review of occurrence, fate, ecotoxicity and removal technologies. *Environmental Pollution*, 241, 1153–1166.

Rodriguez-Mozaz, S., Vaz-Moreira, I., Della Giustina, S. V., Llorca, M., Barceló, D., Schubert, S., Berendonk, T. U., Michael-Kordatou, I., Fatta-Kassinos, D. & Martinez, J. L. 2020. Antibiotic residues in final effluents of European wastewater treatment plants and their impact on the aquatic environment. *Environment International*, 140, 105733.

Rosman, N., Salleh, W., Mohamed, M. A., Jaafar, J., Ismail, A. & Harun, Z. 2018. Hybrid membrane filtration-advanced oxidation processes for removal of pharmaceutical residue. *Journal of Colloid and Interface Science*, 532, 236–260.

Shao, S., Hu, Y., Cheng, J. & Chen, Y. 2018. Research progress on distribution, migration, transformation of antibiotics and antibiotic resistance genes (ARGs) in aquatic environment. *Critical Reviews in Biotechnology*, 38, 1195–1208.

Sharma, K. & Kaushik, G. 2021. Pharmaceuticals: An emerging problem of environment and its removal through biodegradation. *Environmental Microbiology and Biotechnology*, 267–292.

Sharp, P. F., Gemmell, H. G., Murray, A. D. & Sharp, P. F. 2005. *Practical nuclear medicine*. Springer.

Shrader-Frechette, K. S. 1993. *Burying uncertainty: Risk and the case against geological disposal of nuclear waste*. University of California Press.

Siegel, J. A. & Sparks, R. B. 2002. Radioactivity appearing at landfills in household trash of nuclear medicine patients: Much ado about nothing? *Health Physics*, 82, 367–372.

Singh, A., Gogoi, A., Saikia, P., Karunanidhi, D. & Kumar, M. 2021. Integrated use of inverse and biotic ligand modelling for lake water quality resilience estimation: A case of Ramsar wetland, (Deepor Beel), Assam, India. *Environmental Research*, 200, 111397.

Spellberg, B., Bartlett, J. G. & Gilbert, D. N. 2013. The future of antibiotics and resistance. *New England Journal of Medicine*, 368, 299–302.

Stuart, M., Lapworth, D., Crane, E. & Hart, A. 2012. Review of risk from potential emerging contaminants in UK groundwater. *Science of the Total Environment*, 416, 1–21.

Togola, A. & Budzinski, H. 2008. Multi-residue analysis of pharmaceutical compounds in aqueous samples. *Journal of Chromatography A*, 1177, 150–158.

United States Nuclear Regulatory Commission. 1996. *Radioactive waste: Production, storage, disposal*. The Commission.

Yu, X., Sui, Q., Lyu, S., Zhao, W., Cao, X., Wang, J. & Yu, G. 2020. Do high levels of PPCPs in landfill leachates influence the water environment in the vicinity of landfills? A case study of the largest landfill in China. *Environment International*, 135, 105404.

Ziessman, H. A., O'Malley, J. P. & Thrall, J. H. 2013. *Nuclear medicine: The requisites e-book*. Elsevier Health Sciences.

Zuccato, E., Castiglioni, S., Fanelli, R., Reitano, G., Bagnati, R., Chiabrando, C., Pomati, F., Rossetti, C. & Calamari, D. 2006. Pharmaceuticals in the environment in Italy: Causes, occurrence, effects and control. *Environmental Science and Pollution Research*, 13, 15–21.

3 Fate and Transportation of Pharmaceutical Residues and PPCPs in the Aquatic System
Physiological Effects and Hazards

Mario Alberto Burgos-Aceves, César Arturo Ilizaliturri-Hernández, and Caterina Faggio

CONTENTS

3.1 INTRODUCTION

Regarding the United States Environmental Protection Agency (US EPA), pharmaceuticals and personal care products (PPCPs) are any manufactured items used for personal health, cosmetic reasons or used by agribusiness to enhance the growth or health of livestock (Daughton, 2004; Cizmas et al., 2015). This definition encompasses prescription and non-prescription human, illegal, and veterinary drugs (Boxall et al., 2012), from which 12 PPCPs, eight pharmaceuticals and four personal care products (Table 3.1), have been identified as a priority since they pose a risk of toxicity in aquatic environments, including freshwater and marine environments (Srain et al., 2021).

The pharmaceutical industry is responsible for developing, manufacturing, and marketing PPCPs. It is a market that, worldwide, has increased significantly, generating around 1.27 trillion dollars by 2020, where the European and the USA companies are the leaders in marketing. However, Brazil, India, Russia, Colombia, and Egypt are now playing an essential role in the pharmaceutical industry (Figure 3.1). Meanwhile, in Latin American countries, despite market advances, their production is still incipient. Instead, the Chinese pharmaceutical industry has shown significant progress in recent years (Liu & Wong, 2013; Mikulic, 2020; González Peña et al., 2021).

Like the production, PPCPs are extensively and increasingly used in human and veterinary medicine worldwide, resulting in a continuous discharge of PPCPs and their metabolites into the environment (Nikolaou et al., 2007), perhaps due to inadequate management, treatment, and disposal. Many of these PPCPs have been detected worldwide; some can persist for several months or even years (Monteiro & Boxall, 2009, 2010). Other PPCPs are not as persistent and undergo several degradation

DOI: 10.1201/9781003361091-3

TABLE 3.1
NORMAN List of Principal Categories of Pharmaceutical and Personal Care Products (PPCPs) as Likely Emerging Pollutants

Pharmaceuticals	Personal care products	Priority PPCPs
Anaesthetics	Anti-microbial agents	*Pharmaceuticals*
Analgesics	Antioxidants	Carbamazepine
Anorexics	Fragrances	Erythromycin
Anthelmintics	Insect repellents	Fluoxetine
Anti-inflammatories	Moth repellents	Metoprolol
Antibacterials	Parabens	Naproxen
Antibiotics	Sunscreen agents	Ofloxacin
Anticonvulsants	Siloxanes	Sertraline
Antidiabetics	Other	Sulfamethoxazole
Antiemetics		
Antihistaminics		*Personal care*
Antihypertensives		Bisphenol A
Anti-microbial agents		Linear alkylbenzene sulfonate
Antineoplastics		Nonylphenol
Antipruritics		Triclosan
Antipsychotics		
Antispasmodics—muscolotropics		
Antiulceratives		
Antivirals		
Anxiolytics		
Beta-blockers		
Blood viscosity agents		
Bronchodilators		
Diuretics		
Eco-labels		
Human medicinal product active substances		
Lipid regulators		
Muscle relaxants		
NSAIDs		
Pharmaceuticals transformation products		
Psychiatric drugs		
Sedatives, hypnotics		
Steroids and hormones		
Stimulants		
Veterinary medicinal product active substances		
X-ray contrast media		
Other		
Drugs of abuse		
Designer drugs		
Opiates, opioids, and metabolites		
Sympathomimetics		
Synthetic cannabinoids and psychoactive compounds		

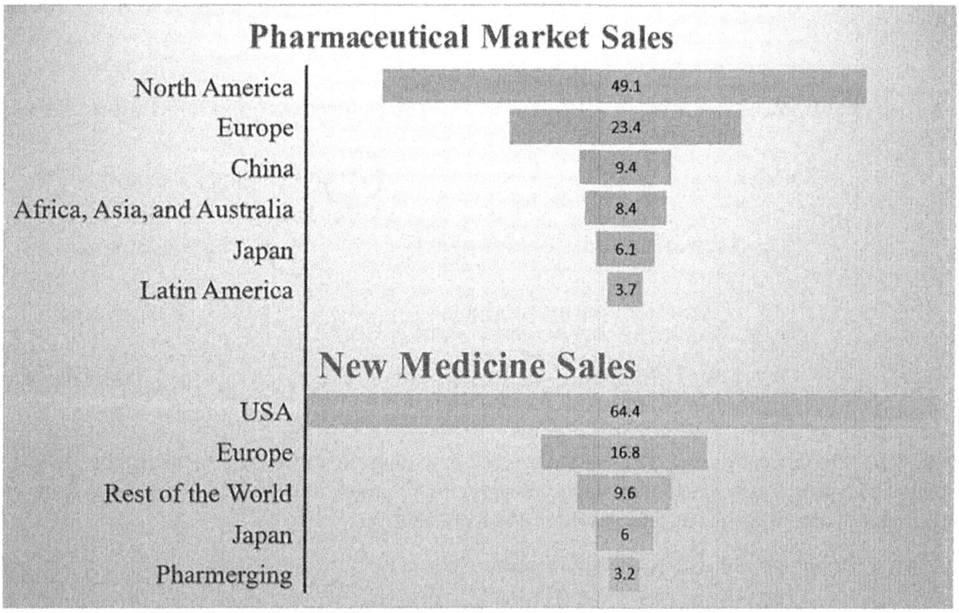

FIGURE 3.1 Breakdown of the world pharmaceutical sales market in 2021, including selling new drugs launched during 2016–2021.

Notes: Pharmaceutical Market Sales: North America includes the USA and Canada; Europe includes Belarus, Turkey, Russia, and Ukraine; Africa, Asia, and Australia do not include China and Japan. New Medicine Sales: Europe (Top 5) comprises France, Germany, Italy, Spain, and the United Kingdom; Pharmerging comprises 21 countries, Algeria, Argentina, Bangladesh, Brazil, Colombia, Chile, China, Egypt, India, Indonesia, Kazakhstan, Mexico, Nigeria, Pakistan, Philippines, Poland, Russia, Saudi Arabia, South Africa, Turkey, and Vietnam.

Source: EFPIA (2022).

processes, but given these chemicals continuous consumption and introduction into the environment, they can become pseudo-persistent (Barceló & Petrović, 2007; Ebele et al., 2017; Richmond et al., 2017). As PPCPs are biologically active compounds with unique properties, they may induce physiological effects even at low doses, making them substances capable of altering biological processes in different organisms (Reyes et al., 2021). Therefore, they have been recognized as environmental pollutants of emerging global concern (Figure 3.2) and, more recently, as agents of global change (Monteiro & Boxall, 2010; Bernhardt et al., 2017). The quantity introduced into aquatic systems is poorly characterized (Schwarzenbach et al., 2006). However, some studies have identified the incidence of PPCPs as trace contaminants present in the waste, aquatic systems, or finished drinking water (Sharma et al., 2009, 2013; Tölgyesi et al., 2010; Verlicchi et al., 2010; Anquandah et al., 2013; Michael et al., 2013; Petrovic et al., 2014; Kuzmanovic et al., 2015; Qin et al., 2015), suggesting that the contamination by PPCPs may be widespread, thus being a multifaceted problem that has become an area of great concern in environmental chemistry (González Peña et al., 2021). Despite the general lack of toxic effects at environmentally relevant concentrations, PPCPs are capable of causing various sublethal ecological effects in many components of aquatic ecosystems; therefore, emphasis should be placed on assessing the potentially harmful effects of these substances on ecosystems and human health (Richmond et al., 2017). For this reason, the use of animals as model organisms is used even more to track the health of environmental systems worldwide. They can be used to understand biological processes, gain insight into the environment's health status, and better understand the effects of PPCPs on organisms (Porretti et al., 2022; Zicarelli et al., 2022).

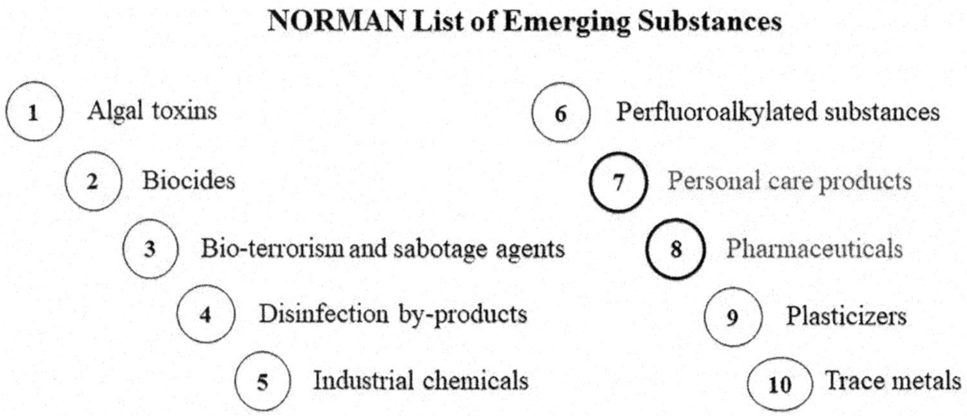

FIGURE 3.2 The currently most frequently discussed emerging substances and pollutants by NORMAN. Item (2) includes their transformation products and agricultural protection products. (5) includes flame retardants, lubricants, anti-microbial agents, gasoline, and food additives.

Source: www.norman-network.net/?q=node/81#sub29.

The first evidence related to the presence of PPCP in the environment was carried out in the 1970s (Garrison et al., 1976; Hignite & Azarnoff, 1977). Nevertheless, it has only been in the last two decades that a significant quantity of research has been done to establish the occurrence, fate, effects, and risks of PPCPs in the environment, primarily in aquatic organisms (Kolpin et al., 2002; Cunningham et al., 2006; Fent et al., 2006; Barceló & Petrović, 2007; Fu et al., 2018; Junaid et al., 2019; Papageorgiou 2019; Liu et al., 2020a, 2020b, 2020c; Yuan et al., 2020; Kar et al., 2020). Likewise, research and policy formulation strategies related to the risk assessment of environmental exposure to PPCPs have been developed (Boxall et al., 2012). However, despite this growing interest in the effects of PPCPs on aquatic organisms, knowledge gaps remain due to the relatively small number of chemicals tested at different sample matrices. Most studies focus on quantifying pre-selected or targeted analytes rather than identifying all the compounds present in the samples, despite advances in instrumentation and analytical capabilities that allow the detection of low concentrations (as low as picograms per litre, pg/L) of various PPCPs in wastewater, surface water, groundwater, drinking water, soil, and aquatic organisms (Christian et al., 2003; Arpin-Pont et al., 2016; Cantwell et al., 2017; Richardson & Ternes, 2018; Snow et al., 2020; Adeleye et al., 2022). In order to identify the priority compounds for developing environmental management guidelines and identify the gaps that require further exploration, it is necessary to synthesize the available information from previously conducted research (Reyes et al., 2021; Adeleye et al., 2022). In addition, there is an urgent need to develop methods to remove these emerging pollutants from aquatic environments (Oluwole et al., 2020).

3.2 FATE AND TRANSPORTATION OF PHARMACEUTICAL RESIDUES AND PERSONAL CARE PRODUCTS

Water resources are increasingly limited, and the quality of water bodies is seriously endangered by the presence of different pollutants that risk human health and aquatic environments (Balusamy et al., 2020; Wu et al., 2020). Organic compounds can enter the environment depending on their usage and application mode. This can happen through municipal, industrial, and agricultural waste disposal, pharmaceutical product excretion, and accidental spills. Once in the environment, they may be widely distributed from production, use, and disposal (Farré et al., 2008). World consumption

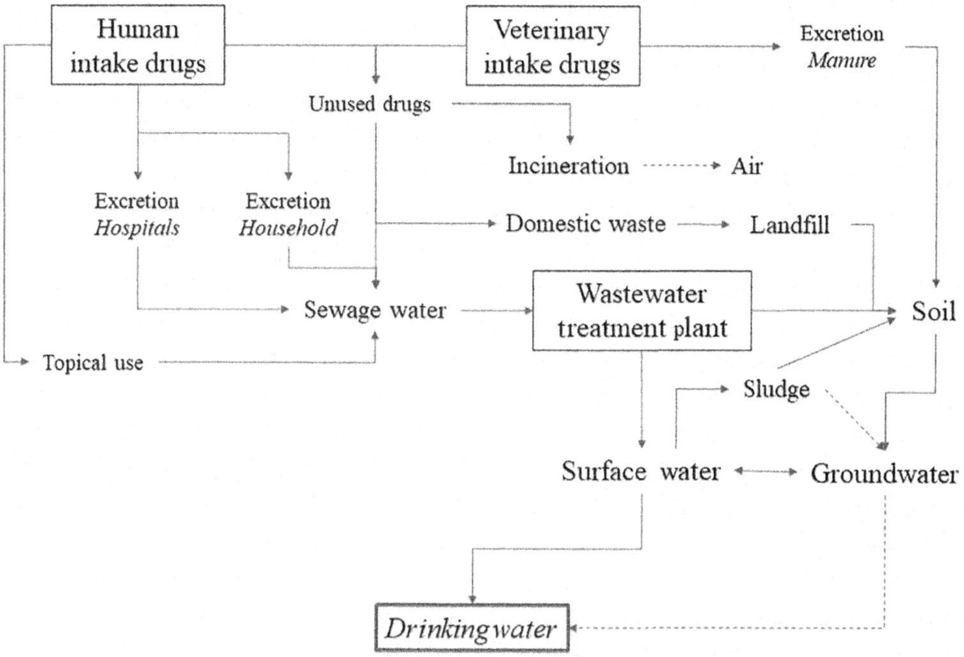

FIGURE 3.3 Potential pathways by which drugs and other PPCPs enter the environment. They can be both point and non-point sources.

of PPCPs has been widely increasing, and this emerging pollutant has been reported in the aquatic environment worldwide. PPCPs can enter surrounding water bodies primarily from wastewater effluent discharges from wastewater treatment plants and animal farms (Ebele et al., 2017). Because many pharmaceuticals and supplements contain concentrations above the human or animal body's ability to process, these chemicals not used by the body are excreted as waste or washed down the shower or sink drains, reaching the urban sewage network and even the wastewater treatment plants (Figure 3.3). Improper expired or unused medications disposal directly through the toilet may be another source of contamination (Ávila & García, 2015).

Once inside the environment, these emerging contaminants are subject to degradation by chemical and photochemical processes that contribute to their elimination. However, this will depend on the area in which they are present, whether they are in groundwater, surface water, or sediment, and whether they come from sewage treatment plants or drinking water facilities, what different transformations can take place, sometimes producing products that can differ in their environmental behaviour and ecotoxicological profile (Farré et al., 2008). Thus, depending on the physicochemical characteristics of PPCPs and their transformation products, they can reach groundwater and contaminate aquifers or remain retained in the soil and eventually accumulate (Figure 3.2). However, in all of these areas, studying the fate of emerging pollutants as well as the use of an efficient treatment process to treat municipal wastewater is crucial to eliminate or minimize the presence of PPCPs and their metabolites, to prevent possible adverse effects on the aquatic biota (Farré et al., 2008; Ávila & García, 2015).

3.2.1 Fate and Transportation in the Freshwater Environment

To date, PPCPs have been detected in many surface water lakes, rivers, and streams, ranging from ng/L to mg/L (Chen et al., 2013), but are more frequently detected in rivers than lakes (Adeleye et al., 2022). Since then, several routes of freshwater contamination with PPCP residues have been identified (Daughton & Ternes, 1999). Based on previous remarks, concentrations of common PPCPs are higher in freshwater than in groundwater and saline waters, where antibiotics and analgesics are the most frequently detected PPCPs in freshwater (Adeleye et al., 2022). The water released from wastewater treatment plants represents the primary source of PPCPs for the freshwater environment since many of them are not effectively eliminated by treatment systems (Diamantini et al., 2019). Absorption of PPCPs by the body after therapeutic use, followed by excretion and release into the sewage system or septic tank, appears to be one of the main routes. Therefore, post-use human drugs may enter aquatic systems after ingestion and subsequent excretion as unmetabolized parent compounds or metabolites through wastewater treatment plants (Pérez & Barceló, 2007). The drugs and their metabolites that make it through the wastewater treatment plant can enter rivers or streams or reach groundwater after leaching. Nevertheless, the concentrations of PPCPs in freshwater are generally lower than in raw or treated wastewater due to dilution effects (Adeleye et al., 2022).

Furthermore, pharmaceuticals can reach surface waters through runoff from fields treated with digested sludge for agricultural purposes (Nikolaou et al., 2007; Farré et al., 2008). These human and veterinary drugs and their metabolites pollute the soil and could enter the food chain. Consequently, agricultural runoff containing human and veterinary drug waste can enter freshwater environments and percolate into groundwater by leaching and runoff from livestock slurries (Farré et al., 2008). Another source of PPCP freshwater contamination is wastewater from drug or personal care product factories, which goes directly to wastewater treatment plants (Fick et al., 2009). The liquid effluent is discharged directly into freshwater, and the sludge is deposited on the ground as fertilizer. Also, through leaching, PPCPs can reach groundwater from the soil, which could threaten drinking water. Externally applied PPCPs, on the other hand, are primarily discharged through shower, bathroom, pool, and sink drains that go directly to wastewater treatment plants and, ultimately, to freshwater bodies (Peck, 2006). Therefore, domestic, industrial, and hospital wastewater are the main contributors of PPCPs to the urban wastewater system. The wastewater treatment plant is the receiving end, becoming a pool of PPCP products, where urban rivers are the primary receptors (Ghesti Pivetta & do Carmo Cauduro Gastaldini, 2019).

The scope of PPCPs depends on the compound's physicochemical properties and the receipt medium's characteristics. The distribution of PPCPs generally occurs through the aqueous medium because they are hydrophilic with high polarity and low volatility, in addition to their dispersion within the food chain (Caliman & Gavrilescu, 2009). The transport of PPCPs is associated with the absorption behaviour of the compound in the wastewater treatment plant, the soil, and the water-sediment system (Boxall, 2004). Several PPCP residues have been found in sludge samples from wastewater treatment plants, which can be released into the environment when these sludges are applied directly to agricultural land as fertilizers (Van Wieren et al., 2012). It has been reported that PPCPs can be transported to groundwater when these sludges are applied to agricultural land (Heberer, 2002) and fields irrigated with treated wastewater (Pedersen et al., 2005). This may result in the uptake of PPCPs by crops, thus creating a potential route of exposure to PPCPs through dietary intake in humans (Wu et al., 2014, 2015). Runoff of PPCP-containing biosolids, whether from landfills or applied to agricultural land, can be carried into surrounding surface waters or seep into groundwater (Kleywegt et al., 2007), posing a risk to aquatic life and public health. Sorption in sediments is another mechanism through which PPCPs are transported to freshwater. The sediment acts as a drain and accumulates these emerging pollutants, which can be released into the body of water (Zhao et al., 2013). Several studies have shown that some PPCPs are more persistent in sediment than in water, such as sulfamethoxazole, carbamazepine, Triclosan, and ciprofloxacin (Chenxi et al., 2008; Conkle et al., 2012). Osenbrück et al. (2007) identified that the presence of carbamazepine, galaxolide, and bisphenol A in urban groundwater results from a mixture of local river water infiltration, sewer exfiltration, and urban stormwater recharge. Therefore, a much better understanding of attenuation processes, their dependence on local conditions, and their temporal and spatial variations are essential for a robust risk assessment of PPCP debris in groundwater. This may be particularly important in light of possible long-term effects due to the high retention of pollutants in groundwater and, of course, if urban groundwater is used for drinking water.

Nevertheless, the reduction in bioavailability or toxicity of PPCPs is not influenced by absorption into sediments or suspended solids in receiving waters. Several studies have reported that PPCPs can accumulate in different environmental compartments, including sediments (Azzouz & Ballesteros, 2012; Silva et al., 2013; Chen & Zhou, 2014). Hence, these chemicals compounds' continuous and gradual release from the sediments to the overlying water is possible. Therefore, it seems imperative to better understand the toxicological impacts of PPCPs on freshwater sediments, as sediments act as a sink for these chemicals discharged into rivers and lakes, estuaries, and groundwater, and many studies have used this retention function to assess historical pollutant loads, decomposition or transformation, and other remobilization processes (Ebele et al., 2017; Tong et al., 2017, Cundy & Croudace, 2017).

3.2.2 Fate and Transportation in the Marine Environment

Despite the increasing pollution of the marine environment by PPCPs, few studies have been conducted on their presence and fate in marine biota (Arpin-Pont et al., 2016). Marine (coastal) and estuarine environments are considered the primary recipients of wastewater discharges since populated areas are concentrated on the world's coasts. These discharges are contained in many compounds, among which the PPCPs stand out (Maranho et al., 2015; Kallenborn et al., 2018). PPCPs can indirectly enter the marine environment through submarine or sewage treatment plant outfalls (Fenet et al., 2014) or runoff via rivers and streams (Farré et al., 2008). Recreational activities such as bathing are another source of PPCPs entering the marine environment, especially concerning sunscreen agents (Langford & Thomas, 2008; Bachelot et al., 2012; Fisch et al., 2017; Ren et al., 2021). Aquaculture activities such as fish farming are an alternative source of PPCP pollution in seawater, mainly antibiotics (Zou et al., 2011) and antiparasitic drugs (Rico & Van den Brink, 2014). However, data on the presence of PPCPs in the marine environment is limited because few studies have been conducted. Factors such as dilution and diffusion seem to play an essential role in the levels of PPCPs contained in seawater. Another factor could be the complexity of marine environment hydrodynamics in coastal areas (Weigel et al., 2002; Arpin-Pont at el., 2016). In a study by Vidal-Dorsch et al. (2012), several PPCPs were detected in marine receiving water of sewage treatment plant effluent discharge at submarine outfalls, highlighting nonylphenol, naproxen, gemfibrozil, atenolol, and tris(1-chloro-2-propyl)phosphate in concentrations below 1 μg/L, lower than those presently known for chronic toxicity thresholds. For example, nonylphenol, a critical chemical polluting many habitats (Raju et al., 2018), was found in this study at an average concentration of 0.23 μg/L in seawater and at 1.4 μg/L in the effluents. Conferring to the US Environmental Protection Agency, this value is lower than the aquatic life objective of 1.7 μg/L for seawater (Brooke & Thursby, 2005). Therefore, it is still challenging to determine the environmental significance of low levels of PPCPs in seawater at this time because toxicity thresholds for aquatic life have been developed for only a few. Hence, it is essential to assess the range of contamination by identifying vulnerable marine sites; considering the persistence and mobility of contaminants, interactions with other compounds, and exposure conditions that could lead to bioaccumulation and sublethal toxicity to improve the assessment exposure of aquatic organisms (Vidal-Dorsch et al., 2012; Arpin-Pont at el., 2016). Alternatively, it is also essential to consider that increasing evidence shows ocean acidification can directly and indirectly affect marine organisms. Its combined effects with other stressors, such as PPCPs, must be considered. Some data suggest that ocean acidification will increase the toxicity of some pharmaceutical substances for marine organisms, which has clear implications for coastal benthic ecosystems suffering from chronic pollution (Freitas et al., 2016, 2020). Unfortunately, this is an aspect that has received very little attention to date.

The presence of PPCPs in marine sediment is also concerning. Sediments are natural reservoirs of many chemicals in the water column and can act as a sink or source for various xenobiotics that can persist in the environment for an extended period (Daughton, 2001) since its chemical degradation can be inhibited when the compound is fixed to the sediment (Pan & Xing, 2011). Allowing them to bioaccumulate, becoming incorporated into the food chain and affecting people via food, or being toxic to one or more links in the food chain, thus disturbing ecosystems more broadly. In coastal marine environments, sedimentation increases in areas where freshwater meets seawater since an increase in salinity causes the precipitation of dissolved substances in low-salinity water (Beretta et al., 2014). As far as PPCPs are concerned, their environmental behaviour is believed to be controlled for the most part by their interaction with soil and sediment particles, and the presence of PPCPs in marine sediments is mainly of the lipophilic type (Richardson et al., 2005; Amine et al., 2012; Arpin-Pont at el., 2016). Despite this, few extensive studies have been conducted on the distribution and effects of spiked sediments with PPCPs in the marine environment (Pan & Xing, 2011; Beretta et al., 2014; Maranho et al., 2015; Pusceddu et al., 2018). These few publications showed that residues of PPCPs such as antibiotics (chloramphenicol, erythromycin, sulfathiazole), antiepileptics (carbamazepine), non-steroidal anti-inflammatory drugs (diclofenac, ibuprofen), anti-microbials (Triclosan), musks (AHTN, HHCB), and UV-filters (EHMC, O.C., OD-PABA) could be found in marine sediment (Zeng et al., 2008; Cantwell et al., 2010; Amine et al., 2012; Liang et al., 2013; Na et al., 2013; Beretta et al., 2014; Pei et al., 2022). However, the detection frequencies could not be estimated, so it is difficult to highlight a general tendency due to the lack of data on marine sediments. Nonetheless, some studies highlighted, as expected, that the highest concentrations were reported mainly in areas affected by anthropogenic activities (Arpin-Pont et al., 2016). According to studies, concentrations of PPCPs in marine sediments vary from very low to non-detectable, although some may occur in relatively high concentrations (Pei et al., 2022). Long et al. (2013) evaluated the presence of 119 PPCPs in marine sediments. The results showed that only 14 of the 119 compounds were quantifiable in sediments. The most frequently detected compounds were diphenhydramine (an antihistamine), which was detected at 87.5%, and triclocarban (an antibacterial) in 35.0% of the samples, with a maximum concentration of 4.81 ng/g and 16.6 ng/g dry weight, respectively. Meanwhile, in a study conducted by Xie et al. (2019), of the 34 PPCPs, 21 were detected in sediments from a mariculture area in the Pearl River Delta at concentrations ranging from 0.002 to 11 ng/g d.w. The antibiotics norfloxacin, cefotaxime sodium, isochlortetracyclin, roxithromycin, oxytetracycline, spectinomycin, and the anti-inflammatory ketoprofen were present in all sediment samples. The migration of PPCPs in sediments poses a potential risk to the surface water and groundwater environment. Understanding the spatial distribution and vertical profile of PPCPs in sediments on a regional scale is valuable for comprehensive PPCP risk prevention (Xie et al., 2022). Thus, investigating the ecological risk of PPCPs in marine sediments can allow us to create a scientific basis to establish a centralized management system and control the use of PPCPs and risk prevention of PPCPs to protect the environment and ensure the safety of the sea (Pei et al., 2022).

3.3 PHYSIOLOGICAL EFFECTS ON AQUATIC ORGANISMS

The incidence of PPCPs in the aquatic environment is usually reported in the lowest concentrations, fluctuating from ng/L to μg/L, due to inadequate elimination during conventional wastewater treatment processes (Sui et al., 2015; Osorio et al., 2016) and ecotoxicological studies have agreed that these compounds are not jeopardous to aquatic organisms (Franzellitti et al., 2013). However, recent reports have shown biological effects at such low concentrations (Table 3.2) and given their ubiquity in the environment.

Nowadays, scientists are concerned about the possible toxicological effects of these substances on ecosystems and public health (Table 3.2). PPCPs in aquatic environments may affect aquatic life through persistence, bioaccumulation, and toxicity (World Health Organization, 2012). For this reason, in recent years, the possible toxicological effects related to the appearance of PPCPs in aquatic organisms have been evaluated (Fent et al., 2006; Sehonova et al., 2018; Srain et al., 2021).

TABLE 3.2
Potential Effects of Pharmaceutical and Personal Care Products on Humans and Wildlife

Class of compound	Representative concentration	Potential effects on human or biota health	Examples
Pharmaceuticals	Up to 1 µg/L	Endocrine disrupting	Antibiotics, painkillers, caffeine, birth-control pill, antiepileptics
Personal care products	Up to 1 µg/L	Bioaccumulative, endocrine disrupting	Soaps, fragrances, Triclosan

Lethal effects of some PPCPs in aquatic species have been reported in concentrations ranging from 0.00053 to > 3000 mg/L and exposure periods from 5 min to 50 days. Among these, the most commonly found in the aquatic environment are anti-inflammatories (e.g., diclofenac or ibuprofen), antibiotics (e.g., erythromycin, azithromycin), beta-blockers (metoprolol), lipid regulators (gemfibrozil), anti-depressants (fluoxetine), antiepileptics (carbamazepine), diuretics, antidiabetics, synthetic hormones (e.g., alpha oestradiol), and others (Aliko et al., 2021; Srain et al., 2021). Toxicological concerns regarding the environmental release of PPCPs include the inducement of abnormal physiological processes and reproductive impairment, increased incidences of cancer, change of antibiotic-resistant bacteria, and the potential for augmented toxicities when chemical mixtures occur in the environment (Richardson et al., 2005; Sehonova et al., 2018).

Little research is available in the scientific literature on PPCPs accumulation in biota (Ali et al., 2018). PPCPs and their metabolites can potentially bioaccumulate in individuals at different trophic levels (Mackay & Barnthouse, 2010). Therefore, it is necessary to comprehend the bioaccumulation process. We can point out that it is a process by which contaminants enter the food chain from all possible exposure routes, be it water, sediments, soil, air, or intake, and accumulate in the biological tissues of aquatic organisms in concentrations higher than water column or sediments (Walker et al., 2013; Walker & Grant, 2015; Wilkinson et al., 2018). Recent literature on concentration levels of PPCPs (focused on drugs of abuse and their metabolites), cocaine, and its metabolite benzoylecgonine are most frequently detected in some wild aquatic organisms, which are consistent with the widespread incidence in surface water and poor removal efficiency of these compounds (Chen et al., 2021). On the other hand, among public studies on the biota, reports on the uptake in fish tissues predominate (Brooks et al., 2005; Valdersnes et al., 2006; Mottaleb et al., 2009; Subedi et al., 2012; Gao et al., 2015; Mottaleb et al., 2015). This may be because fish play a key role in aquatic food webs and are also an important food source for humans (Arnnok et al., 2017; Sehonova et al., 2017). However, the presence of these contaminants in the tissues of other aquatic species, such as algae, aquatic plants, crustaceans, and molluscs, has also been reported (Coogan et al., 2007; Ali et al., 2018; Wilkinson et al., 2018; Wang et al., 2019; Ruan et al., 2020; Yang et al., 2020; Álvarez-Ruiz et al., 2021; Chen et al., 2021; Xin et al., 2021; Xu et al., 2022). The bioaccumulation factor will depend on the species and the compound but can range from 0 to > 50,000 orders of magnitude (Liu et al., 2020a, 2020b, 2020c). The evidence so far tells us that the likelihood of drug bioaccumulation is from primary producers to top predators in aquatic food webs, leading to potential acute or chronic toxic effects in aquatic organisms through absorption and transformation and potentially biomagnifying as they are transferred through food webs (Chen et al., 2021).

Consequently, only a few PPCPs are subjected to environmental risk assessments where appropriate ecotoxicity tests are included (Ferrari et al., 2004; Srain et al., 2021). Then, further evaluation is required for potentially toxic and toxicological effects on aquatic organisms (Daughton, 2001; Martin-Diaz et al., 2009; Chen et al., 2021). Since, for many PPCPs, the potential effects on humans and aquatic organisms are not clearly understood. A prime reason for this lies in analytical capabilities to detect PPCPs in the environment. Currently, suitable protocols are deficient for identifying and measuring PPCPs without using costly technological methods. Although PPCPs are found at relatively low levels in the environment, their potential risk to aquatic organisms is primarily related to chronic exposure. Therefore, these compounds' natural ecological risk is difficult to understand (Arpin-Pont et al., 2016).

3.3.1 Physiological Effects on Freshwater Biota

The occurrence of PPCPs in the freshwater ecosystem at environmentally relevant trace concentrations potentially elicits the ecotoxicological effects on various aquatic species under chronic or acute exposure. In a study conducted by Lagesson et al. (2016), the impact of five drugs (diphenhydramine, oxazepam, trimethoprim, diclofenac, and hydroxyzine) was assessed on the European perch and four species of invertebrates (damselfly larvae, mayfly larvae, waterlouse, and ramshorn snail). The results suggest that the bioaccumulation and absorption capacity will depend on the type of drug and the species. It is in the benthic species (ramshorn snail and waterlouse) where the highest concentrations were reported. Data results suggest that bottom-dwelling organisms in lower trophic positions are the primary recipients of PPCPs (Vernouillet et al., 2010; Du et al., 2014; Ding et al., 2015; Ruhí et al., 2016). However, other field studies have found low or no bioaccumulation and trophic biomagnification potential of pharmaceuticals in freshwater food webs (Xie et al., 2015). Xu et al. (2022) investigated the concentrations of 35 typical PPCPs in freshwater molluscs, *Hyriopsis cumingii*, *Unio douglasiae*, *Sinanodonta woodiana*, *Lamprotula leai*, and *Corbicula fluminea*. Sixteen of these compounds were found in all mollusc soft tissues. The results show differences in concentration and presence between compounds, species, and organs. PPCPs concentrations ranged between 736.1 and 4801 ng/g d.w., highlighting ketoprofen (mean: 390.8 ng/g; range: 42.5–1207 ng/g), ibuprofen (mean: 347.4 ng/g; range: 44.9–992.7 ng/g), Triclosan (mean: 193.4 ng/g; range: ND-995.4 ng/g), gemfibrozil (mean: 190.1 ng/g; range: 40.8–1224 ng/g) and triclocarban (mean: 121.6 ng/g; range: 0.94–647.9 ng/g) for presenting the highest average concentrations. Meanwhile, the mollusc species that presented the highest average concentration of PPCPs were *C. fluminea* (3.18 ± 1.13 μg/g) and *U. douglasiae* (2.84 ± 1.22 μg/g), while *S. woodiana* presented the lowest concentration of PPCPs (1.01 ± 0.21 μg/g). Regarding the concentrations of PPCPs in organs, differences were also reported. The gonad was the main accumulator organ of PPCPs, with concentrations ranging between 0.3 (*S. woodiana*) and 36.6 (*H. cumingii*) μg/g. This interspecies difference in PPCPs accumulation may be influenced by factors such as body size, metabolic capacity, and food chain structure (Fu et al., 2019). Thus, redistribution of organ-specific PPCP residues can arise to the difference in assimilation rates and clearance of each organ (Xu et al., 2022). Therefore, the distribution of organic compounds in aquatic organisms may be closely related to the lipid content in various organs (Elskus et al., 2005).

One of the crucial aspects to point out about pharmaceutical compounds is their high effectiveness at low doses; that is, they are explicitly designed to maximize their biological activity at low concentrations and have a specific effect on metabolic, enzymatic, or cell signalling mechanisms (Ebele et al., 2017). Fluoxetine, the active ingredient of the anti-depressant Prozac®, is known to increase serotonin levels in the synaptic space of neurons, acting as selective serotonin (5-hydroxytryptamine, 5-HT) reuptake inhibitor in treating depression and other mood disorders (Hiemke & Härtter, 2000). Moreover, 5-HT is a high-level physiological controller in aquatic organisms, as it involves hormonal and neural mechanisms and plays a crucial role in regulating food intake, metabolism, and reproductive success in invertebrates (Tierney, 2001; Fabbri & Capuzzo, 2010). According to data, fluoxetine can potentially impair relevant physiological functions in invertebrates (Péry et al., 2008; Franzellitti et al., 2013; Fong & Ford, 2014; Ford et al., 2018). This active anti-depressant ingredient has been detected in freshwater environments between 0.012 and 0.54 μg/L (Kolpin et al., 2002; Chen et al., 2006; Gardner et al., 2012). According to results obtained by Ford et al. (2018), fluoxetine significantly could cause foot detachment only at higher concentrations (1 mg/L) in acute (4 h) exposure in the great pond snail *Lymnaea stagnalis*. Whilst, in a trial where *Daphnia magna* adults were acutely (6 days) exposed to fluoxetine in a range of 1–100 μg/L, no apparent effects on the Daphnia physiology were observed. Whereas, when chronically (30 days) exposed to the anti-depressant at 36 μg/L, adults significantly enhanced fertility (Flaherty & Dodson, 2005). In another study, reproduction in *D. magna* was reduced considerably by chronic (21 days) exposure to the anti-depressant at 31 μg/L (Péry et al., 2008). The reproductive capacity was also reduced in *Ceriodaphnia dubia* chronically (45 days) exposed to fluoxetine at 112 μg/L, but not when *C. dubia* were exposed at 57 μg/L, since fecundity increased (Brooks et al., 2003). Henry et al. (2004) also found a reproduction decrease for the same species exposed at 89 μg/L and 447 μg/L. According to other data, fluoxetine affections can be found at low exposure concentrations. The freshwater planarian *Schmidtea mediterranea*, exposed to fluoxetine from 1.0 μg/L for nine days, began to show an increment in locomotor activity and a reduction in feeding activity. A decreased reproduction and a significant increase in DNA damage began to be observed at 10 μg/L and 0.1 μg/L, respectively. All these endpoints consistently showed a linear increase or decrease (whatever the case) with increasing fluoxetine concentration (Ofoegbu et al., 2019). Thus, fluoxetine presents a dose-dependent effect (Nentwig, 2007; Ofoegbu et al., 2019). The effects of fluoxetine at low concentrations (around 10 μg/L) were also observed in a study with *D. magna* and the snail *Potamopyrgus antipodarum*, especially on parthenogenetic reproduction, causing a significant decrease (Péry et al., 2008). In a multigenerational study with the exposure of newborns of *D. magna* individuals previously exposed to the anti-depressant, the results showed that the length of the newborns was affected by fluoxetine, with second-generation exposed individuals showing much more pronounced effects than the first one, exposed to 8.9 μg/L (Péry et al., 2008). Although, regarding the chronic effects, we can say that the data published so far give us contradictory information about the effects of fluoxetine on invertebrates. A review of existing data on anti-depressants' impact on various endpoints in freshwater species reveals a wide variability of results. While some reports indicate adverse effects on aquatic organisms at levels at least one order of magnitude higher than those reported in municipal wastewater concentrations, others point to an anti-depressant effect at lower concentrations. We can then point out that the aquatic toxicity of individual PPCPs will depend on the model species, experimental designs and endpoints, the moment and duration of exposure, contaminant concentrations, and the developmental stage of organisms when exposure occurs (Srain et al., 2021).

Algae play an essential role in the aquatic ecosystem; they form the basis of aquatic food webs with a vital function in transferring energy and nutrients to species of higher trophic levels. So the influence of algae can affect the life of higher trophic organisms. They are sensitive to environmental disturbances and pollution and can accumulate many water contaminants that can be transferred to a higher trophic level species—aspects for which some algae can even be used as environmental indicators. This sensitivity was demonstrated in the work carried out by Bi et al. (2018). In this study, exposure to Triclosan, fluoxetine, and the mixture of both PPCPs had an inhibitory effect on the growth of seven species of algae: *Chlorella pyrenoidosa*, *C. ellipsoidea*, *Scenedesmus obliquus*, *S. quadricauda*, *Dunaliella salina*, *D. parva*, and *Chlamydomonas microsphaera*. The results showed that the inhibitory effect of Triclosan on algal growth varied by a 50-fold difference between the seven algal species, with *D. parva* being the most sensitive while *C. ellipsoidea* was the most minor susceptible. Meanwhile, fluoxetine had slightly higher toxicity than Triclosan in the seven algae species evaluated. The mixture of Triclosan and fluoxetine did not consistently affect the seven algae species. Thus, the toxicity of PPCPs is highly dependent on the algal species, as the sensitivity of various species in response to the same compound results in species-dependent toxicity. The findings so far showed that PPCPs could cause unexpected effects on algae and their communities, with Chlorophyta and Diatoms specifically being the most accessible and sensitive groups to PPCPs (Xin et al., 2021).

Secondly, Dar et al. (2022) point out that Triclosan, at acute exposures, causes alterations at the biomolecular level, modifying existing proteins or synthesizing new ones to respond to the stress caused by the drug, as observed in embryos of the four species of freshwater fish *Cyprinus carpio*, *Ctenopharyngodon idella*, *Labeo rohita*, and *Cirrhinus mrigala*. In another study, this anti-microbial agent can be highly toxic, as observed in the larvae of *L. rohita* after 96 h of exposure. Triclosan can significantly cause biochemical and transcriptomic alterations in the fish larvae that result in oxidative stress, impairment of metabolic processes, and dysfunction of the liver, kidney, and digestive system (Sharma et al., 2021). Whilst chronic exposure to tramadol hydrochloride (0.2, 2, 20, 200, and 600 μg/L) for 28 days did not affect either mortality or growth in juvenile zebrafish (*Danio rerio*), over time, it can cause an imbalance between free radicals and antioxidants (Plhalova et al., 2020).

3.3.2 Physiological Effects on Seawater Biota

Studies on the effects of PPCPs on marine and estuarine organisms are scarce, perhaps due to the high polarity of some of them (Arpin-Pont et al., 2016; Prichard & Granek, 2016). However, these compounds have been detected at 7112 ng/g, as reported for UV-filters in two mussel species from French coastal regions. The high UV-filter concentrations corresponded to the sites with the most increased tourist activity and, thus, recreational pressure (Bachelot et al., 2012). The occurrence and bioaccumulation of PPCPs in macroalgae, barnacles, and fish from contaminated coastal waters of the Saudi Red Sea were also reported (Ali et al., 2018). Eleven of the 20 target PPCPs were detected in the barnacles at concentrations between the limit of quantification (L.O.Q.) and 17.9 ng/g d.w. While in fish species studied, 17 of the 20 target PPCPs were detected with levels in the < L.O.Q.–93.5 ng/g d.w. range. In the marine macroalgae, eight of the 20 target PPCPs were detected at concentrations between 1.7 ng/g and 44.3 ng/g d.w. The authors reveal that the appearance and accumulation of PPCPs in macroalgae and barnacles could indicate a new way to absorb these chemicals by marine biota, specifically in polluted waters with a continuous supply of non-persistent pollutants such as PPCPs for long-term exposure of local benthic organisms (Ali et al., 2018). Thus, PPCPs can bioaccumulate within the organisms (Bachelot et al., 2012). An aspect of PPCPs contamination in the marine environment has to be considered (Arpin-Pont et al., 2016). Based on the studies, antibiotics have been the main represented PPCP class with more than ten detected molecules (Bachelot et al., 2012; Nakata et al., 2012; Na et al., 2013; Ali et al., 2018), perhaps due to the extensive use of antibiotics in human therapies, animal husbandry, agricultural activities, and aquaculture (Ávila & García, 2015). However, musk fragrances are ubiquitous, persistent and bioaccumulative pollutants in marine organisms that are sometimes highly toxic (Nakata, 2005; Aminot et al., 2021; Arruda et al., 2022).

Most studies have focused on the exploratory PPCPs effects at the cellular or subcellular level, and few studies examine organismal- and community-level effects (Prichard & Granek, 2016). In Mediterranean mussels, *Mytilus galloprovincialis*, chronically exposed to the anti-inflammatory acetylsalicylic acid (10 and 100 μg/L) for 10 to 20 days, no alterations in the cell viability of haemocytes and digestive gland cells were reported. However, there were alterations in the digestive gland's physiological mechanisms of volume regulation. Still, a time-dose reaction to acetylsalicylic acid in the gills and digestive gland showed numerous alterations, such as lipofuscin deposits and haemocyte infiltration (Pagano et al., 2022). Pusceddu et al. (2018) evaluated chronic exposure to ibuprofen and Triclosan in marine sediments in the sea urchins *Lytechinus variegatus* and the bivalve *Perna perna*, using sub-individual and developmental endpoints. Bioassays exhibited an alteration in the embryo-larval development of both species, causing a significant reduction in the lysosomes of the haemocytes. Ibuprofen and Triclosan appear to act directly on cell membranes by membranotropic effects, causing rupture of lysosomal membranes, thus showing a marked cytotoxic effect (Cortez et al., 2012; Pusceddu et al., 2018). Whereas the marine polychaetes *Hediste diversicolor* exposed to the carbamazepine, ibuprofen, and propranolol-spiked sediments elicited an anti-inflammatory action, as demonstrated by inhibiting the cyclooxygenase (COX) activity (Maranho et al., 2015). In a study with the Mediterranean mussels, exposure to the anti-depressant fluoxetine at 0.3 ng/L caused alterations in serotonin synthesis after seven days of exposure, resulting in adverse effects on preproduction, metabolism, and locomotion at concentrations close to or even lower than environmental levels (Franzellitti et al., 2013). In the sea snail *Gibbula umbilicalis*, fluoxetine can cause foot detachment only at high concentrations (1 mg/L), as reported by Ford et al. (2018).

However, fluoxetine, also present in the sediments, did not show anti-inflammatory properties or cause changes in the energy status in the marine polychaetes *H. diversicolor* after 14 days of laboratory exposure (Maranho et al., 2015). Fluoxetine can significantly alter camouflage efficiencies on the uniform and sandy backgrounds in the young cuttlefish *Sepia officinalis*. Hatchlings exposed to 1 ng/L of fluoxetine exhibited a diminution in uniform camouflage efficiency in a dose-dependent way. They also showed a significant increase in the frequency of sand-digging behaviours, which might make them highly visible to predators (Di Poi et al., 2014).

Exposure to dissolved caffeine and carbamazepine, alone or in combination, induced oxidative stress in *H. diversicolor* after 28 days of exposure to 0.3 μg/L (Pires et al., 2016a). The impacts of carbamazepine and caffeine were also evidenced in the polychaete species *Diopatra neapolitana*. After 28 days of exposure to caffeine or carbamazepine, the regenerative capacity of *D. neapolitana* was reduced dose-dependently (Pires et al., 2016b). Oxidative stress through promoting lipid peroxidation in the brachyuran crab *Carcinus maenas* exposed to increasing concentrations of carbamazepine and novobiocin for 28 days was also observed (Aguirre-Martínez et al., 2013). In the rainbow trout, *Oncorhynchus mykiss*, fenofibrate, diclofenac, fluoxetine, clofibrate, sulfamethoxazole, carbamazepine gadolinium chloride, and propranolol can cause a specific Ethoxyresorufin-O-Deethylase (EROD) inhibition in primary hepatocytes after exposure to a concentration ranging from 0.92–118.13 μg/L. However, clofibrate, fenofibrate, and fluoxetine showed a more significant cytotoxic effect and induced oxidative stress at sublethal concentrations (Laville et al., 2004). In another study, acute toxicity of carbamazepine in juvenile rainbow trout was observed, causing inhibition of all antioxidant activities, mainly in the gills. Changes in the haematological profile were also reported, with increased erythrocyte count, haemoglobin, mean corpuscular haemoglobin concentration, monocytes, and neutrophil granulocytes (Li et al., 2011). In the harbour seal *Phoca vitulina*, carbamazepine can have an immunosuppressive effect by inhibiting the proliferation of peripheral blood mononuclear cells after exposure to a concentration ranging from 0.50–100 mg/L (Kleinert et al. 2018).

In conclusion, it is required to point out that in recent years, considerable heterogeneity of PPCP types has been detected in seawater and sediments in marine ecosystems worldwide. Among them are pollutants that the Stockholm Convention banned 20 years ago (Avellan et al., 2022). From the point of view of environmental diagnosis, we can say that some of these substances can still be found in very high concentrations in both seawater and sediment samples. However, more studies investigating the effects of PPCPs in the various taxonomic groups are still lacking. Most studies to date indicate little short-term and long-term toxicity, and therefore the main issues of concern with PPCPs are their ability to bioaccumulate to high levels and their propensity to cause oestrogenic and endocrine effects (Brausch & Rand, 2011).

3.4 SUMMARY AND CONCLUSIONS

This chapter reviewed the literature to detail as clearly and concisely as possible the fate, transport, and effects of the various PPCPs detected in aquatic environments, whether marine or freshwater. In the first instance, we can say that wastewater treatment plants have been established as the primary point sources responsible for the release of PPCPs in aquatic environments. This is a consequence of failures in the elimination and purification of water during treatment in the most urbanized areas and in developing countries where a more significant presence of PPCPs in aquaculture environments is reported. Therefore, it is necessary to implement monitoring strategies that are most effective concerning the use, disposal, occurrence, and impacts in the different stages of the PPCP life cycle to control their release into the environment. Therefore, it is essential to acquire more accurate data on the fate of PPCPs in marine and freshwater ecosystems and their effects of PPCPs on aquatic organisms.

The reports published so far indicate that several aquatic species are affected by PPCPs, both sublethal and lethal exposures in environmentally relevant concentrations that can alter ecological processes. However, the various aquatic species affected by PPCPs and the variety of sublethal and lethal effects produced by their exposure to PPCPs remain to be identified. These reports may have provided a deep understanding of the toxicity of PPCPs to aquatic organisms, even when present at the lowest concentrations, such as sublethal effects. In addition, the studies will help support future research to determine the range and magnitude of PPCP concentrations and effects in aquatic environments and to help reduce the sources of PPCPs released into the environment. Current concentrations of PPCPs in the environment are lower than the threshold effect levels for traditional acute ecotoxicology tests. However, aquatic species are exposed to low concentrations throughout their lives. Chronic toxicity effects accumulate over time from exposure to the environmental presence of PPCPs. Studies on the chronic toxicity of PPCPs in aquatic species are increasing. However, there are still gaps in information on the chronic effects of many PPCPs on species. For this, it is also necessary to carry out future research on chronic toxicity to better understand and manage the toxic effects produced by PPCPs in aquatic environments. Chronic toxicity studies are required to investigate the sublethal impacts of understudied PPCPs in understudied aquatic species. Therefore, it will be possible to effectively assess the ecological hazards of PPCPs for aquatic ecosystems with the completion of further studies on chronic toxicity.

Ecological effects may be augmented by concurrent exposures to multiple PPCPs commonly identified in aquatic environments, although PPCPs are typically tested for ecotoxicological effects as individual compounds and rarely as mixtures. However, environmental PPCPs are present as mixtures that can raise toxicity and produce synergetic effects. Thus, toxicity may be the sum of individual PPCPs concentrations, and ecotoxicity effects may arise even when individual PPCPs exist at no-observed-effect concentration levels. Mixtures of compounds with different toxic modes of action can injure aquatic organisms due to their combined toxicity. Further research on the combined toxicity of PPCP mixtures to aquatic organisms is warranted.

REFERENCES

Adeleye, A.S., Xue, J., Zhao, Y., Taylor, A.A., Zenobio, J.E., Sun, Y., Han, Z., Salawu, O.A., Zhu, Y. 2022. Abundance, fate, and effects of pharmaceuticals and personal care products in aquatic environments. *Journal of Hazardous Materials* 424(Pt B):127284.

Aguirre-Martínez, G.V., Del Valls, T.A., Martín-Díaz, M.L. 2013. Early responses measured in the brachyuran crab *Carcinus maenas* exposed to carbamazepine and novobiocin: application of a 2-tier approach. *Ecotoxicology and Environmental Safety* 97:47–58.

Ali, A.M., Rønning, H.T., Sydnes, L.K., Alarif, W.M., Kallenborn, R., Al-Lihaibi, S.S. 2018. Detection of PPCPs in marine organisms from contaminated coastal waters of the Saudi Red Sea. *Science of the Total Environment* 621:654–662.

Aliko, V., Korriku, R.S., Pagano, M., Faggio, C. 2021. Double-edged sword: fluoxetine and Ibuprofen as development jeopardizers and apoptosis' inducers in common toad, *Bufo bufo*, tadpoles. *Science of the Total Environment* 776:145945.

Álvarez-Ruiz, R., Picó, Y., Campo, J. 2021. Bioaccumulation of emerging contaminants in mussel (*Mytilus galloprovincialis*): influence of microplastics. *Science of the Total Environment* 796:149006.

Amine, H., Gomez, E., Halwani, J., Casellas, C., Fenet, H. 2012. U.V. filters, ethylhexyl methoxycinnamate, octocrylene and ethylhexyl dimethyl PABA from untreated wastewater in sediment from eastern Mediterranean river transition and coastal zones. *Marine Pollution Bulletin* 64:2435–2442.

Aminot, Y., Munschy, C., Héas-Moisan, K., Pollono, C., Tixier, C. 2021. Levels and trends of synthetic musks in marine bivalves from French coastal areas. *Chemosphere* 268:129312.

Anquandah, G.A.K., Sharma, V.K., Panditi, V.R., Gardinali, P.R., Kim, H., Oturan, M.A. 2013. Ferrate(VI) oxidation of propranolol: kinetics and products. *Chemosphere* 91:105–109.

Arnnok, P., Singh, R.R., Burakham, R., Pérez-Fuentetaja, A., Aga, D.S. 2017. Selective uptake and bioaccumulation of anti-depressants in fish from effluent impacted Niagara River. *Environmental Science & Technology* 51(18):10652–10662.

Arpin-Pont, L., Martinez Bueno, M.J., Gomez, E., Fenet, H. 2016. Occurrence of PPCPs in the marine environment: a review. *Environmental Science and Pollution Research* 23(6):4978–4991.

Arruda, V., Simões, M., Gomes, I.B. 2022. Synthetic musk fragrances in water systems and their impact on microbial communities. *Water* 14:69.

Avellan, A., Duarte, A., Rocha-Santos, T. 2022. Organic contaminants in marine sediments and seawater: a review for drawing environmental diagnostics and searching for informative predictors. *Science of the Total Environment* 808:152012.

Ávila, C., García, J. 2015. Pharmaceuticals and personal care products (PPCPS) in the environment and their removal from wastewater through constructed wetlands. *Comprehensive Analytical Chemistry* 67:195–244.

Azzouz, A., Ballesteros, E. (2012). Combined microwave-assisted extraction and continuous solid-phase extraction prior to gas chromatography-mass spectrometry determination of pharmaceuticals, personal care products and hormones in soils, sediments and sludge. *Science of the Total Environment* 419:208–215.

Bachelot, M., Li, Z., Munaron, D., Le Gall, P., Casellas, C., Fenet, H., Gomez, E. 2012. Organic U.V. filter concentrations in marine mussels from French coastal regions. *Science of the Total Environment* 420:273–279.

Balusamy, B., Senthamizhan, A., Uyar, T. 2020. Functionalized electrospun nanofibers as a versatile platform for colorimetric detection of heavy metal ions in water: a review. *Materials* 13(10):2421.

Barceló, D., Petrović, M. 2007. Pharmaceuticals and personal care products (PPCPs) in the environment. *Analytical and Bioanalytical Chemistry* 387:1141–1142.

Beretta, M., Britto, V., Tavares, T.M., da Silva, S.M.T., Lowe Pletsch, A. 2014. Occurrence of pharmaceutical and personal care products (PPCPs) in marine sediments in the Todos os Santos Bay and the north coast of Salvador, Bahia, Brazil. *Journal of Soils and Sediments* 14:1278–1286.

Bernhardt, E.S., Rosi, E.J., Gessne, M.O. 2017. Synthetic chemicals as agents of global change. *Frontiers in Ecology and the Environment* 15(2):84–90.

Bi, R., Zeng, X., Mu, L., Hou, L., Liu, W., Li, P., Chen, H., Dan Li, D., Bouchez, A., Tang, J., Xie, L. 2018. Sensitivities of seven algal species to Triclosan, fluoxetine and their mixtures. *Scientific Reports* 8:15361.

Boxall, A.B.A. 2004. The environmental side effects of medication. *EMBO Reports* 5:1110–1116.

Boxall, A.B.A., Rudd, M.A., Brooks, B.W., Caldwell, D.J., Choi, K., Hickmann, S., Innes, E., Ostapyk, K., Staveley, J.P., Verslycke, T., Ankley, G.T., Beazley, K.F., Belanger, S.E., Berninger, J.P., Carriquiriborde, P., Coors, A., Deleo, P.C., Dyer, S.D., Ericson, J.F., Gagné, F., Van Der Kraak, G. 2012. Pharmaceuticals and personal care products in the environment: what are the big questions? *Environmental Health Perspectives* 120(9):1221–1229.

Brausch, J.M., Rand, G.M. 2011. A review of personal care products in the aquatic environment: environmental concentrations and toxicity. *Chemosphere* 82(11):1518–1532.

Brooke, L.T., Thursby, G. 2005. *Ambient aquatic life water quality criteria for nonylphenol. EPA-822-R-05–005.* Washington, DC: U.S. Environmental Protection Agency, Office of Water, Office of Science and Technology. www.epa.gov/sites/default/files/2019-03/documents/ambient-wqc-nonylphenol-final.pdf

Brooks, B.W., Chambliss, C.K., Stanley, J.K., Ramirez, A., Banks, K.E., Johnson, R.D., Lewis, R.J. 2005. Determination of select anti-depressants in fish from an effluent-dominated stream. *Environmental Toxicology and Chemistry* 24(2):464–469.

Brooks, B.W., Turner, P.K., Stanley, J.K., Weston, J.J., Glidewell, E.A., Foran, C.M., Slattery, M., La Point, T.W., Huggett, D.B. 2003. Waterborne and sediment toxicity of fluoxetine to select organisms. *Chemosphere* 52:135–142.

Caliman, F.A., Gavrilescu, M. 2009. Pharmaceuticals, personal care products and endocrine disrupting agents in the environment—a review. *Clean Soil Air Water* 37:277–303.

Cantwell, M.G., Katz, D.R., Sullivan, J.C., Ho, K., Burgess, R.M. 2017. Temporal and spatial behavior of pharmaceuticals in Narragansett Bay, Rhode Island, United States. *Environmental Toxicology and Chemistry* 36(7):1846–1855.

Cantwell, M.G., Wilson, B.A., Zhu, J., Wallace, G.T., King, J.W., Olsen, C.R., Burgess, R.M., Smith, J.P. 2010. Temporal trends of Triclosan contamination in dated sediment cores from four urbanized estuaries: evidence of preservation and accumulation. *Chemosphere* 78:347–352.

Chen, K., Zhou, J.L. 2014. Occurrence and behavior of antibiotics in water and sediments from the Huangpu River, Shanghai, China. *Chemosphere* 95:604–612.

Chen, L., Guo, C., Sun, Z., Xu, J. 2021. Occurrence, bioaccumulation and toxicological effect of drugs of abuse in aquatic ecosystem: a review. *Environmental Research* 200:111362.

Chen, M., Ohman, K., Metcalfe, C., Ikonomou, M.G., Amatya, P.L., Wilson, J. 2006 Pharmaceuticals and endocrine disruptors in wastewater treatment effluents and in the water supply system of Calgary, Alberta, Canada. *Water Quality Research Journal of Canada* 41:351–364.
Chen, W., Xu, J., Lu, S., Jiao, W., Wu, L., Chang, A.C. 2013. Fates and transport of PPCPs in soil receiving reclaimed water irrigation. *Chemosphere* 93(10):2621–2630.
Chenxi, W., Spongberg, A.L., Witter, J.D. 2008. Determination of the persistence of pharmaceuticals in biosolids using liquid-chromatography tandem mass spectrometry. *Chemosphere* 73(4):511–518.
Christian, T., Schneider, R.J., Färber, H.A., Skutlarek, D., Meyer, M.T., Goldbach, H.E. 2003. Determination of antibiotic residues in manure, soil, and surface waters. *Acta Hydrochimica et Hydrobiologica* 31(1):36–44.
Cizmas, L., Sharma, V.K., Gray, C.M., McDonald, T.J. 2015. Pharmaceuticals and personal care products in waters: occurrence, toxicity, and risk. *Environmental Chemistry Letters* 13(4):381–394.
Conkle, J.L., Gan, J., Anderson, M.A. 2012. Degradation and sorption of commonly detected PPCPs in wetland sediments under aerobic and anaerobic conditions. *Journal of Soils and Sediments* 12:1164–1173.
Coogan, M.A., Edziyie, R.E., La Point, T.W., Venables, B.J. 2007. Algal bioaccumulation of triclocarban, Triclosan, and methyl-Triclosan in a North Texas wastewater treatment plant receiving stream. *Chemosphere* 67(10):1911–1918.
Cortez, F.S., Seabra Pereira, C.D., Santos, A.R., Cesar, A., Choueri, R.B., Martini, G.D.A., Bohrer-Morel, M.B. 2012. Biological effects of environmentally relevant concentrations of the pharmaceutical Triclosan in the marine mussel *Perna perna* (Linnaeus, 1758). *Environmental Pollution* 168:145–150.
Cundy, A.B., Croudace, I.W. 2017. The fate of contaminants and stable Pb isotopes in a changing estuarine environment: 20 years on. *Environmental Science & Technology* 51(17):9488–9497.
Cunningham, V.L., Buzby, M., Hutchinson, T., Mastrocco, F., Parke, N., Roden, N. 2006. Effects of human pharmaceuticals on aquatic life: next steps. *Environmental Science & Technology* 40(11):3456–3462.
Dar, O.I., Aslam, R., Sharma, S., Jia, A.Q., Kaur, A., Faggio, C. 2022. Biomolecular alterations in the early life stages of four food fish following acute exposure of Triclosan. *Environmental Toxicology and Pharmacology* 91:103820.
Daughton, C.G. 2001. Pharmaceuticals and personal care products in the environment: overarching issues and overview. *In* Daughton, C.G., Jones-Lepp, T.L. (Eds.) *Pharmaceuticals and personal care products in the environment: scientific and regulatory issues* (A.C.S. Symposium Series 791). Washington, DC: American Chemical Society, pp. 2–38. http://pubs.acs.org/doi/pdfplus/10.1021/bk-2001-0791.ch001
Daughton, C.G. 2004. Pharmaceuticals and personal care products (PPCPs) as environmental pollutants: pollution from personal actions. Presented at Regional Science Liaison/Hazardous Substances Technical Liaison Meeting, Las Vegas, NV, February 26.
Daughton, C.G., Ternes, T.A. 1999. Pharmaceuticals and personal care products in the environment: agents of subtle change? *Environmental Health Perspectives* 107(6):907–938.
Diamantini, E., Mallucci, E., Bellin, A. 2019. A parsimonious transport model of emerging contaminants at the river network scale. *Hydrology and Earth System Sciences* 23:573–593.
Ding, J., Lu, G., Li, S., Nie, Y., Liu, J. 2015. Biological fate and effects of propranolol in an experimental aquatic food chain. *Science of the Total Environment* 532:31–39.
Di Poi, C., Bidel, F., Dickel, L., Bellanger, C. 2014. Cryptic and biochemical responses of young cuttlefish *Sepia officinalis* exposed to environmentally relevant concentrations of fluoxetine. *Aquatic Toxicology* 151:36–45.
Du, B., Haddad, S.P., Luek, A., Scott, W.C., Saari, G.N., Kristofco, L.A., Connors, K.A., Rash, C., Rasmussen, J.B., Chambliss, C.K., Brooks, B.W. 2014. Bioaccumulation and trophic dilution of human pharmaceuticals across trophic positions of an effluent-dependent wadeable stream. *Philosophical Transactions of the Royal Society* 369:20140058.
Ebele, A.J., Abdallah, M.A.-E., Harrad, S. 2017. Pharmaceuticals and personal care products (PPCPs) in the freshwater aquatic environment. *Emerging Contaminants* 3:1–16.
EFPIA (The European Federation of Pharmaceutical Industries and Associations). 2022. The pharmaceutical industry in figures key data 2022. www.efpia.eu/media/637143/the-pharmaceutical-industry-in-figures-2022.pdf [Accessed August 14, 2022].
Elskus, A.A., Collier, T.K., Monosson, E. 2005. Interactions between lipids and persistent organic pollutants in fish. *Biochemistry and Molecular Biology of Fishes* 6:119–152.
Fabbri, E., Capuzzo, A., 2010. Cyclic A.M.P. signaling in bivalve molluscs: an overview. *Journal of Experimental Zoology A* 313:179–200.

Farré, M., Pérez, S., Kantiani, L., Barceló, D. 2008. Fate and toxicity of emerging pollutants, their metabolites and transformation products in the aquatic environment. *Trends in Analytical Chemistry* 27:991–1007.

Fenet, H., Arpin-Pont, L., Vanhoutte-Brunier, A., Munaron, D., Fiandrino, A., Martínez Bueno, M.-J., Boillot, C., Casellas, C., Mathieu, O., Gomez, E. 2014. Reducing P.E.C. uncertainty in coastal zones: a case study on carbamazepine, oxcarbazepine and their metabolites. *Environment International* 68:177–184.

Fent, K., Weston, A.A., Caminada, D., 2006. Ecotoxicology of human pharmaceuticals. *Aquatic Toxicology* 76(2):122–159.

Ferrari, B., Mons, R., Vollat, B., Fraysse, B., Paxéus, N., Lo Giudice, R., Pollio, A., Garric, J. 2004. Environmental risk assessment of six human pharmaceuticals: are the current environmental risk assessment procedures sufficient for the protection of the aquatic environment? *Environmental Toxicology and Chemistry* 23(5):1344–1354.

Fick, J., Söderström, H., Lindberg, R.H., Phan, C., Tysklind, M., Larsson, D.G. 2009. Contamination of surface, ground, and drinking water from pharmaceutical production. *Environmental Toxicology and Chemistry* 28(12):2522–2527.

Fisch, K., Waniek, J.J., Schulz-Bull, D.E. 2017. Occurrence of pharmaceuticals and UV-filters in riverine runoffs and waters of the German Baltic Sea. *Marine Pollution Bulletin* 124(1):388–399.

Flaherty, C.M., Dodson, S.I. 2005. Effects of pharmaceuticals on Daphnia survival, growth, and reproduction. *Chemosphere* 61(2):200–207.

Fong, P.P., Ford, A.T. 2014. The biological effects of anti-depressants on the molluscs and crustaceans: a review. *Aquatic Toxicology* 151:4–13.

Ford, A.T., Hyett, B., Cassidy, D., Malyon, G. 2018. The effects of fluoxetine on attachment and righting behaviours in marine (*Gibbula unbilicalis*) and freshwater (*Lymnea stagnalis*) gastropods. *Ecotoxicology* 27(4):477–484.

Franzellitti, S., Buratti, S., Valbonesi, P., Fabbri, E. 2013. The mode of action (M.O.A.) approach reveals interactive effects of environmental pharmaceuticals on *Mytilus galloprovincialis*. *Aquatic Toxicology* 140–141:249–256.

Freitas, R., Almeida, Â., Calisto, V., Velez, C., Moreira, A., Schneider, R.J., Esteves, V.I., Wrona, F.J., Figueira, E., Soares, A. 2016. The impacts of pharmaceutical drugs under ocean acidification: new data on single and combined long-term effects of carbamazepine on *Scrobicularia plana*. *Science of the Total Environment* 541:977–985.

Freitas, R., Silvestro, S., Pagano, M., Coppola, F., Meucci, V., Battaglia, F., Intorre, L., Soares, A., Pretti, C., Faggio, C. 2020. Impacts of salicylic acid in *Mytilus galloprovincialis* exposed to warming conditions. *Environmental Toxicology and Pharmacology* 80:103448.

Fu, Q., Malchi, T., Carter, L.J., Li, H., Gan, J., Chefetz, B. 2019. Pharmaceutical and personal care products: from wastewater treatment into Agro-Food systems. *Environmental Science & Technology* 53(24):14083–14090.

Fu, W., Fu, J., Li, X., Li, B., Wang, X. 2018. Occurrence and fate of PPCPs in typical drinking water treatment plants in China. *Environmental Geochemistry and Health* 41:5–15.

Gao, L., Yuan, T., Cheng, P., Bai, Q., Zhou, C., Ao, J., Wang, W., Zhang, H. 2015. Effects of Triclosan and triclocarban on the growth inhibition, cell viability, genotoxicity and multixenobiotic resistance responses of *Tetrahymena thermophila*. *Chemosphere* 139:434–440.

Gardner, M., Comber, S., Scrimshaw, M.D., Cartmell, E., Lester, J., Ellor, B. 2012. The significance of hazardous chemicals in wastewater treatment works effluents. *Science of the Total Environment* 437:363–372.

Garrison, A.W., Pope, J.D., Allen, F.R. 1976. GC/MS analysis of organic compounds in domestic waste waters. *In* Keith, C.H. (Ed.) *Identification and analysis of organic pollutants in water*. Ann Arbor, MI: Science Publishers, pp. 517–556.

González Peña, O.I., López Zavala, M.Á., Cabral Ruelas, H. 2021. Pharmaceuticals market, consumption trends and disease incidence are not driving the pharmaceutical research on water and wastewater. *International Journal of Environmental Research and Public Health* 18(5):2532.

Heberer, T. 2002. Occurrence, fate, and removal of pharmaceutical residues in the aquatic environment: a review of recent research data. *Toxicology Letters* 131:5–17.

Henry, T.B., Kwon, J.W., Armbrust, K.L., Black, M.C. 2004. Acute and chronic toxicity of five selective serotonin reuptake inhibitors in *Ceriodaphnia dubia*. *Environmental Toxicology and Chemistry* 23(9):2229–2233.

Hiemke, C., Härtter, S. 2000. Pharmacokinetics of selective serotonin reuptake inhibitors. *Pharmacology & Therapeutics* 85:11–28.

Hignite, C., Azarnoff, D.L. 1977. Drugs and drug metabolites as environmental contaminants: chlorophenoxyisobutyrate and salicylic acid in sewage water effluent. *Life Sciences* 20:337–341.

Junaid, M., Wang, Y., Hamid, N., Deng, S., Li, W.-G., Pei, D.-S. 2019. Prioritizing selected PPCPs on the basis of environmental and toxicogenetic concerns: a toxicity estimation to confirmation approach. *Journal of Hazardous Materials* 380:120828.

Kallenborn, R., Brorström-Lundén, E., Reiersen, L.O., Wilson, S. 2018. Pharmaceuticals and personal care products (PPCPs) in Arctic environments: indicator contaminants for assessing local and remote anthropogenic sources in a pristine ecosystem in change. *Environmental Science and Pollution Research* 25(33):33001–33013.

Kar, S., Sanderson, H., Roy, K., Benfenati, E., Leszczynski, J. 2020. Ecotoxicological assessment of pharmaceuticals and personal care products using predictive toxicology approaches. *Green Chemistry* 22:148–1516.

Kleinert, C., Lacaze, E., Fortier, M., Hammill, M., De Guise, S., Fournier, M. 2018. T lymphocyte-proliferative responses of harbor seal (*Phoca vitulina*) peripheral blood mononuclear cells (PBMCs) exposed to pharmaceuticals *in vitro*. *Marine Pollution Bulletin* 127:225–234.

Kleywegt, S., Smyth, S.-A., Parrott, J., Schaefer, K., Lagacé, E., Payne, M., Topp, E., Beck, A., McLaughlin, A., Ostapyk, K. 2007. Pharmaceuticals and personal care products in the Canadian environment: research and policy directions. NWRI Scientific Assessment Report Series No.8. 53 p.

Kolpin, D.W., Furlong, E.T., Meyer, M.T., Thurman, E.M., Zaugg, S.D., Barber, L.B., Buxton, H.T. 2002. Pharmaceuticals, hormones, and other organic wastewater contaminants in U.S. streams, 1999–2000: a national reconnaissance. *Environmental Science & Technology* 36(6):1202–1211.

Kuzmanovic, M., Ginebreda, A., Petrovic, M., Barceló, D. 2015. Risk assessment based prioritization of 200 organic micropollutants in 4 Iberian rivers. *Science of the Total Environment* 503–504:289–299.

Lagesson, A., Fahlman, J., Brodin, T., Fick, J., Jonsson, M., Byström, P., Klaminder, J. 2016. Bioaccumulation of five pharmaceuticals at multiple trophic levels in an aquatic food web—insights from a field experiment. *Science of the Total Environment* 568:208–215.

Langford, K.H., Thomas, K.V. 2008. Inputs of chemicals from recreational activities into the Norwegian coastal zone. *Journal of Environmental Monitoring* 10:894–898.

Laville, N., Aït-Aïssa, S., Gomez, E., Casellas, C., Porcher, J.M. 2004. Effects of human pharmaceuticals on cytotoxicity, EROD activity and ROS production in fish hepatocytes. *Toxicology* 196(1–2):41–55.

Li, Z.H., Zlabek, V., Velisek, J., Grabic, R., Machova, J., Kolarova, J., Li, P., Randak, T. 2011. Acute toxicity of carbamazepine to juvenile rainbow trout (*Oncorhynchus mykiss*): effects on antioxidant responses, hematological parameters and hepatic EROD. *Ecotoxicology and Environmental Safety* 74(3):319–327.

Liang, X., Chen, B., Nie, X., Shi, Z., Huang, X., Li, X. 2013. The distribution and partitioning of common antibiotics in water and sediment of the Pearl River Estuary, South China. *Chemosphere* 92(11):1410–1416.

Liu, J.-L., Wong, M.-H. 2013. Pharmaceuticals and personal care products (PPCPs): a review on environmental contamination in China. *Environment International* 59:208–224.

Liu, N., Jin, X., Feng, C., Wang, Z., Wu, F., Johnson, A.C., Xiao, H., Hollert, H., Giesy, J.P. 2020a. Ecological risk assessment of fifty pharmaceuticals and personal care products (PPCPs) in Chinese surface waters: a proposed multiple-level system. *Environment International* 136:105454.

Liu, N., Jin, X., Yan, Z., Luo, Y., Feng, C., Fu, Z., Tang, Z., Wu, F., Giesy, J.P. 2020b. Occurrence and multiple-level ecological risk assessment of pharmaceuticals and personal care products (PPCPs) in two shallow lakes of China. *Environmental Sciences Europe* 32:69.

Liu, S., Wang, C., Wang, P., Chen, J., Wang, X., Yuan, Q. 2020c. Anthropogenic disturbances on distribution and sources of pharmaceuticals and personal care products throughout the Jinsha River Basin, China. *Environmental Research* 198:110449.

Long, E.R., Dutch, M., Weakland, S., Chandramouli, B., Benskin, J.P. 2013. Quantification of pharmaceuticals, personal care products, and perfluoroalkyl substances in the marine sediments of Puget Sound, Washington, U.S.A. *Environmental Toxicology and Chemistry* 32(8):1701–1710.

Mackay, D., Barnthouse, L. 2010. Integrated risk assessment of household chemicals and consumer products: addressing concerns about Triclosan. *Integrated Environmental Assessment and Management* 6(3):390–392.

Maranho, L.A., André, C., DelValls, T.A., Gagné, F., Martín-Díaz, M.L. 2015. Toxicological evaluation of sediment samples spiked with human pharmaceutical products: energy status and neuroendocrine effects in marine polychaetes *Hediste diversicolor*. *Ecotoxicology and Environmental Safety* 118:27–36.

Martin-Diaz, L., Franzellitti, S., Buratti, S., Valbonesi, P., Capuzzo, A., Fabbri, E. 2009. Effects of environmental concentrations of the antiepilectic drug carbamazepine on biomarkers and cAMP-mediated cell signaling in the mussel *Mytilus galloprovincialis*. *Aquatic Toxicology* 94(3):177–185.

Michael, I., Rizzo, L., McArdell, C.S., Manaia, C.M., Merlin, C., Schwartz, T., Dagot, C., Fatta Kassinos, D. 2013. Urban wastewater treatment plants as hotspots for the release of antibiotics in the environment: a review. *Water Research* 47:957–995.

Mikulic, M. 2020. *Statistics & facts, global pharmaceutical, industry*. Hamburg: Statista.

Monteiro, S.C., Boxall, A.B.A. 2009. Factors affecting the degradation of pharmaceuticals in agricultural soils. *Environmental Toxicology and Chemistry* 28:2546–2554.

Monteiro, S.C., Boxall, A.B.A., 2010. Occurrence and fate of human pharmaceuticals in the environment. *In* Whitacre, D.M. (Ed.) *Reviews of environmental contamination and toxicology*. New York, NY: Springer, pp. 53–154.

Mottaleb, M.A., Meziani, M.J., Matin, M.A., Arafat, M.M., Wahab, M.A. 2015. Emerging micro-pollutants pharmaceuticals and personal care products (PPCPs) contamination concerns in aquatic organisms—LC/MS and GC/MS analysis. *In* Kurwadkar, S., Zhang, X., Ramirez, D., Mitchell, F.L. (Eds.) *Emerging micro-pollutants in the environment: occurrence, fate, and distribution; A.C.S. Symposium Series*. Washington, DC: American Chemical Society.

Mottaleb, M.A., Usenko, S., O'Donnell, J.G., Ramirez, A.J., Brooks, B.W., Chambliss, C.K. 2009. Gas chromatography-mass spectrometry screening methods for select U.V. filters, synthetic musks, alkylphenols, an anti-microbial agent, and an insect repellent in fish. *Journal of Chromatography A* 1216:815–823.

Na, G., Fang, X., Cai, Y., Ge, L., Zong, H., Yuan, X., Yao, Z., Zhang, Z. 2013. Occurrence, distribution, and bioaccumulation of antibiotics in coastal environment of Dalian, China. *Marine Pollution Bulletin* 69(1):233–237.

Nakata, H. 2005. Occurrence of synthetic musk fragrances in marine mammals and sharks from Japanese coastal waters. *Environmental Science & Technology* 39(10):3430–3434.

Nakata, H., Shinohara, R.-I., Nakazawa, Y., Isobe, T., Sudaryanto, A., Subramanian, A., Tanabe, S., Zakaria, M.P., Zheng, G.J., Lam, P.K.S., Kim, E.Y., Min, B.-Y., We, S.-U., Viet, P.H., Tana, T.S., Prudente, M., Frank, D., Lauenstein, G., Kannan, K. 2012. Asia-Pacific mussel watch for emerging pollutants: distribution of synthetic musks and benzotriazole U.V. stabilizers in Asian and U.S. coastal waters. *Marine Pollution Bulletin* 64:2211–2218.

Nentwig, G. 2007. Effects of pharmaceuticals on aquatic invertebrates. Part II: the anti-depressant drug fluoxetine. *Archives of Environmental Contamination and Toxicology* 52:163–170.

Nikolaou, A., Meric, S., Fatta, D. 2007. Occurrence patterns of pharmaceuticals in water and wastewater environments. *Analytical and Bioanalytical Chemistry* 387:1225–1234.

Ofoegbu, P.U., Lourenço, J., Mendo, S., Soares, A., Pestana, J. 2019. Effects of low concentrations of psychiatric drugs (carbamazepine and fluoxetine) on the freshwater planarian, *Schmidtea mediterranea*. *Chemosphere* 217:542–549.

Oluwole, A.O., Omotola, E.O., Olatunji, O.S. 2020. Pharmaceuticals and personal care products in water and wastewater: a review of treatment processes and use of photocatalyst immobilized on functionalized carbon in A.O.P. degradation. *BMC Chemistry* 14(1):62.

Osenbrück, K., Gläser, H.-R., Knöller, K., Weise, S.M., Möder, M., Wennrich, R., Schirmer, M., Reinstorf, F., Busch, W., Strauch, G. 2007. Sources and transport of selected organic micropollutants in urban groundwater underlying the city of Halle (Saale), Germany. *Water Research* 41(15):3259–3270.

Osorio, V., Larrañaga, A., Aceña, J., Pérez, S., Barceló, D. 2016. Concentration and risk of pharmaceuticals in freshwater systems are related to the population density and the livestock units in Iberian Rivers. *Science of the Total Environment* 540:267–277.

Pagano, M., Savoca, S., Impellitteri, F., Albano, M., Capillo, G., Faggio, C. 2022. Toxicological evaluation of acetylsalicylic acid in non-target organisms: chronic exposure on *Mytilus galloprovincialis* (Lamarck, 1819). *Frontiers in Physiology* 13:920952.

Pan, B., Xing, B. 2011. Pharmaceuticals and personal care products in soils and sediments. *In* Xing, B., Senesi, N., Huang, P. (Eds.) *Biophysicochemical processes of anthropogenic organic compounds in environmental systems*. Medford, MA: Wiley, pp. 185–213.

Papageorgiou, M., Zioris, I., Danis, T., Bikiaris, D., Lambropoulou, D. 2019. Comprehensive investigation of a wide range of pharmaceuticals and personal care products in urban and hospital wastewaters in Greece. *Science of the Total Environment* 694:133565.

Peck, A.M. 2006. Analytical methods for the determination of persistent ingredients of personal care products in environmental matrices. *Analytical and Bioanalytical Chemistry* 386:907–939.

Pedersen, J.A., Soliman, M., Suffet, I.H. 2005. Human pharmaceuticals, hormones, and personal care product ingredients in runoff from agricultural fields irrigated with treated wastewater. *Journal of Agricultural and Food Chemistry* 53(5):1625–1632.

Pei, S., Li, B., Wang, B., Liu, J., Song, X. 2022. Distribution and ecological risk assessment of pharmaceuticals and personal care products in sediments of North Canal, China. *Water* 14:1999.

Pérez, S., Barceló, D. 2007. Application of advanced MS techniques to analysis and identification of human and microbial metabolites of pharmaceuticals in the aquatic environment. *Trends in Analytical Chemistry* 26(6):494–514.

Péry, A.R., Gust, M., Vollat, B., Mons, R., Ramil, M., Fink, G., Ternes, T., Garric, J. 2008. Fluoxetine effects assessment on the life cycle of aquatic invertebrates. *Chemosphere* 73(3):300–304.

Petrovic, M., Škrbic, B., Živancev, J., Ferrando-Climent, L., Barcelo, D. 2014. Determination of 81 pharmaceutical drugs by high performance liquid chromatography coupled to mass spectrometry with hybrid triple quadrupole-linear ion trap in different types of water in Serbia. *Science of the Total Environment* 468–469:415–4128.

Pires, A., Almeida, Â., Calisto, V., Schneider, R.J., Esteves, V.I., Wrona, F.J., Soares, A.M., Figueira, E., Freitas, R. 2016a. *Hediste diversicolor* as bioindicator of pharmaceutical pollution: results from single and combined exposure to carbamazepine and caffeine. *Comparative Biochemistry and Physiology C: Toxicology & Pharmacology* 188:30–38.

Pires, A., Almeida, Â., Correia, J., Calisto, V., Schneider, R.J., Esteves, V.I., Soares, A.M., Figueira, E., Freitas, R. 2016b. Long-term exposure to caffeine and carbamazepine: impacts on the regenerative capacity of the polychaete *Diopatra neapolitana*. *Chemosphere* 146:565–573.

Pivetta, G., do Carmo Cauduro Gastaldini, M. 2019. Presence of emerging contaminants in urban water bodies in southern Brazil. *Journal of Water and Health* 17(2):329–337.

Plhalova, L., Sehonova, P., Blahova, J., Doubkova, V., Tichy, F., Faggio, C., Berankova, P., Svobodova, Z. 2020. Evaluation of tramadol hydrochloride toxicity to juvenile zebrafish—morphological, antioxidant and histological responses. *Applied Sciences* 10:2349.

Porretti, M., Arrigo, F., Di Bella, G., Faggio, C. 2022. Impact of pharmaceutical products on zebrafish: an effective tool to assess aquatic pollution. *Comparative Biochemistry and Physiology C: Toxicology & Pharmacology* 261:109439.

Prichard, E., Granek, E.F. 2016. Effects of pharmaceuticals and personal care products on marine organisms: from single-species studies to an ecosystem-based approach. *Environmental Science and Pollution Research* 23:22365–22384.

Pusceddu, F.H., Choueri, R.B., Pereira, C.D.S., Cortez, F.S., Santos, D.R.A., Moreno, B.B., Santos, A.R., Rogero, J.R., Cesar, A. 2018. Environmental risk assessment of Triclosan and ibuprofen in marine sediments using individual and sub-individual endpoints. *Environmental Pollution* 232:274–283.

Qin, Q., Chen, X., Zhuang, J. 2015. The fate and impact of pharmaceuticals and personal care products in agricultural soils irrigated with reclaimed water. *Critical Reviews in Environmental Science and Technology* 45:1379–1408.

Raju, S., Sivamurugan, M., Gunasagaran, K., Subramani, T., Natesan, M. 2018. Preliminary studies on the occurrence of nonylphenol in the marine environments, Chennai-a case study. *JoBAZ* 79:52.

Ren, B., Geng, J., Wang, Y., Wang, P. 2021. Emission and ecological risk of pharmaceuticals and personal care products affected by tourism in Sanya City, China. *Environmental Geochemistry and Health* 43(8):3083–3097.

Reyes, N.J.D.G., Geronimo, F.K.F., Yano, K.A.V., Guerra, H.B., Kim, L.-H. 2021. Pharmaceutical and personal care products in different matrices: occurrence, pathways, and treatment processes. *Water* 13:1159.

Richardson, B.J., Lam, P.K.S., Martin, M. 2005. Emerging chemicals of concern: pharmaceuticals and personal care products (PPCPs) in Asia, with particular reference to Southern China. *Marine Pollution Bulletin* 50(9):913–920.

Richardson, S.D., Ternes, T.A. 2018. Water analysis: emerging contaminants and current issues. *Analytical Chemistry* 90(1):398–428.

Richmond, E.K., Grace, M.R., Kelly, J.J., Reisinger, A.J., Rosi, E.J., Walters, D.M. 2017. Pharmaceuticals and personal care products (PPCPs) are ecological disrupting compounds (EcoDC). *Elementa: Science of the Anthropocene* 5:66.

Rico, A., Van den Brink, P.J. 2014. Probabilistic risk assessment of veterinary medicines applied to four major aquaculture species produced in Asia. *Science of the Total Environment* 468–469:630–641.

Ruan, Y., Lin, H., Zhang, X., Wu, R., Zhang, K., Leung, K.M.Y., Lam, J.C.W.P., Lam, K.S. 2020. Enantiomer-specific bioaccumulation and distribution of chiral pharmaceuticals in a subtropical marine food web. *Journal of Hazardous Materials* 394:122589.

Ruhí, A., Acuña, V., Barceló, D., Huerta, B., Mor, J.R., Rodríguez-Mozaz, S., Sabater, S. 2016. Bioaccumulation and trophic magnification of pharmaceuticals and endocrine disruptors in a Mediterranean river food web. *Science of the Total Environment* 540:250–259.

Schwarzenbach, R.P., Escher, B.I., Fenner, K., Hofstetter, T.B., Johnson, C.A., von Gunten, U., Wehrli, B. 2006. The challenge of micropollutants in aquatic systems. *Science* 313:1072–1077.

Sehonova, P., Plhalova, L., Blahova, J., Doubkova, V., Prokes, M., Tichy, F., Fiorino, E., Faggio, C., Svobodova, Z. 2017. Toxicity of naproxen sodium and its mixture with tramadol hydrochloride on fish early life stages. *Chemosphere* 188:414–423.

Sehonova, P., Svobodova, Z., Dolezelova, P., Vosmerova, P., Faggio, C. 2018. Effects of waterborne antidepressants on non-target animals living in the aquatic environment: a review. *Science of the Total Environment* 631–632:789–794.

Sharma, S., Dar, O.I., Singh, K., Kaur, A., Faggio, C. 2021. Triclosan elicited biochemical and transcriptomic alterations in *Labeo rohita* larvae. *Environmental Toxicology and Pharmacology* 88:103748.

Sharma, V.K., Anquandah, G.A.K., Yngard R.A., Kim, H., Fekete, J., Bouzek, K., Ray, A.K., Golovko, D. 2009. Nonylphenol, octylphenol, and bisphenol-A in the aquatic environment: a review on occurrence, fate, and treatment. *Journal of Environmental Science and Health, A: Toxic/Hazardous Substances and Environmental Engineering* 44:423–442.

Sharma, V.K., Liu, F., Tolan, S., Sohn, M., Kim, H., Oturan, M.A. 2013. Oxidation of β-lactam antibiotics by ferrate(VI). *Chemical Engineering Journal* 221:446–451.

Silva, A.K., Amador, J., Cherchi, C., Miller, S.M., Morse, A.N. Pellegrin, M.-L. Wells, M.J.M. 2013. Emerging pollutants—part I: occurrence, fate and transport. *Water Environment Research* 85(10):1978–2021.

Snow, D.D., Cassada, D.A., Biswas, S., Malakar, A., D'Alessio, M., Marshall, A.H.L., Sallach, J.B. 2020. Detection occurrence, and fate of emerging contaminants in agricultural environments. *Water Environment Research* 92(10):1741–1750.

Srain, H.S., Beazley, K.F., Walker, T.R. 2021. Pharmaceuticals and personal care products and their sublethal and lethal effects in aquatic organisms. *Environmental Research* 29(2):142–181.

Subedi, B., Du, B., Chambliss, C.K., Koschorreck, J., Rüdel, H., Quack, M., Brooks, B.W., Usenko, S. 2012. Occurrence of pharmaceuticals and personal care products in German fish tissue: a national study. *Environmental Science & Technology* 46(16):9047–9054.

Sui, Q., Cao, X., Lu, S., Zhao, W., Qiu, Z., Yu, G. 2015. Occurrence, sources and fate of pharmaceuticals and personal care products in the groundwater: a review. *Emerging Contaminants* 1:14–24.

Tierney, A.J. 2001. Structure and function of invertebrate 5-HT receptors: a review. *Comparative Biochemistry and Physiology A* 128:791–804.

Tölgyesi, A., Verebey, Z., Sharma, V.K., Kovacsics, L., Fekete, J. 2010. Simultaneous determination of corticosteroids, androgens, and progesterone in river water by liquid chromatography-tandem mass spectrometry. *Chemosphere* 78:972–979.

Tong, L., Qin, L., Xie, C., Liu, H., Wang, Y., Guan, C., Huang, S. 2017. Distribution of antibiotics in alluvial sediment near animal breeding areas at the Jianghan Plain, Central China. *Chemosphere* 186:100–107.

Valdersnes, S., Kallenborn, R., Sydnes, L.K. 2006. Identification of several Tonalide (R) transformation products in the environment. *International Journal of Environmental Analytical Chemistry* 86(7):461–471.

Van Wieren, E.M., Seymour, M.D., Peterson, J.W. 2012. Interaction of the fluoroquinolone antibiotic, ofloxacin, with titanium oxide nanoparticles in water: adsorption and breakdown. *Science of the Total Environment* 441:1–9.

Verlicchi, P., Galletti, A., Petrovic, M., Barceló, D. 2010. Hospital effluents as a source of emerging pollutants: an overview of micropollutants and sustainable treatment options. *Journal of Hydrology* 389(3):416–428.

Vernouillet, G., Eullaffroy, P., Lajeunesse, A., Blaise, C., Gagné, F., Juneau, P. 2010. Toxic effects and bioaccumulation of carbamazepine evaluated by biomarkers measured in organisms of different trophic levels. *Chemosphere* 80(9):1062–1068.

Vidal-Dorsch, D.E., Bay, S.M., Maruya, K., Snyder, S.A., Trenholm, R.A., Vanderford, B.J. 2012. Contaminants of emerging concern in municipal wastewater effluents and marine receiving water. *Environmental Toxicology and Chemistry* 31(12):2674–2682.

Walker, T.R., Grant, J. 2015. Metal(loid)s in sediment, lobster and mussel tissues near historical gold mine sites. *Marine Pollution Bulletin* 101(1):404–408.

Walker, T.R., MacAskill, D., Weaver, P. 2013. Legacy contaminant bioaccumulation in rock crabs in Sydney Harbour during remediation of the Sydney Tar Ponds, Nova Scotia, Canada. *Marine Pollution Bulletin* 77(1–2):412–417.

Wang, Y., Yin, T., Kelly, C., Gin, K.Y.-H. 2019. Bioaccumulation behaviour of pharmaceuticals and personal care products in a constructed wetland. *Chemosphere* 222:275–285.

Weigel, S., Kuhlmann, J., Huhnerfuss, H., 2002. Drugs and personal care products as ubiquitous pollutants: occurrence and distribution of clofibric acid, caffeine and DEET in the North Sea. *Science of the Total Environment* 295:131–141.

Wilkinson, J.L., Hooda, P.S., Swinden, J., Barker, J., Barton, S. 2018. Spatial (bio)accumulation of pharmaceuticals, illicit drugs, plasticisers, perfluorinated compounds and metabolites in river sediment, aquatic plants and benthic organisms. *Environmental Pollution* 234:864–875.

World Health Organization. 2012. Pharmaceuticals in drinking-water. www.who.int/publications/i/item/9789241502085 [Accessed August 2, 2022].

Wu, H., Gai, Z., Guo, Y., Li, Y., Hao, Y., Lu, Z.N. 2020. Does environmental pollution inhibit urbanization in China? A new perspective through Residents' medical and health costs. *Environmental Research* 182:109128.

Wu, X., Conkle, J.L., Ernst, F., Gan, J. 2014. Treated wastewater irrigation: uptake of pharmaceutical and personal care products by common vegetables under field conditions. *Environmental Science & Technology* 48(19):11286–11293.

Wu, X., Dodgen, L.K., Conkle, J.L., Gan, J. 2015. Plant uptake of pharmaceutical and personal care products from recycled water and biosolids: a review. *Science of the Total Environment* 536:655–666.

Xie, H., Hao, H., Xu, N., Liang, X., Gao, D., Xu, Y., Gao, Y., Tao, H., Wong, M. 2019. Pharmaceuticals and personal care products in water, sediments, aquatic organisms, and fish feeds in the Pearl River Delta: occurrence, distribution, potential sources, and health risk assessment. *Science of the Total Environment* 659:230–239.

Xie, J., Liu, Y., Wu, Y., Li, L., Fang, J., Lu, X. 2022. Occurrence, distribution and risk of pharmaceutical and personal care products in the Haihe River sediments, China. *Chemosphere* 302:134874.

Xie, Z., Lu, G., Liu, J., Yan, Z., Ma, B., Zhang, Z., Chen, W. 2015. Occurrence, bioaccumulation, and trophic magnification of pharmaceutically active compounds in Taihu Lake, China. *Chemosphere* 138:140–147.

Xin, X., Huang, G., Zhang, B. 2021. Review of aquatic toxicity of pharmaceuticals and personal care products to algae. *Journal of Hazardous Materials* 410:124619.

Xu, X., Xu, Y., Xu, Nan, Pan, B., Ni, J. 2022. Pharmaceuticals and personal care products (PPCPs) in water, sediment and freshwater mollusks of the Dongting Lake downstream the Three Gorges Dam. *Chemosphere* 301:134721.

Yang, H., Lu, G., Yan, Z., Liu, J., Dong, H., Bao, X., Zhang, X., Sun, Y. 2020. Residues, bioaccumulation, and trophic transfer of pharmaceuticals and personal care products in highly urbanized rivers affected by water diversion. *Journal of Hazardous Materials* 391:122245.

Yuan, X., Hu, J., Li, S., Yu, M. 2020. Occurrence, fate, and mass balance of selected pharmaceutical and personal care products (PPCPs) in an urbanized river. *Environmental Pollution* 266:115340.

Zeng, X., Mai, B., Sheng, G., Luo, X., Shao, W., An, T., Fu, J., 2008. Distribution of polycyclic musks in surface sediments from the Pearl River Delta and Macao coastal region, South China. *Environmental Toxicology and Chemistry* 27:18–23.

Zhao, J.L., Zhang, Q.Q., Chen, F., Wang, L., Ying, G.G., Liu, Y.S., Yang, B., Zhou, L.J., Liu, S., Su, H.C., Zhang, R.Q. 2013. Evaluation of Triclosan and triclocarban at river basin scale using monitoring and modeling tools: implications for controlling of urban domestic sewage discharge. *Water Research* 47(1):395–405.

Zicarelli, G., Multisanti, C.R., Falco, F., Faggio, C. 2022. Evaluation of toxicity of Personal Care Products (PCPs) in freshwaters: Zebrafish as a model. *Environmental Toxicology and Pharmacology* 94:103923.

Zou, S., Xu, W., Zhang, R., Tang, J., Chen, Y., Zhang, G., 2011. Occurrence and distribution of antibiotics in coastal water of the Bohai Bay, China: impacts of river discharge and aquaculture activities. *Environmental Pollution* 159:2913–2920.

4 Presence and Distribution of Pharmaceuticals Residue in Food Products and the Food Chain

Pavla Lakdawala, Jana Blahova, and Caterina Faggio

CONTENTS

4.1 INTRODUCTION

In the last twenty years, thousands of ecotoxicological studies have reported pharmaceutical products being released and distributed in oceans, rivers, lakes, streams, and underground water around the world at concentrations ranging from ng/L to µg/L with typically higher values in freshwaters compared to marine waters (Mezzelani et al., 2018). Presence of these micropollutants has been well documented in soils and sludges. They are ubiquitous in the aquatic environment mainly due to their extensive use in human medicine, agricultural practice and animal husbandry, and inefficient removal by wastewater treatment plants (WWTPs). Conventional WWTPs have limited technologies and therefore removal has a large variation with a range from 0 to 100%. Thus, high concentrations of these growing environmental pollutants are not only commonly detected in influent water to wastewater treatment plants, but can also be analysed in effluents at harmful concentrations (Miller et al., 2018; Patel et al., 2019). Alarming are also recent findings of pharmaceutical residues in areas of Antarctica, which is acknowledged as one of the few remaining untouched and uncontaminated places in the world (Emnet et al., 2015; Hermandez et al., 2019; Olalla et al., 2020). Pharmaceutical fate is influenced by both physicochemical properties of the compound (e.g. water solubility, lipophilicity) as well as external factors, such as water temperature, sunlight, or pH. It can be clearly stated that pharmaceutical residues in the aquatic ecosystem are nowadays evaluated as emerging contaminants of great environmental concern (Zenker et al., 2014).

Pharmacological residues are not only commonly found in almost all abiotic matrices of aquatic ecosystems, but are also abundant in many biotic matrices. Since they are often designed to cross biological membranes, rate of uptake and internal body concentration are critical to monitor (Miller et al., 2018). Aquatic organisms in contaminated environments, as non-target species, are exposed to bioactive pharmacological substances via uptake from the water by gill or/and skin, via uptake

DOI: 10.1201/9781003361091-4

of suspended particles (ingestion), via the contaminated diet or the combined uptake (e.g. via all routes). However, compared to traditional persistent organic pollutants (e.g. dioxins, polychlorinated biphenyls) they are more water soluble and therefore have lower predisposition to accumulate. Their increasing lipophilicity also increases the likelihood to accumulate in non-target aquatic organisms. After taken up, they may store in organisms, especially in fatty tissues and thy concentrations decrease by depuration mechanisms such as gill and gastrointestinal elimination, organism growth and biotransformation. On the other hand, many pharmaceuticals are fully excreted after removal of the source of contamination (Van der Oost et al., 2003; Zenker et al., 2014).

4.2 INPUT OF PHARMACEUTICALS INTO THE AQUATIC FOOD CHAIN

In the aquatic ecosystem, organisms are intricately interconnected through food chains and food webs which describe the structure of communities and present interactions between species. The food chain presents a linear sequence of aquatic organisms through which nutrients and energy pass as one organism eats another (Figure 4.1) (Arnot and Gobas, 2004). Water and aquatic sediments act as an important sink for various pharmaceuticals and might also be significant sources of contamination in aquatic food chain. In general, chemicals reach higher concentrations in aquatic organisms that are higher up the food chain (Xie et al., 2017; Boström, 2019).

At the base of the food chain lie the primary producers. These autotrophic organisms (such as micro/macroalgae, plants) obtain nutrition from inorganic materials and sunlight energy. They provide materials and energy for higher trophic levels. Thus, they are an essential foundation of aquatic food webs. Even small disruptions to autotrophic organisms may lead to tremendous effects on the whole ecosystem. In addition, autotrophic organisms play a major role in the global carbon cycle and alleviating climate change. Based on the fact that they play a vital role in aquatic organisms, unexpected effects of pharmaceuticals, or other contaminants, would affect the life of higher trophic organisms through the food web (Xin et al., 2021). The organisms that eat the primary producers are called primary consumers, and they are usually herbivores (e.g. zooplankton, snails, mussels). They are also often exposed to sediment contaminated with pharmaceuticals via ingestion of sediment particles. Zooplankton plays a crucial ecological role in aquatic food webs, linking basal resources

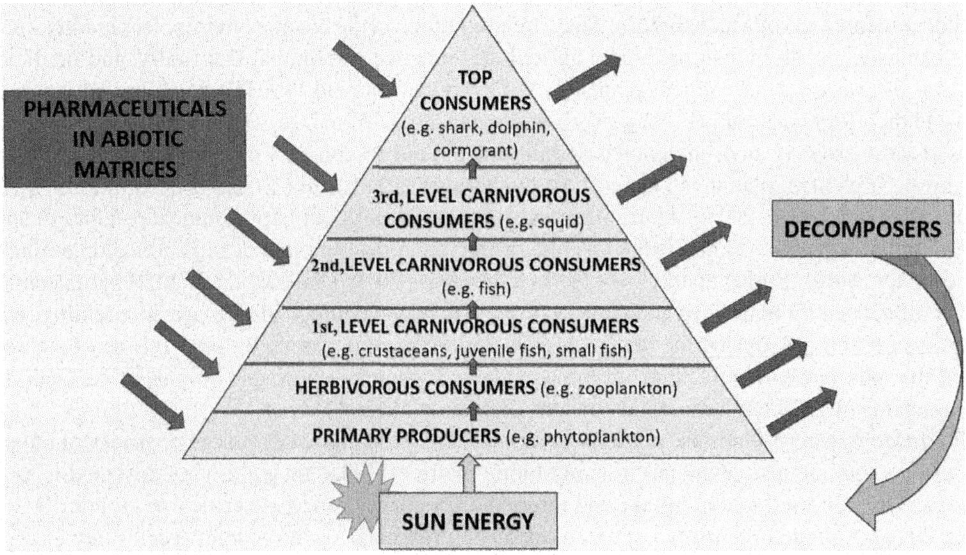

FIGURE 4.1 Input of pharmaceuticals into the aquatic food chain.

to higher-level consumers. Therefore, zooplankton can serve as a pathway for transferring contaminants to fish and other top predators. At the next trophic level, there are secondary consumers. These secondary consumers are generally carnivores (e.g. crustaceans, fish, frog, turtle). Organisms at the very top of an aquatic food chain, large predators, are called apex consumers (e.g. tuna, dolphins, sharks, seals, walruses, fish-eating birds like cormorants or pelicans). As prey consumption can be an important route of exposure to pollutants, organisms at the top of the food chain are a highly exposed group and elevated concentrations of pollutants can be found in their tissues (Brodin et al., 2014; Grabicova et al., 2015; Ding et al., 2016; Xie et al., 2017).

In general, bioaccumulation potential of pharmaceuticals can be expressed as several endpoints, including bioconcentration factor (BCF), bioaccumulation factor (BAF), or biomagnification factor (BMF) (Zenker et al., 2014). The BCF of a compound is defined as the ratio of the concentration of the chemical in the organism and in water. Uptake of chemicals in organisms from contaminated water probably follows a passive diffusion mechanism. The BAF includes the net uptake from an aquatic ecosystem by all possible routes from any sources. It can be calculated on a total organism basis or normalised to the lipid content in tissues. Biomagnification can be defined as a condition where the chemical concentration of various contaminants in an organism exceeds the concentration of its food when the major exposure route occurs from the organism's diet (Drouillard, 2008). Factors BCF, BAF, and BMF greater than 1 indicate that the concentration of pollutant in the aquatic organism is greater than that of the water, concentration in the organism is greater than that of environment from which the pharmaceutical was taken (e.g. water, diet, and sediment), or concentration in the tissue is greater than the tissue content of its prey, respectively (van der Oost et al., 2003). Some ecotoxicological studies also use other parameters such as biota-sediment accumulation factor (BSAF) or trophic magnification factor (TMF). The biota-sediment accumulation factor (BSAF) is a useful parameter for understanding the partitioning of pharmaceutical contamination from sediment to benthic organisms. However, field-based BSAF data for pharmaceuticals are still limited (Xie et al., 2017). The extent of biomagnification for a given contaminant in a food web can be quantified by establishing the trophic magnification factor (TMF). This index is calculated from the slope of a regression of the logarithmically transformed normalised contaminant concentrations against trophic levels of the organisms. Compared with other bioaccumulation metrics such as BCF or BAF, the TMF is potentially more robust to errors because it is not prone to uncertainty in measurement of chemical concentrations in abiotic media (McLeod et al., 2015; Xie et al., 2017).

The bioaccumulation capacity of pharmaceuticals in aquatic organisms displays positive relationships with the chemical properties of chemicals, such as $\log K_{ow}$, $\log D_{ow}$ (pH-dependent K_{ow}), and $\log D_{lipw}$ (liposome—water distribution coefficient) (Yang et al., 2020). It seems that for compounds with higher hydrophobicity, their bioaccumulation profiles are related to $\log K_{ow}$ and $\log D_{ow}$ values, while for most pharmaceuticals with polarity, their bioaccumulation potential is greatly affected by pK_a values. However, for some pharmaceuticals, their bioaccumulation could be affected by other environmental factors, such as photodegradation and biodegradation, which is in no obvious correlation with their physicochemical properties.

The transfer of pharmaceuticals from primary producers through zooplankton and first-level carnivorous consumers (such as crustaceans or early-life stages of fish) to higher trophic levels of carnivorous consumers has been hardly described in natural conditions; however, there are various experimental studies describing the potential of bioconcentration of various groups of pharmaceuticals. Various studies have reported that aquatic organisms at lower trophic positions (e.g. algae) seem to bioaccumulate pharmaceuticals to a greater extent than organisms at higher trophic positions. Possibly, the metabolism of pharmaceuticals at high trophic levels is the main reason for not observing biomagnification. The compounds that are slowly metabolised by organisms exhibit the highest trophic magnification in the aquatic environment. Also, other factors, such as habitat use or feeding patterns might significantly influence the bioaccumulation of pharmaceuticals (Tang et al., 2022). The trophic transfer risks are usually determined by trophic magnification factors which are calculated from the slope of logarithmically transformed concentrations of chemicals versus the trophic level of organisms in the food web (Fu et al., 2022).

Bioaccumulation potential in itself poses a significant hazard criterion because the effects of chronic exposure, at low environmentally relevant doses, may only be recognised in a later phase of life, may only be manifested in higher trophic value or have possible multi-generation effects. Bioaccumulation data can be also used to support both ecological and human health risk assessment (van der Oost et al., 2003; Arguello-Pérez et al., 2020). The most appropriate way to perform bioaccumulation studies in aquatic organisms, especially in fish, to determine BCF and BMF are toxicity tests based on OECD guideline 305 Bioaccumulation in Fish: Aqueous and Dietary Exposure (Organisation for Economic Co-operation and Development, 2017). Bioconcentration or bioaccumulation of pharmaceuticals or their metabolites under controlled laboratory conditions is documented in various organisms such as algae, zooplankton, or fish (Vernouillet et al., 2010; Valdés et al., 2016; Garcia-Galan et al., 2017; Vaclavik et al., 2020). Similarly, a number of field studies carried out around the world also indicate that these compounds tend to accumulate in the biota. Compared to classical persistent organic pollutants, available data on bioaccumulation and possible trophic transfer are still limited. The difficulty of characterising drugs in terms of bioaccumulation is compounded by the fact that they are a relatively important heterogeneous group of chemical compounds (Grabicova et al., 2015; Ruhí et al., 2016; Miller et al., 2019; Pemberthy et al., 2020; Yang et al., 2020).

4.3 PHARMACEUTICAL RESIDUES IN ORGANISMS OF LOWER TROPHIC LEVEL

Organisms of lower trophic levels are dominantly represented by primary producents and benthic organisms. Accumulation of bioactive pharmaceuticals in sediments can result in elevated contaminant concentrations in these organisms. Moreover, various benthic aquatic organisms such as snails and amphipod crustaceans are known to feed on primary producers such as aquatic plants, substrate algae, and periphyton which may contain trace amounts of potentially bioactive contaminants. Thus, benthic organisms, including plants, periphyton, various molluscs and crustaceans, are exposed to pharmaceuticals from contaminated water, sediments, and even diet (Wilkinson et al., 2018). This makes them be a possible sink for pharmaceutically active ingredients in the aquatic environment and a potential source of contamination for higher trophic level organisms. The bioaccumulation potential of individual drugs in lower trophic level aquatic organisms is very different and depends mainly on their physicochemical properties and of course on the type of aquatic organism and its position in the trophic chain. In controlled laboratory conditions, aquatic organisms such as algae or invertebrates seem to bioaccumulate pharmaceuticals to a greater extent than organisms occupying higher trophic positions (e.g. fish) (Lagesson et al., 2016; Xie et al., 2017; Mezzelani et al., 2018). Occurrence of drugs in various species of wild freshwater and marine organisms at a lower trophic level has been confirmed worldwide (Tables 4.1 and 4.2).

4.4 PHARMACEUTICAL RESIDUES IN FISH

Similarly to organisms at lower trophic levels, residues of pharmaceuticals have been widely reported in fish (Tables 4.3 and 4.4). The most commonly detected pharmaceutical substances include antibiotics, antidepressants, anti-inflammatories and analgesics, antiepileptics. In addition to the parent compounds, their metabolites can also be detected. A typical example is the widely detected norfluoxetine—the primary active metabolite of the antidepressant fluoxetine. The detected concentrations of pharmaceuticals are variable, varying between fish species and individual tissues. The highest concentrations of these contaminants are most often found in liver, where they reach up to tens of $ng.g^{-1}$. From the point of view of food safety, the occurrence of these contaminants in muscle is also an important issue, where concentrations in the order of $ng.g^{-1}$ are mainly analysed. Many toxicological studies confirmed occurrence of drugs in the kidney, brain, or bile (Du et al.,

TABLE 4.1
Occurrence of selected pharmaceutical residues in invertebrates and primary producents (Part 1)

Group	Compound	Concentration	Species	Country	Reference
Anaesthetics	Ketamine	<LOQ to 22.5 ng.g^{-1}	*Gammarus pulex*	United Kingdom	Miller et al. (2019)
Antibiotics	Clarithromycin	<LOQ to 3.9 ng.g^{-1}	Benthic organisms	Czech Republic	Grabicova et al. (2022)
	Sulfamethazine	0–430 ng.g^{-1}	*Mytilus* spp.	USA	Dodder et al. (2014)
	Oxytetracycline	11.76–16.23 µg.kg^{-1}	*Carcinus maenas*	Portugal	Fonseca et al. (2021)
	Tetracycline	8.23–8.36 µg.kg^{-1}	*Hediste diversicolor*	Portugal	Fonseca et al. (2021)
Anticonvulsants	Carbamazepine	<LOQ to 31.5 ng.g^{-1}	*Gammarus pulex*	United Kingdom	Miller et al. (2019)
	Carbamazepine	<LOQ to 34.3 ng.g^{-1}	*Asellus aquaticus*	United Kingdom	Miller et al. (2019)
	Carbamazepine	1.3±0.22 ng.g^{-1}	Perihyton	USA	Du et al. (2014)
	Carbamazepine	1.2±0.28 ng.g^{-1}	*Planorbis* sp.	USA	Du et al. (2014)
	Carbamazepine	1.3±5.3 ng.g^{-1}	*Geukensia demissa*	USA	Klosterhaus et al. (2013)
Antipsychotic and antidepressants	Citalopram	<LOQ to 42.4 ng.g^{-1}	*Gammarus pulex*	United Kingdom	Miller et al. (2019)
	Citalopram	0.13¬–6.1 ng.g^{-1}	Benthic organisms	Czech Republic	Grabicova et al. (2022)
	Risperidone	<LOQ to 22.5 ng.g^{-1}	*Gammarus pulex*	United Kingdom	Miller et al. (2019)
	Sertraline	0–5.5 ng.g^{-1}	*Mytilus* spp.	USA	Dodder et al. (2014)
	Sertraline	<LOQ to 5.6 ± 0.8 ng.g^{-1}	*Erpodella octoculata*	Czech Republic	Grabicova et al. (2015)
	Fluoxetine	2.25–9.93 ng.g^{-1}	*Mytilus galloprovincialis*	Portugal	Silva et al. (2017)

LOQ – limit of quantification

TABLE 4.2
Occurrence of selected pharmaceutical residues in aquatic invertebrates and primary producents (Part 2)

Group	Compound	Concentration	Species	Country	Reference
Cardiovascular	Metoprolol	<LOQ to 5.4 $ng.g^{-1}$	*Dreissena polymorpha*	Czech Republic	Grabicova et al. (2022)
	Propranolol	<LOQ to 45.5 $ng.g^{-1}$	*Gammarus pulex*	United Kingdom	Miller et al. (2019)
	Propranolol	31.0±2.0 $ng.g^{-1}$	*Lemna gibba*	Argentina	Mastrángelo et al. (2022)
	Telmisartan	0.67–31 $ng.g^{-1}$	Benthic organisms	Czech Republic	Grabicova et al. (2022)
	Valsartan	<LOQ to 2.3 ± $ng.g^{-1}$	*Erpodella octoculata*	Czech Republic	Grabicova et al. (2015)
	Diltiazem	0.21±0.11 $ng.g^{-1}$	*Hyalella azteca*	USA	Du et al. (2014)
	Diltiazem	0.37±0.07 $ng.g^{-1}$	*Planorbis* sp.	USA	Du et al. (2014)
Nonsteroidal anti-inflammatory drug; Analgesics	Diclofenac	1–4.6 $ng.g^{-1}$	*Mytilus galloprovincialis*	United Kingdom	Capolupo et al. (2017)
	Diclofenac	2.09 $ng.g^{-1}$ (mean value)	Periphyton	United Kingdom	Wilkinson et al. (2018)
	Diclofenac	1.4 $ng.g^{-1}$ (mean value)	*Bithynia tentaculata*	United Kingdom	Wilkinson et al. (2018)
	Diclofenac	33±13 $ng.g^{-1}$	*Erpodella octoculata*	Czech Republic	Grabicova et al. (2015)
	Paracetamol	$<$1.86–2.61 $ng.g^{-1}$	*Potamogeton sp.*	United Kingdom	Wilkinson et al. (2018)
	Ibuprofen	<LOQ to 9.39 $ng.g^{-1}$	*Mytilus gallaprovincialis*	Adriatic coast	Mezzelani et al. (2016)
	Celecoxib	26±14 $ng.g^{-1}$	*Corbicula fluminea*	USA	Du et al. (2014)
	Celecoxib	25±19 $ng.g^{-1}$	*Hyalella azteca*	USA	Du et al. (2014)

LOQ – limit of quantification

TABLE 4.3
Occurrence of selected pharmaceutical residues in various tissues of fish (Part 1)

Group	Compound	Concentration (tissue)	Species	Country	Reference
Anaesthetics	Lidocaine	3.9±0.1 µg.kg^{-1} (muscle)	*Cyprinus carpio*	Germany	Boulard et al. (2020)
	Lidocaine	3.4±0.3 µg.kg^{-1} (liver)	*Cyprinus carpio*	Germany	Boulard et al. (2020)
Antibiotics	Ciprofloxacin	28.5–96.22 µg.kg^{-1} (liver)	*Halobatrachus didactylus*	Portugal	Fonseca et al. (2021)
	Enrofloxacin	5.41–11.25 µg.kg^{-1} (liver)	*Dicentrarchus labrax* juveniles	Portugal	Fonseca et al. (2021)
	Ofloxacin	10.63–22.50 µg.kg^{-1} (liver)	*Dicentrarchus labrax* adults	Portugal	Fonseca et al. (2021)
	Tetracycline	12.05–32.39 µg.kg^{-1} (liver)	*Diplodus bellottii*	Portugal	Fonseca et al. (2021)
	Sulfamethoxazole	1.87 µg.kg^{-1} (muscle)	*Hoplias lacerdae*	Argentina	Ondarza et al. (2019)
	Erythromycin	2.9 µg.kg^{-1} (liver)	*Rhamdia quelen*	Argentina	Ondarza et al. (2019)
	Erythromycin	1.4 µg.kg^{-1} (muscle)	*Rhamdia quelen*	Argentina	Ondarza et al. (2019)
Anticonvulsants	Carbamazepine	1.8±0.27 µg.kg^{-1} (muscle)	*Gambusia affinis*	USA	Du et al. (2014)
	Carbamazepine	1.69±0.51 µg.kg^{-1} (muscle)	*Prochilodus lineatus*	Argentina	Rojo et al. (2021)
	Levomepromazine	3.1±2.2 µg.kg^{-1} (kidney)	*Salmo trutta* m. *fario*	Czech Republic	Grabicova et al. (2017)
	Levomepromazine	11±11 µg.kg^{-1} (liver)	*Salmo trutta* m. *fario*	Czech Republic	Grabicova et al. (2017)
Antipsychotic and antidepressants	Fluoxetine	36.80 µg.kg^{-1} (liver)	*Diplodus bellottii*	Portugal	Fonseca et al. (2021)
	Fluoxetine	70.31–195.40 µg.kg^{-1} (liver)	*Dicentrarchus labrax* juveniles	Portugal	Fonseca et al. (2021)
	Norfluoxetine	<LOQ–3.22 (whole body)	*Pimephales promelas*	USA	Metcalfe et al. (2010)
	Venlafaxine	Up to 1.6 µg.kg^{-1} (muscle)	various fish species	Uruguay	Rojo et al. (2019)
	Norfluoxetine	13±2.3 µg.kg^{-1} (muscle)	*Gambusia affinis*	USA	Du et al. (2014)
	Sertraline	14±4.6 µg.kg^{-1} (muscle)	*Gambusia affinis*	USA	Du et al. (2014)

LOQ – limit of quantification

TABLE 4.4
Occurrence of selected pharmaceutical residues in various tissues of fish (Part 2)

Group	Compound	Concentration (tissue)	Species	Country	Reference
	Metoprolol	Up to 2.4 μg.kg^{-1} (muscle)	various fish species	Uruguay	Rojo et. Al. (2019)
	Metoprolol	<LOQ to 0.7 μg.kg^{-1} (muscle)	*Liza aurata*	Spain	Moreno-González et al. (2016)
	Diltiazem	0.14±0.08 μg.kg^{-1} (muscle)	*Gambusia affinis*	USA	Du et al. (2014)
	Nadolol	<LOQ to 0.7 μg.kg^{-1} (muscle)	*Liza aurata*	Spain	Moreno-González et al. (2016)
Cardiovascular	Propranolol	1.6–7.1 μg.kg^{-1} (muscle)	*Carassius auratus*	China	Xie et al. (2017)
	Propranolol	3.2–17 μg.kg^{-1} (brain)	*Carassius auratus*	China	Xie et al. (2017)
	Propranolol	1.9–4.6 μg.kg^{-1} (liver)	*Carassius auratus*	China	Xie et al. (2017)
	Atenolol	53.6±12.0 μg.kg^{-1} (muscle)	*Salminus brasiliensis*	Argentina	Rojo et al. (2021)
	Enalapril	10.4±9.3 μg.kg^{-1} (muscle)	*Megaleporinus obtusidens*	Argentina	Rojo et al. (2021)
	Diclofenac	0.1–0.2 μg.kg^{-1} (muscle)	*Carassius auratus*	China	Liu et al. (2015)
	Diclofenac	1–1.6 μg.kg^{-1} (brain)	*Carassius auratus*	China	Liu et al. (2015)
	Diclofenac	8.7–29.0 μg.kg^{-1} (liver)	*Carassius auratus*	China	Liu et al. (2015)
	Diclofenac	1.8–5.9 μg.kg^{-1} (muscle)	*Hypophthalmichthys molitrix*	China	Xie et al. (2017)
Nonsteroidal	Diclofenac	5.6–16 μg.kg^{-1} (brain)	*Hypophthalmichthys molitrix*	China	Xie et al. (2017)
anti-inflammatory drug;	Diclofenac	10–27 μg.kg^{-1} (liver)	*Hypophthalmichthys molitrix*	China	Xie et al. (2017)
Analgesics	Ibuprofen	2.8–40 μg.kg^{-1} (muscle)	*Cyprinus carpio*	China	Xie et al. (2017)
	Ibuprofen	12–51 μg.kg^{-1} (brain)	*Cyprinus carpio*	China	Xie et al. (2017)
	Ibuprofen	10–48 μg.kg^{-1} (liver)	*Cyprinus carpio*	China	Xie et al. (2017)
	Tramadol	0.43–0.31 μg.kg^{-1} (kidney)	*Salmo trutta* m. *fario*	Czech Republic	Grabicova et al. (2017)
	Tramadol	2.6–0.8 μg.kg^{-1} (liver)	*Salmo trutta* m. *fario*	Czech Republic	Grabicova et al. (2017)

LOQ – limit of quantification

2014; Mezzelani et al., 2018 Ondarza et al., 2019; Pemberthy et al., 2020; Rojo et al., 2021). The relative frequency of finding different pharmaceutically active compounds vary depending upon the country, region, area consumption pattern, and manufacturing industry locations as well. Also surprising are the frequent findings of pharmacological residues in ecologically protected areas. These observations suggest that these pollutants would be released and transported into these areas from the surrounding locations (Ondarza et al., 2019).

Bioaccumulation of pharmaceuticals in fish is not only affecting organisms that are directly exposed to the chemicals but also threatens their predators and even humans. Although the majority of the previous studies report that biomagnification of most pharmaceuticals through the trophic web is not overall significant, it still seems to be species- and compound-specific (Rojo et al., 2019).

Fish primarily accumulate ionisable base pharmaceuticals by inhalation as well as via the dietary route of exposure. Therefore, it is necessary to understand bioaccumulation dynamics of pharmaceuticals among fish species and other biota across trophic levels (Grabicova et al., 2022). For example, ibuprofen and diclofenac are hydrophilic (log K_{ow}=1.90 for diclofenac and 2.48 for ibuprofen) (Scheytt et al., 2005), which does not favour their bioaccumulation in fish. It is important to mention that bioaccumulation is the result of the competing process of all routes of uptake and elimination, which means that contaminants can be directly accumulated by the liver but not by muscle (Lahti et al., 2011). Therefore, some pollutants can be directly accumulated by the liver and be partially or completely metabolised instead of be accumulated by muscles (Escarrone et al., 2016).

4.5 PHARMACEUTICAL RESIDUES IN ANIMALS VITALLY CONNECTED WITH AQUATIC ENVIRONMENT OR SPECIES FEEDING ON AQUATIC BIOTA

Aquatic insects connect aquatic and terrestrial food webs as their life cycle includes aquatic and terrestrial life stages, linking aquatic and terrestrial habitat for energy and nutrient flow. However, unfortunately, this linkage also enables contaminant transfer from the aquatic to the terrestrial environment (Veseli et al., 2022). Consequently, even amphibians, reptiles, birds, or mammals might be exposed to pharmaceuticals coming from contaminated food sources originating in the aquatic habitat (e.g. fish, insects).

Among the riparian and terrestrial organisms, some species might be more at risk than others. For example, due to the negative effects of anthropogenic activities (e.g. habitat destruction, pollution), amphibians are the most globally threatened group of vertebrates (Egea-Serrano et al., 2012). Or, in case of mammals, bats may be even more susceptible than other mammals to the effects of low doses of bioaccumulative contaminants due to their annual life cycles requiring significant fat deposition followed by extreme fat depletion during hibernation or migration, at which time contaminants may be mobilised into the brain and other tissues (Clark and Shore, 2001).

Data on pharmaceutical residue concentrations in riparian and terrestrial organisms connected with aquatic environment are presented in Table 4.5.

4.6 SUMMARY AND CONCLUSIONS

Pharmaceutical residues have not only been reported in various water bodies globally, but also in soils, sediments, and even aquatic biota. Even though the concentrations of pharmaceuticals seem to be low, they have actually been developed to be biologically active in small doses. Unfortunately, aquatic organisms are exposed to pharmaceutically active ingredients through a combination of uptake from water (through gills), uptake of suspended particles (ingestion), and also through contaminated diet. As a result, pharmaceutical residues with various modes of action have been found in aquatic primary producents, zooplankton, fish, and even riparian and terrestrial organisms that are vitally connected with the aquatic environment. In an aquatic ecosystem, organisms are interconnected through food chains and food webs which structure the communities and present interactions between species. Therefore,

TABLE 4.5
Pharmaceuticals reported in animals vitally connected with aquatic environment or species feeding on aquatic biota

Group	Name	Concentration	Species	Country	Reference
Antibiotics	Trimethoprim	~ 3000 ng.g^{-1}	*Hydropsychidae* spp.	Australia	Richmond et al. (2018)
Anticoagulants	Warfarin	57.6 ng.g^{-1} (homogenized tissues)	Various bat species (*Myotis lucifugus, M. sodalis, M. septentrionalis, Eptesicus fuscus*)	USA	Secord et al. (2015)
Antidepressants	Citalopram	~ 6000 ng.g^{-1}	*Hydropsychidae* spp.	Australia	Richmond et al. (2018)
	Fluoxetine	~ 2000 ng.g^{-1}	*Hydropsychidae* spp.	Australia	Richmond et al. (2018)
	Sertraline	<LOQ–5.5 ± 1.6 ng.g^{-1}	*Hydropsyche* sp.	Czech Republic	Grabicova et al. (2015)
	Venlafaxine	~ 6000 ng.g^{-1}	*Hydropsychidae* spp.	Australia	Richmond et al. (2018)
Heart and bloodstream	Diltiazem	0.56–8.63 ng.ml^{-1} (plasma)	*Pandion haliaetus*	USA	Lazarus et al. (2015)
	Metoprolol	~ 2500 ng.g^{-1}	*Hydropsychidae* spp.	Australia	Richmond et al. (2018)
Nonsteroidal and anti-inflammatory drugs, Analgesics	Diclofenac	3.6–21.1 ng.g^{-1} (feathers)	*Ichtyaetus melanocephalus*	Italy	Distefano et al. (2022)
	Diclofenac	<LOD–3370 ng.l^{-1} (plasma)	*Pandion haliaetus*	USA	Bean et al. (2018)
	Ibuprofen	16.69–20.42 ng.g^{-1} (liver)	*Accipiter gentilis*	Germany	Badry et al. (2021)
	Ibuprofen	21.53–74.59 ng.g^{-1} (liver)	*Haliaeetus albicilla*	Germany	Badry et al. (2021)
	Ibuprofen	<LOQ–113.9 ng.g^{-1} (feathers)	*Ichtyaetus melanocephalus*	Italy	Distefano et al. (2022)
	Ketoprofen	40.378 ng.g^{-1} (mean value)	*Odonata*	Croatia	Veseli et al. (2022)
	Naproxen	23.866 ng.g^{-1} (mean value)	*Odonata*	Croatia	Veseli et al. (2022)
	Paracetamol	<LOD–3950 ng.l^{-1} (plasma)	*Pandion haliaetus*	USA	Bean et al. (2018)

LOQ – limit of quantification

it is necessary to understand bioaccumulation dynamics of pharmaceuticals across all trophic levels of the aquatic food chain. Bioaccumulation poses a significant hazard because the effects of chronic exposure, at low environmentally relevant doses, may only be recognised in later phases of life, may manifest only in higher trophic levels, or may have possible multi-generation effects.

The bioaccumulation of pharmaceuticals seems to be highly species- and compound-specific, with persistent pharmaceuticals accumulating in top consumers via trophic transfer, although higher bioaccumulation factors are generally found in benthic and invertebrate primary consumers. It means that, for example, aquatic species at lower trophic positions often bioaccumulate pharmaceuticals to a greater extent than fish occupying higher trophic levels. However, compounds that are slowly metabolised by organisms exhibit more significant trophic magnification. Also, other factors, such as habitat use or feeding patterns, might influence the bioaccumulation of pharmaceuticals.

REFERENCES

Arguello-Pérez, M.A., Ramírez-Ayala, E., Mendoza-Pérez, J.A., Monroy-Mendieta, M.M., Vazquez-Guevara, M., Lezama-Cervantes, C., Godinez-Dominguez, E., Silva-Batiz, F., Tintos-Gomez, A. 2020. Determination of the bioaccumulative potential risk of emerging contaminants in fish muscle as an environmental quality indicator in coastal lagoons of the Central Mexican Pacific. *Water* 12:2721.

Arnot, J.A., Gobas, F.A.P.C. 2004. A food web bioaccumulation model for organic chemicals in aquatic ecosystem. *Environmental Toxicology and Chemistry* 23:2343–2355.

Badry, A., Schenke, D., Treu, G., Krone, O. 2021. Linking landscape composition and biological factors with exposure levels of rodenticides and agrochemicals in avian apex predators from Germany. *Environmental Research* 193:110602.

Bean, T.G., Rattner, B.A., Lazarus, R.S., Day, D.D., Burket, S.R., Brooks, B.W., Haddad, S.P., Bowerman, W.W. 2018. Pharmaceuticals in water, fish and osprey nestlings in Delaware River and Bay. *Environmental Pollution* 232:533–545.

Boström, M.L. 2019. *Uptake and bioaccumulation of ionizable pharmaceuticals in aquatic organisms.* MediaTryck, Lund University. https://lucris.lub.lu.se/ws/portalfiles/portal/71376456/0_Spikfil_Marja_B.pdf

Boulard, L., Parrhysius, P., Jacobs, B., Dierkes, G., Wick, A., Buchmeier, G., Koschorreck, J., Ternes, T.A. 2020. Development of an analytical method to quantify pharmaceuticals in fish tissues by liquid chromatography-tandem mass spectrometry detection and application to environmental samples. *Journal of Chromatography A* 1633:461612.

Brodin, T., Piovano, S., Fick, J., Klaminder, J., Heynen, M., Jonsson, M. 2014. Ecological effects of pharmaceuticals in aquatic systems—impact through behavioural alterations. *Philosophical Transactions of the Royal Society* 369:20130580.

Capolupo, M., Franzellitti, S., Kiwan, A., Valbonesi, P., Dinelli, E., Pignotti, E., Birke, M., Fabbri, E. 2017. A comprehensive evaluation of the environmental quality of a coastal lagoon (Ravenna, Italy): integrating chemical and physiological analyses in mussels as a biomonitoring strategy. *Science of The Total Environment* 598:146–159.

Clark, D.R., Shore, R.F. 2001. Chiroptera. In: *Ecotoxicology of wild mammals*, ed. R.F. Shore, and B.A. Rattner, 159–214. London: Wiley.

Ding, J., Lu, G., Liu, J., Yang, H., Li, Y. 2016. Uptake, depuration, and bioconcentration of two pharmaceuticals, roxithromycin and propranolol, in *Daphnia magna. Ecotoxicology and Environmental Safety* 126:85–93.

Distefano, G.G., Zangrando, R., Basso, M., Panzarin, L., Gambaro, A., Ghirardini, A.V., Picone, M. 2022. Assessing the exposure to human and veterinary pharmaceuticals in waterbirds: the use of feathers for monitoring antidepressants and nonsteroidal anti-inflammatory drugs. *Science of The Total Environment* 821:153473.

Dodder, N.G., Maruya, K.A., Ferguson, P.L., Grace, R., Kiosterhaus, S., La Guardia, M.J., Lauenstein, G.G., Ramirez, J. 2014. Occurrence of contaminants of emerging concern in mussels (*Mytilus spp.*) along the California coast and the influence of land use, storm water discharge, and treated wastewater effluent. *Marine Pollution Bulletin* 81:340–346.

Drouillard, K.G. 2008. *Encyclopedia of ecology* (ed. B. Fath, 2nd Edition). Amsterdam: Elsevier, 353–358.

Du, B., Haddad, S.P., Luek, A., Scott, W.C., Saari, G.N., Kristofco, L.A., Connors, K.A., Rash, C., Rasmussen, J.B., Chambliss, C.K., Brooks, B.W. 2014. Bioaccumulation and trophic dilution of human pharmaceuticals across trophic positions of an effluent-dependent wadeable stream. *Philosophical Transactions of the Royal Society B: Biological Sciences* 369:20140058.

Egea-Serrano, A., Relyea, R.A., Tejedo, M., Torralva, M. 2012. Understanding of the impact of chemicals on amphibians: a meta-analytic review. *Ecology and Evolution* 2:1382–1397.

Emnet, P., Gaw, S., Northcott, G., Storey, B., Graham, L. 2015. Personal care products and steroid hormones in the Antarctic coastal environment associated with two Antarctic research stations, McMurdo Station and Scott Base. *Environmental Research* 136:331–342.

Escarrone, A.L.V., Caldas, S.S., Primel, E.G., Martins, S.E., Nery, L.E.M. 2016. Uptake, tissue distribution and depuration of triclosan in the guppy *Poecilia vivipara* acclimated to freshwater. *Science of The Total Environment* 560–561:218–224.

Fonseca, V.F., Duarte, I.A., Duarte, B., Freitas, A., Vila Pouca, A.S., Barbosa, J., Gillanders, B.M., Reis-Santos, P. 2021. Environmental risk assessment and bioaccumulation of pharmaceuticals in a large urbanized estuary. *Science of the Total Environment* 783:147021.

Fu, Q., Meyer, C., Patrick, M., Kosfeld, V., Ruedel, H., Koschorreck, J., Hollender, J. 2022. Comprehensive screening of polar emerging organic contaminants including PFASs and evaluation of the trophic transfer behavior in a freshwater food web. *Water Research* 218:118514.

Garcia-Galan, M.J., Sordet, M., Bulete, A., Garric, J., Vulliet, E. 2017. Evaluation of the influence of surfactants in the bioaccumulation kinetics of sulfamethoxazole and oxazepam in benthic invertebrates. *Science of the Total Environment* 592:554–564.

Grabicova, K., Grabic, R., Blaha, M., Kumar, V., Cerveny, D., Fedorova, G., Randak, T. 2015. Presence of pharmaceuticals in benthic fauna living in a small stream affected by effluent from a municipal sewage treatment plant. *Water Research* 72:145–153.

Grabicova, K., Grabic, R., Fedorova, G., Fick, J., Cerveny, D., Kolarova, J., Turek, J., Zlabek, V., Randak, T. 2017. Bioaccumulation of psychoactive pharmaceuticals in fish in an effluent dominated stream. *Water Research* 124:654–662.

Grabicova, K., Vojs Stanova, A., Svecova, H., Novakova, P., Kodes, V., Leontovycova, D., Brooks, B.W., Grabic, R. 2022. Invertebrates differentially bioaccumulate pharmaceuticals: implications for routine biomonitoring. *Environmental Pollution* 309:119715.

Hermandez, F., Calisto-Ulloa, N., Gomez-Fuentes, C., Gomez, M., Ferrer, J., Gonzalez-Rocha, G., Bello-Toledo, H., Botero-Coy, A.M., Boix, C., Ibanez, M., Montory, M. 2019. Occurrence of antibiotics and bacterial resistance in wastewater and sea water from the Antarctic. *Journal of Hazardous Materials* 363:447–456.

Klosterhaus, S.L., Grace, R., Hamilton, M.C., Yee, D. 2013. Method validation and reconnaissance of pharmaceuticals, personal care products, and alkylphenols in surface waters, sediments, and mussels in an urban estuary. *Environmental International* 54:92–99.

Lagesson, A., Fahlman, J., Brodin, T., Fick, J., Jonsson, M., Bystrom, P., Klaminder, J. 2016. Bioaccumulation of five pharmaceuticals at multiple trophic levels in an aquatic food web—insights from a field experiment. *Science of the Total Environment* 568:208–215.

Lahti, M., Brozinski, J.-M., Jylha, A., Kronberg, L., Oikari, A. 2011. Uptake from water, biotransformation, and biliary excretion of pharmaceuticals by rainbow trout. *Environmental Toxicology and Chemistry* 30:1403–1411.

Lazarus, R.S., Rattner, B.A., Brooks, B.W., Du, B., McGowan, P.C., Blazer, V.S., Ottinger, M.A. 2015. Exposure and food web transfer of pharmaceuticals in ospreys (*Pandion haliaetus*): predictive model and empirical data. *Integrated Environmental Assessment and Management* 11:118–129.

Liu, J., Lu, G., Xie, Z., Zhang, Z., Li, S., Yan, Z. 2015. Occurrence, bioaccumulation and risk assessment of lipophilic pharmaceutically active compounds in the downstream rivers of sewage treatment plants. *Science of the Total Environment* 511:54–62.

Mastrángelo, M.M., Valdés, M.E., Eissa, B., Ossana, N.A., Barcelo, D., Sabater, S., Rodriguez-Mozaz, S., Giorgi, A.D.N. 2022. Occurrence and accumulation of pharmaceutical products in water and biota of urban lowland rivers. *Science of the Total Environment* 828:154303.

McLeod, A.M., Arnot, J.A., Borga, K., Selck, H., Kashian, D.R., Krause, A., Paterson, G., Haffner, G.D., Drouillard, K.G. 2015. Quantifying uncertainty in the trophic magnification factor related to spatial movements of organisms in a food web. *Integrated Environmental Assessment and Management* 11:306–318.

Metcalfe, C.D., Chu, S., Judt, C., Li, H., Oakes, K.D., Servos, M.R., Andrews, D.M. 2010. Antidepressants and their metabolites in municipal wastewater, and downstream exposure in an urban watershed. *Environmental toxicology and Chemistry* 29:79–89.

Mezzelani, M., Gorbi, S., Regoli, F. 2018. Pharmaceuticals in the aquatic environments: evidence of emerged threat and future challenges for marine organisms. *Marine Environmental Research* 140:41–60.

Miller, T.H., Bury, N.R., Owen, S.F., MacRae, J.I., Barron, L.P. 2018. A review of the pharmaceutical exposome in aquatic fauna. *Environmental Pollution* 239:129–146.

Miller, T.H., Ng, K.T., Bury, S.T., Bury, S.E., Bury, N.R., Barron, L.P. 2019. Biomonitoring of pesticides, pharmaceuticals and illicit drugs in a freshwater invertebrate to estimate toxic or effect pressure. *Environment International* 129:595–606.

Moreno-González, R., Rodríguez-Mozaz, S., Huerta, B., Barceló, D., León, V.M. 2016. Do pharmaceuticals bioaccumulate in marine molluscs and fish from a coastal lagoon? *Environmental Research* 146:282–298.

Olalla, A., Moreno, L., Valcárcel, Y. 2020. Prioritisation of emerging contaminants in the northern Antarctic Peninsula based on their environmental risk. *Science of the Total Environment* 742:140147.

Ondarza, P.M., Haddad, S.P., Avigliano, E., Miglioranza, K.S.B., Brooks, B.W. 2019. Pharmaceuticals, illicit drugs and their metabolites in fish from Argentina: implications for protected areas influenced by urbanization. *Science of the Total Environment* 649:1029–1037.

Organisation for Economic Co-operation and Development (OECD). 2017. *Guidance document on aspects of OECD TG 305 on fish bioaccumulation. Series on testing and assessment, No. 264. ENV/JM/MONO(2017)16.* Paris: OECD Publishing. www.oecd.org/env/ehs/testing/1-GD-OECD-TG305-2016-04-12.pdf (accessed August 2, 2022).

Patel, M., Kumar, R., Kishor, K., Mlsna, T., Pittman, C.U., Jr., Mohan, D. 2019. Pharmaceuticals of emerging concern in aquatic systems: chemistry, occurrence, effects, and removal methods. *Chemical Reviews* 119:3510–3673.

Pemberthy, D.M., Padilla, Y., Echeverri, A., Peñuela, G.A. 2020. Monitoring pharmaceuticals and personal care products in water and fish from the Gulf of Urabá, Colombia. *Heliyon* 6:e04215.

Richmond, E.K., Rosi, E.J., Walters, D.M., Fick, J., Hamilton, S.K., Brodin, T., Sundelin, A., Grace, M.R. 2018. A diverse suite of pharmaceuticals contaminates stream and riparian food webs. *Nature Communications* 9:4491.

Rojo, M., Álvarez-Muñoz, D., Dománico, A., Foti, R., Rodriguez-Mozaz, S., Barceló, D., Carriquiriborde, P. 2019. Human pharmaceuticals in three major fish species from the Uruguay River (South America) with different feeding habits. *Environmental Pollution* 252(Part A):146–154.

Rojo, M., Cristos, D., Gonzales, P., Lopez-Aca, V., Domanico, A., Carriquiriborde, P. 2021. Accumulation of human pharmaceuticals and activity of biotransformation enzymes in fish from two areas of the lower Rio de la Plata Basin. *Chemosphere* 266:129012.

Ruhí, A., Acuna, V., Barcelo, D., Huerta, B., Mor, J.R., Rodriguez-Mozaz, S., Sabater, S. 2016. Bioaccumulation and trophic magnification of pharmaceuticals and endocrine disruptors in a Mediterranean river food web. *Science of the Total Environment* 540:250–259.

Scheytt, T., Mersmann, P., Lindstädt, R., Heberer, T. 2005. 1-Octanol/water partition coefficients of 5 pharmaceuticals from human medical care: carbamazepine, clofibric acid, diclofenac, ibuprofen, and propyphenazone. *Water, Air and Soil Pollution* 165:3–11.

Secord, A.L., Patnode, K.A., Carter, C., Redman, E., Gefell, D.J., Major, A.R., Sparks, D.W. 2015. Contaminants of emerging concern in bats from the Northeastern United States. *Archives of Environmental Contamination and Toxicology* 69:411–421.

Silva, L.J.G., Pereira, A.M.P.T., Rodrigues, H., Meisel, L.M., Lino, C.M., Pena, A. 2017. SSRIs antidepressants in marine mussels from Atlantic coastal areas and human risk assessment. *Science of the Total Environment* 603–604:118–125.

Tang, J., Zhang, J., Su, L., Jia, Y., Yang, Y. 2022. Bioavailability and trophic magnification of antibiotics in aquatic food webs of Pearl River, China: influence of physicochemical characteristics and biotransformation. *Science of the Total Environment* 820:153285.

Vaclavik, J., Sehonova, P., Medkova, D., Stastny, K., Charvatova, M., Faldyna, M., Mares, J., Svobodova, Z. 2020. High resolution mass spectrometry analysis of the sertraline residues contained in the tissues of rainbow trout reared in model experimental conditions. *Physiological Research* 69:619–625.

Valdés, M.E., Huerta, B., Wunderlin, D.A., Bistoni, M.A., Barceló, D., Rodriguez-Mozaz, S. 2016. Bioaccumulation and bioconcentration of carbamazepine and other pharmaceuticals in fish under field and controlled laboratory experiments. Evidences of carbamazepine metabolization by fish. *Science of the Total Environment* 557–588:58–67.

Van der Oost, R., Beyer, J., Vermeulen, N.P.E. 2003. Fish bioaccumulation and biomarkers in environmental risk assessment: a review. *Environmental Toxicology and Pharmacology* 13:57–149.

Vernouillet, G., Eullaffroy, P., Lajeunesse, A., Blaise, C., Gagne, F., Juneau, P. 2010. Toxic effects and bioaccumulation of carbamazepine evaluated by biomarkers measured in organisms of different trophic levels. *Chemosphere* 80:1062–1068.

Veseli, M., Rožman, M., Vilenica, M., Petrović, M., Previšić, A. 2022. Bioaccumulation and bioamplification of pharmaceuticals and endocrine disruptors in aquatic insects. *Science of the Total Environment* 838:156208.

Wilkinson, J.L., Hooda, P.S., Swinden, J., Barker, J., Barton, S. 2018. Spatial (bio)accumulation of pharmaceuticals, illicit drugs, plasticisers, perfluorinated compounds and metabolites in river sediment, aquatic plants and benthic organisms. *Environmental Pollution* 234:864–875.

Xie, Z., Lu, G., Yan, Z., Liu, J., Wang, P., Wang, Y. 2017. Bioaccumulation and trophic transfer of pharmaceuticals in food webs from a large freshwater lake. *Environmental Pollution* 222:356–366.

Xin, X., Huang, G., Zhang, B. 2021. Review of aquatic toxicity of pharmaceuticals and personal care products to algae. *Journal of Hazardous Materials* 410:124619.

Yang, H., Lu, G., Yan, Z., Liu, J., Dong, H., Bao, X., Zhang, X., Sun, Y. 2020. Residues, bioaccumulation, and trophic transfer of pharmaceuticals and personal care products in highly urbanized rivers affected by water diversion. *Journal of Hazardous Materials* 391:122245.

Zenker, A., Cicero, M.R., Prestinaci, F., Bottoni, P., Carere, M. 2014. Bioaccumulation and biomagnification potential of pharmaceuticals with a focus to the aquatic environment. *Journal of Environmental Management* 133:378–387.

5 Analysis and Detection Techniques for Pharmaceutical Residues in the Environment

Raj Kumari, Meenakshi Sharma, and Renu Daulta

CONTENTS

5.1 INTRODUCTION

Pharmaceutical residue contamination of the environment has recently been a common occurrence that also receives media attention. Pharmaceutical chemical contamination of soil can cause disruptions in soil functionality, or limit plant development. There is currently a strong amount of evidence on the harmful impacts on the environment brought on by manufacturing of medicines (Breton, 2003), which include the growth of resistance of antibiotics, the feminisation of fish and humans (Carmen Lidia Chitescu et al., 2016). However, antibiotic pollution puts everyone at danger, regardless of where they live, as a result of the way that antibiotic manufacturing emissions generate resistance in bacteria

DOI: 10.1201/9781003361091-5

that are already present in the environment, spreading to humans. Given the size of the problem it presents and the coordinated worldwide reaction needed to address it, antimicrobial resistance (AMR) is frequently compared to climate change (Pietro Bruni, 2016). Broad-spectrum antibiotic ciprofloxacin was present in concentrations as high as 31 mg/l, which is a million times higher than the typical values seen in treated municipal sewage effluents and harmful to a variety of organisms (Fick, 2009). These discharges have caused previously unheard of levels of river sediment pollutant and contamination of irrigated soils (Larsson, 2010). If manufacturing discharges are considered as a whole, a greater number of Active Pharmaceutical Ingredients (APIs) should experience unfavourable environmental concentrations. Some medications are digested after administration, while others are left unaltered until they are eliminated. These can reach the aquatic systems through excretion or wastewater disposal (Holm et al., 1995). In reality, it shows that the wastewater exiting the treatment plant contains up to 90% of the drug residue. The direct way pharmaceuticals reach the environment is by the combined excretion of active pharmacological components through urine and faeces (Daughton, 2009).

The new analytical tools can now detect pharmaceuticals at all levels present in nature, which has caused some recent concern. Recently, several medications have been linked to detrimental effects on human health and bad developmental outcomes in aquatic creatures. Furthermore, pharmaceuticals have been discovered worldwide in groundwater and in biota ranging from fish to algae, at varying amounts. However, the situation is presumably different from pro-drugs, and it might also be for the metabolites of other medications, like norfluoxetine (Grabicova et al., 2015).

5.2 PHARMACEUTICAL INFILTRATION IN THE ENVIRONMENT

Pharmaceuticals come into the atmosphere via different channels. The ingestion (excretion) route is typically cited as the most extensive route, but improper disposal of pharmaceutical waste also plays a substantial role. Unwanted animal medication waste, such as uneaten medicated fish feed from industrial aquaculture ponds, can be inappropriately disposed (Richardson and Bowron, 1985). The following are some examples of infiltration:

1. Through manufacture: industrial eluent, solutions, and spills from manufacturer pharma plants (Joakim, 2014).
2. Through consumption: excretion hot spots like hospital sewage, animal excrement, manure, or slurry applied to farmland, and individual excretion into the sewage system.
3. Waste: unused medications dumped in landfills, toilets, or sinks (Eckert, 2020).
4. Residue limitations and monitoring information: neither the statutory nor the private sectors have many systematic monitoring programmes, nor are there many regulatory limits defined for pharmaceutical residues. The Sewage Sludge Directive places restrictions on the number of heavy metals in sewage that is put on farms again, but not on the amount of drug residues.

Pharmaceuticals can be eliminated through waste treatment (see Table 5.1). Significant pollution may still exist in cleaned water and post-treatment sewage sludge without optimised local treatment (Sherer, 2006).

5.3 ANALYTICAL TEST METHODS

Because there are so many possible pollutants, choosing analytical priorities is challenging. Numerous methods for ranking risks have been put out, most notably the Stockholm International Water Institute's classification scheme. Different sample extraction and purification techniques are required for wastewater, surface water, soil, sludge, sewage, sediment, landfill leachate, incinerator smokestacks, ash, and plants, necessitating sample preparation methods. To give excellent sensitivity and selectivity for a number of chemical classes within a single multi-residue test method, modern test methods are based on LC-MSMS, occasionally in conjunction with GC-MS.

TABLE 5.1
Pharmaceutical Drugs Category and Detection in the Environment

S. No.	Pharmaceutical category	Pharmaceutical drugs detected in the environment
1	Antibiotics	Macrolides, sulphonamides, tetracyclines, fluoroquinolones, aminoglycosides
2	Non-steroidal anti-inflammatory drugs (NSAIDs)	Acetylsalicylic acid, ibuprofen, diclofenac, mefenamic acid
3	Anticonvulsants	Carbamazepine, primidone
4	Beta-blockers	Metoprolol, propranolol, betaxolol, bisoprolol, nadolol
5	Beta-agonists	Salbutamol
6	Opioids	Dextropropoxyphene
7	X-ray contrast agents	Iopromide, iopamidol, iohexol, diatrizoate

5.3.1 Development of Analytical Methods

When there are no established techniques, new approaches are being created to assess novel products for the presence of pharmacopoeial or non-pharmacopoeial products, with the goal of reducing cost and time while increasing precision and power. An analytical method uses a specific technique and specific, step-by-step instructions to analyse a sample qualitatively, quantitatively, or structurally for one or more analytes. The two basic categories of analytical procedures are classical methods and instrumental methods. The term "classical method" refers to a method in which the signal is inversely proportional to the analyte's absolute concentration. An instrumental method is one in which the signal is proportional to the concentration of the analytes. The three basic categories of classical approaches are: analyte separation, qualitative analysis, and quantitative analysis, the first three steps. Extraction, distillation, precipitation, and filtering are processes used to separate analytes. Instrumental approaches can be categorised into four primary categories: spectroscopic techniques, electrochemical techniques, chromogenic techniques, and miscellaneous techniques (Ravisankar et al., 2015). The different steps involved in analytical method development are shown in Figure 5.1.

5.3.2 The Need for Method Development

The identification, classification, and calculation, together with their dosage forms and organic fluids, are all part of the evaluation process. During the development and manufacture of a drug, the main goals of analytical strategies are to gather information about efficacy (dose of drug), impurity (related to medication safety), bioavailability (which includes important drug characteristics like crystal type, uniformity of drug release), stability (that shows the degradation product), and to produce parameters to confirm this information, before the method development by newer techniques.

The following are the stages that go into method development:

1. Analyte and standard characteristics: all the information needed on the analyte and its structure is gathered, including its physical and chemical characteristics like solubility and optical isomerism.
    a. The analyte's quality is sufficient to ensure complete purity.
    b. The appropriate storage is constructed (refrigerator, desiccators, and freezer).
2. Technique requirements: linearity, selectivity, specificity, range, accuracy, precision, LOD, LOQ, and others must be described in order to produce all the important analytical graphs or data.
3. System appropriateness: system suitability assessment was initially used by the pharmaceutical industry to determine whether a chromatographic system was suited for a

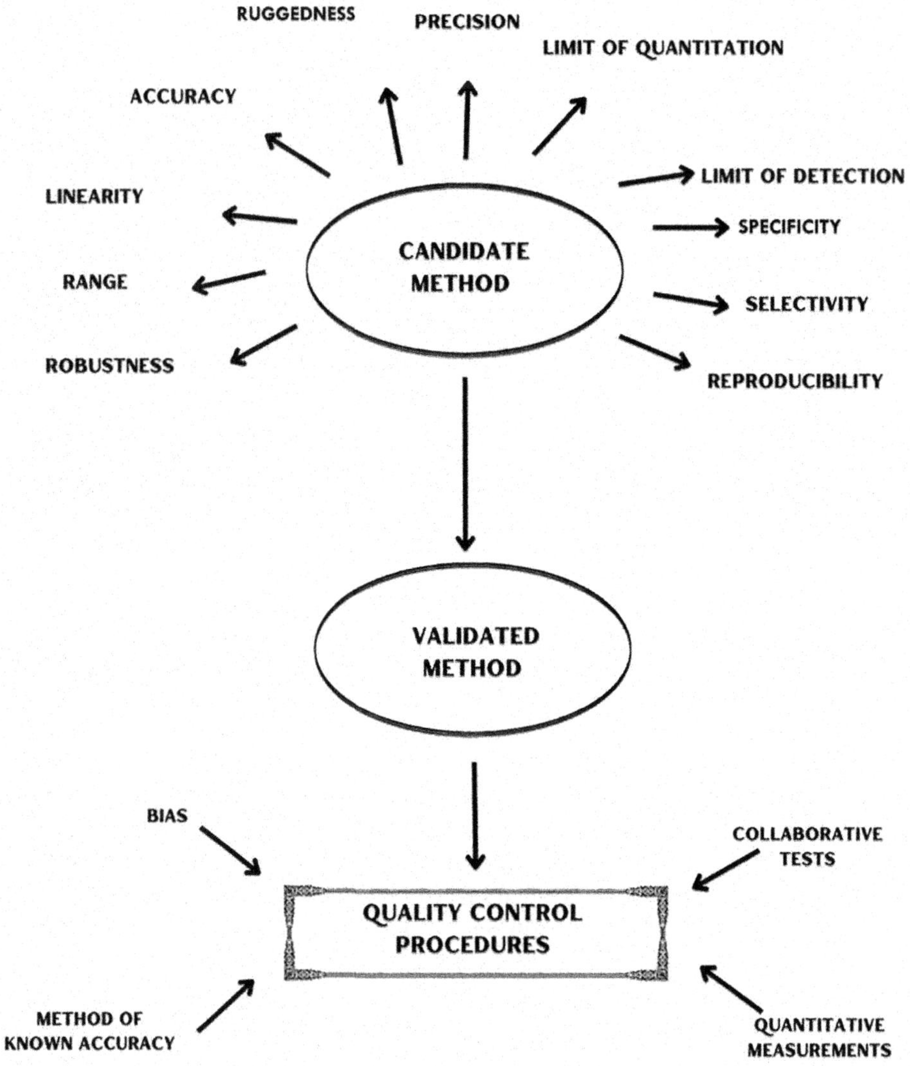

FIGURE 5.1 Different steps involved in analytical method development.

particular analysis and is commonly used today in pharmaceutical laboratories where quality of results is crucial. The following criteria are used in the system suitability tests (SST) report: efficiency (N), capacity factor (K), separation, resolution (Rs), tailing factor (T), and relative variance are examples of theoretical plates (RSD).

The process of documenting that is implied by the approach validation/evaluation. For the specified use, an analytical procedure gives analytical data (The Pharmaceutical, 2018).

5.4 VALIDATION

This concept first emerged in the United States in 1978. Since then, the concept of validation has expanded to encompass a wide range of activities, from analytical methods used for drug standard control to computerised systems for clinical trials, or process control. Validation is based on regulatory specifications, but is not officially supported by them, and is therefore best viewed as an essential component of current good manufacturing practice (cGMP). Any new or modified technique must be validated to ensure that it can produce reliable and consistent results when used by various operators using the same or similar equipment in the same or other facilities. The parameters to be considered are minimum batch product failure; increased efficiency; manufacturing and productivity; assured high quality; time foundation; method optimisation; costs for quality improvement; rejection going down; rising yields (Mahar and Verma, 2014).

1. Accuracy: accuracy is the degree to which test results closely approximate the true value.
2. Precision: the measurement of precision is the degree of agreement between two or more measurements made on the same sample.
   The relative standard deviation is used to express precision.

$$\% \text{RSD} = \text{Standard deviation}/\text{Mean} \times 100$$

3. Linearity: the capacity to produce a response that is equilibrated to the concentration (amount) of analyte in the sample is known as linearity.
   The confidence limit surrounding the slope of the regression line is how linearity is expressed.
4. Limit of detection (LOD): LOD is the smallest concentration of analyte that may be detected or identified, but not quantified, in a sample. A concentration at a specific signal-to-noise ratio—typically 3:1—is how LOD is stated.

$$\text{LOD} = 3.3 \times \text{S/SD}$$

5. Limit of quantitation (LOQ): the lowest amount (concentration) of analyte in a sample that can be quantified. The ICH has suggested the following signal for LOQ: 10:1 noise ratio.

$$\text{LOQ} = 10 \times \text{S/SD}$$

6. Specificity: an analytical method's specificity is its capacity to quantify the analyte precisely in the presence of additional components.

The following conclusions flow from this definition:

1. Identification;
2. Testing for purity;
3. Assay;
4. Range: it is calculated using a linear response curve and is expressed in the same unit as test results;
5. Robustness is the measurement of an analytical procedure's ability to be unaffected by minute alterations in method parameters.

5.5 CHROMATOGRAPHY TECHNIQUE

A physicochemical technique for separating mixtures of substances is chromatography. Chromatography uses two phases, a stationary phase and a mobile phase, to separate a mixture of substances into their constituent components. Chromatography is classified as follows:

Based on interaction of solute to stationary phase: (i) adsorption chromatography; (ii) partition chromatography; (iii) ion-exchange chromatography; (iv) molecular exclusion chromatography.

Based on chromatographic bed shape: (i) column chromatography; (ii) planar chromatography; (iii) paper chromatography; (iv) thin layer chromatography; (v) displacement chromatography.

Techniques by physical state of mobile phase: (i) gas chromatography; (ii) liquid chromatography; (iii) affinity chromatography.

Thin layer chromatography and paper chromatography are used for the (i) separation of mixtures of drugs of chemical or biological origin, plant extracts, etc; (ii) separation of carbohydrates, vitamins, antibiotics, proteins, alkaloids, glycosides, etc; (iii) identification of related compounds in drugs; (iv) detection of the presence of foreign substances in drugs.

Gas chromatography is mainly used for (i) determination of purification of compounds for drugs like clove oil, atropine, sulphate, stearic acid; (ii) quality control and analysis of drug products like antibiotics, general anaesthetics, antivirals etc; (iii) determination of the level of metabolites in body fluids like blood plasma, serum, and urine.

Column chromatography is divided into two types, adsorptions column chromatography and partition column chromatography. Adsorptions column chromatography works on the principle of differences in affinity between the two components towards the stationary phase, thus various components of a sample mixture get separated. (i) Selection of column: the preferred ratio of diameter to the length of the column is 1:20 to 1:100, with the latter having high efficiency of separation. The length of the column used for separation is selected based on the type of adsorbent used for packing; the number of components that are to be separated. (ii) Selection of adsorbents: silica gel for column, activated alumina, starch, charcoal, magnesia, etc. (iii) Selection of solvent system: different mobile phases used are petroleum ether, carbon tetra chloride, ether, toluene, etc. (iv) Packing of column: packing can be performed by two methods, wet packing and dry packing. (v) Elution: isocratic elution and fractional elution. (vi) Detection: coloured components can be identified as well-defined zones in the column. Colourless compound which have the ability to absorb UV visible light can be easily detected by UV-visible detectors. Fluoresce compound can be detected by a variety of detectors. Various instruments based on chromatographic techniques used in pharmaceuticals detection are shown in Figure 5.2.

5.5.1 Ion-Exchange Chromatography

It is a process that allows the separation of ions and polar molecules based on their affinity to the ion exchanger. It works on almost any kind of charged molecule including large proteins, small nucleotides, and amino acids. Cations and anions can be separated using this method. The separations occur by reversible exchange of ions between the ions present in solution and those present in ion exchange resin. It works on almost any kind of charged molecule including large proteins, small nucleotides, and amino acids. The application of ion-exchange chromatography is in: (i) softening and demineralisation of water; (ii) separation of inorganic ions; (iii) separation of sugars, amino acids, and proteins; (iv) purification of solution free from ionic impurities; (v) ion-exchange column in HPLC.

5.5.2 Gel-Filtration Chromatography

The fundamental idea behind this technique is to use materials containing dextran to segregate macromolecules according to how differently their molecules are sized. For various molecular weight ranges, different gels are utilised. The type of solvent employed might either be aqueous or non-aqueous. Small-pored inert molecules make up the stationary phase. The most popular column material is sephadex G type. Dextran, agarose gel, and polyacrylamide are further materials utilised for columns; besides, dextran, agarose gel, polyacrylamide are also used as column materials.

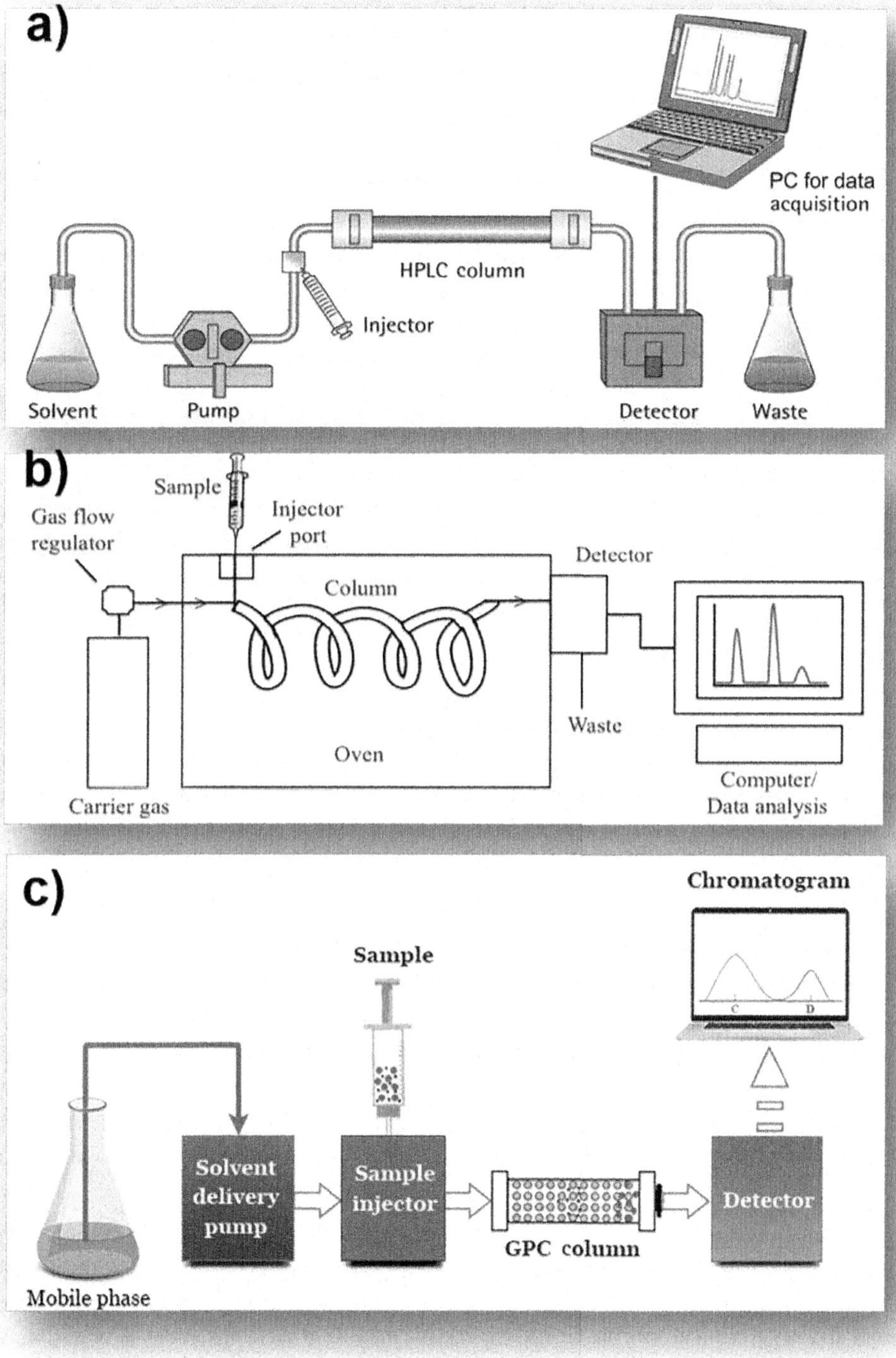

FIGURE 5.2 Photographic image of different type of chromatography instruments used for pharmaceuticals determination: a) high-performance liquid chromatography, b) gas chromatography, and c) gel permeation chromatography.

5.5.3 Ultra-Performance Liquid Chromatography (UPLC)

This sophisticated method of liquid chromatography improves in three key areas: speed, resolution, and sensitivity. When compared to high-performance liquid chromatography, UPLC is suited for particles less than 2 m in diameter and can achieve superior resolution, speed, and sensitivity.

5.5.4 High-Performance Liquid Chromatography (HPLC)

It is one of the most popular analytical procedures. A separation technique including mass transfer between the stationary and mobile phase is what is meant by the term "chromatographic process." To separate the components of a mix, HPLC uses a liquid mobile phase. A liquid or a solid phase is frequently the stationary phase. These components are first made to dissolve in a solvent, after which they are compelled to flow through a chromatographic column under intense pressure. The mixture separates into its component parts in the column. Resolution is important and is influenced by how much the stationary phase and solute components interact. The immovable packing within the column is referred to as the stationary phase. The relationship between the solute and moving and stationary in LC technique has been devised for the identification of amiodarone hydrochloride in tablet and injectable forms. Phases are frequently altered through varied combinations of both solvents and stationary phases. For the detection of amiodarone hydrochloride and its associated components in amiodarone hydrochloride injections, an HPLC method was also created and validated. HPLC has a high level of adaptability that is unmatched by other chromatographic methods and can quickly separate a range of chemical mixtures. HPLC consists of solvent reservoirs, solvent degasser, gradient valve, mixing vessel, high pressure pump, sample injection loop, guard column, analytical column, detectors, and waste collector. HPLC separation using a BDS Hypersil C18 column for high-performance liquid chromatography, solvent used is methanol (25:75, v/v) and its pH-adjusted to using ortho-phosphoric acid using disodium hydrogen phosphate buffer (0.02 M) as the mobile phase. The suggested techniques were effectively employed for determining the researched medicines in tablets in accordance with ICH recommendations. For the determination of gatifloxacin in dosage forms and from human plasma, HPLC and LC/ESI-MS/MS approaches were used.

HPLC in pharmaceutical application: (i) identify active constituents in dosage forms; (ii) evaluate pharmaceutical product shelf-life; (iii) measure dopamine in levodopa.

In environment application: (i) identify diphenhydramine in deposited sample; (ii) bio-monitor pollutants.

In clinical application: (i) analyse antibiotics; (ii) detect endogenous neuropeptides in brain extracellular fluids.

In food and flavouring: (i) ensure soft drink consistency and quality; (ii) analyse natural contamination (e.g. mercury and phenol in sea water).

Carbamazepine (CBZ) assay through HPLC method: material and method: acetonitrile, tartrazine, and carbamazepine powder (HPLC grade) (98%), and all other compounds, were of analytical reagent quality. Samples of carbamazepine (CBZ) were prepared by combining acetonitrile (25%) and distilled water (75%) with carbamazepine (0.1 mg/ml), resulting in a solution with a concentration of 0.1 g/100 mL. This sample stock solution was separated into six distinct concentrations of CBZ in a volumetric flask: 5, 10, 15, 20, 25, and 30 g/ml. Then analysis was done with the help of HPLC.

Method validation: the standard was present at a constant concentration of 10 g/ml in the mobile phase of the RP-HPLC apparatus utilised for the validation procedure, which was composed of 25% acetonitrile and 75% distilled water. A reverse phase column with the specification C(18)150 × 4.6 mm I.D. reversed-phase column was used to achieve the separation with a flow of 1 mL per minute. 10 L was the injection volume. The flow rate was held constant at 1.0 mL/min, with the detection wavelength set at 284 nm, and retention time was 1.86 minutes. The maximum potential run time was 3 minutes. For the drug's stock solution dilution procedures (concentration range: 5–30 g/mL), a calibration curve was created. The results of the average peak area chromatography were compared with International Council on Harmonization's recommendations (ICH). Linearity: over 2.5 to 40 g/mL, the technique established in this study produced linearity of R2 = O.9922 and R2 = O.9912.

Result: the utilised technique complies with acceptable International Councils on Harmonization standards with regards to accuracy, precision, specificity, limit of detection, limit of quantitation, linearity, and range (ICH Guidelines Q2 (RI), 2005). The results of the developed, verified HPLC technique show the presence of carbamazepine in ambient sample.

Solid-phase extraction (SPE) was used to separate the antibiotics ampicillin, amoxicillin, penicillin g, ceftazidime, tetracycline, and doxycycline from aqueous matrices. Ultrasound-assisted extraction (USAE) was used on sediment samples before SPE. The antibiotics were then tested using high-performance liquid chromatography along with a diode array detector or mass spectrometer (HPLC-DAD/MS). These antibiotics were tracked in river waters using the developed SPE/USAE-HPLC-DAD/MS techniques, which were also employed to analyse some sediment samples. The samples under study contained antibiotic residues; to recover antibiotics from sediment samples, 30 minutes were spent extracting 3 grams of dried sediment in an ultrasonic bath with 20 mL of methanol. After 15 minutes of centrifugation at 4000 rpm, the supernatant was collected, and it was then dried under nitrogen. Similar to the river water samples, the residue was reconstituted in 100 mL of distilled water before being extracted using SPE. For the DAD detection, two wavelengths were needed: 197 nm for penicillin and 272 nm for tetracyclines and ceftazidime. The proposed HPLC-DAD approach has low detection and quantification limits (LOD and LOQ) in the g/mL range, high linearity in the 5.21–166.7 g/mL range, correlation coefficients (r) better than 0.999, and good repeatability (measured in three replicates for the 0.85 g/mL concentration) (Virginia Coman, 2017).

5.5.5 Automated Development in HPTLC

High-performance thin layer chromatography (HPTLC) is an improved version of thin layer chromatography. The fundamental technique of TLC can be improved with ways to automate the various procedures, increase the resolution attained, and enable more precise quantitative measurements. When a sample is put manually to a TLC plate, there is a chance that the droplet size and position will be unpredictable; due to its benefits of accuracy in analyte quantification at the micro and even nanogram levels and cost effectiveness, HPTLC has nowadays become a standard analytical technique.

5.5.6 Development of RP-HPLC

A non-polar stationary phase and an aqueous, moderately polar mobile phase are characteristics of reversed-phase HPLC (RP-HPLC or RPC). For the simultaneous detection of 'camylofin' dihydrochloride and diclofenac potassium utilising methylparaben as an indoor standard, a straightforward, quick, and accurate RP-HPLC method has been devised. Using an Inertsil C18 column (250 mm × 4.6 mm, 5 m) as the stationary phase and a mobile phase made up of 0.05 M KH_2PO_4 in water: methanol (35:65, v/v) at a flow rate of 1.5 mL per minute, 27°C for the column, and 220 nm for UV detection, effective chromatographic separation was accomplished for the simultaneous measurement of amoxicillin trihydrate and bromhexine hydrochloride from oily suspension, a relatively easy, quick, and accurate RP-HPLC approach was established. Mobile phase methanol and glacial acetic acid (50:50 v/v) were utilised with an ODS C18 (250 × 4.5 mm ID), 5 particle size. For the simultaneous measurement of tranexamic acid and mefenamic acid in combination tablet dose form, an enhanced compounds RP-HPLC technique with PDA detection has been created and validated. The tactic's short analysis run time of 10 minutes demonstrates its suitability to routine quality control of the medicine.

5.5.7 LC-MS Method

Because of their sensitivity, selectivity, study speed, and cost-effectiveness, LC-MS procedures are applied in a variety of pharmaceutically relevant substances. These analytical functions have continuously advanced, making instruments more user-friendly and trustworthy. These advancements came at the right time and matched the aforementioned changes in the pharmaceutical sector. Analytical method for classifying chemical compounds by mass by employing electric and magnetic fields to separate gaseous ions. The technique is frequently used to determine the masses and relative abundances of different isotopes, study the results of liquid or gas chromatographic separation, examine the vacuum integrity of high-vacuum equipment, and determine the geological age of materials.

5.5.7.1 Applications of LC-MS Method in Clinical Samples

Seven anti-HIV drugs in plasma of HIV-infected patients were quantified simultaneously using the LC-MS/MS method. For the simultaneous quantification of tolmetin (TMT) and MED5 in human plasma, a sensitive, quick assay technique using tandem mass spectrometry and electrospray ionisation in the positive-ion mode has been developed and validated. TMT, MED5, and mycophenolic acid (internal standard, IS) were extracted from human plasma using a simple solid-phase extraction procedure. The LC-MS/MS method for paclitaxel measurement in human plasma was created and verified. Amlodipine Desolate, Olmesartan Medoxomil, and Hydrochlorothiazide in Tablet Dosage Form: Quantitative Estimation.

Histone deacetylase inhibitors (HDIs) and DNA methyltransferase (DNMT) inhibitors are often utilised compounds in pre-clinical and clinical anti-cancer investigations because epigenetic regulators have swiftly become one of the most extensively explored therapeutic agents for a wide range of disorders. For measuring the levels of phenol in mouse plasma, a precise HPLC method was developed and assessed. Purified phenol, Moutan Cortex decoction, and Rhubarbmoutan decoction were all given orally at equivalent doses of 10mg/kg phenol to 180 KM male mice weighing 22–28 g. After being administered intravenously, nimesulide had an 89.42% bioavailability. These pharmacokinetic data imply that intramuscular administration of nimesulide may also be beneficial for treating illness conditions in cattle. In order to measure bupropion and its metabolite hydroxybupropion in human plasma, a lidocaine-based LC-MS/MS assay was developed. The absorption of bupropion was, however, greatly boosted by the hyperlipidaemic meal. The low drug passing through the buccal mucosa is the most crucial component of transbuccal medication administration. The idea of prolonged rapamycin delivery via ReGel is a potential method to stop SMC proliferation and avoid arteriovenous graft stenosis in haemodialysis patients. Multiple components can be determined simultaneously. For the simultaneous measurement of Atorvastatin, Ezetimibe, and Fenofibrate in their ternary mixture of affordable pharmaceutical formulations, a simple, quick, and accurate reversed-phase liquid chromatographic approach is established. Tenatoprazole underwent extensive stress testing as per ICH guideline Q1A (R2). Tenatoprazole was put through a series of stressful conditions, including photolysis, hydrolysis, oxidation, and neutral breakdown. In acidic, neutral, and oxidative environments, extensive deterioration was discovered to take place (Ravali et al., 2011).

5.6 SPECTROSCOPY TECHNIQUE

Spectroscopy is a scientific field that studies how samples or materials interact with light or other electromagnetic radiations (EMR). In other words, the changes in rotational, vibrational, and/or electronic energies are measured by spectroscopy. The interaction of electromagnetic radiation with matter is the basis for spectroscopy research. When an electromagnetic beam strikes a sample, one of four things happens: it absorbs, reflects, transmits, or scatters. As a result, a spectrum is produced and examined. Various spectroscopic instruments that are used in determination of pharmaceuticals are shown in Figure 5.3.

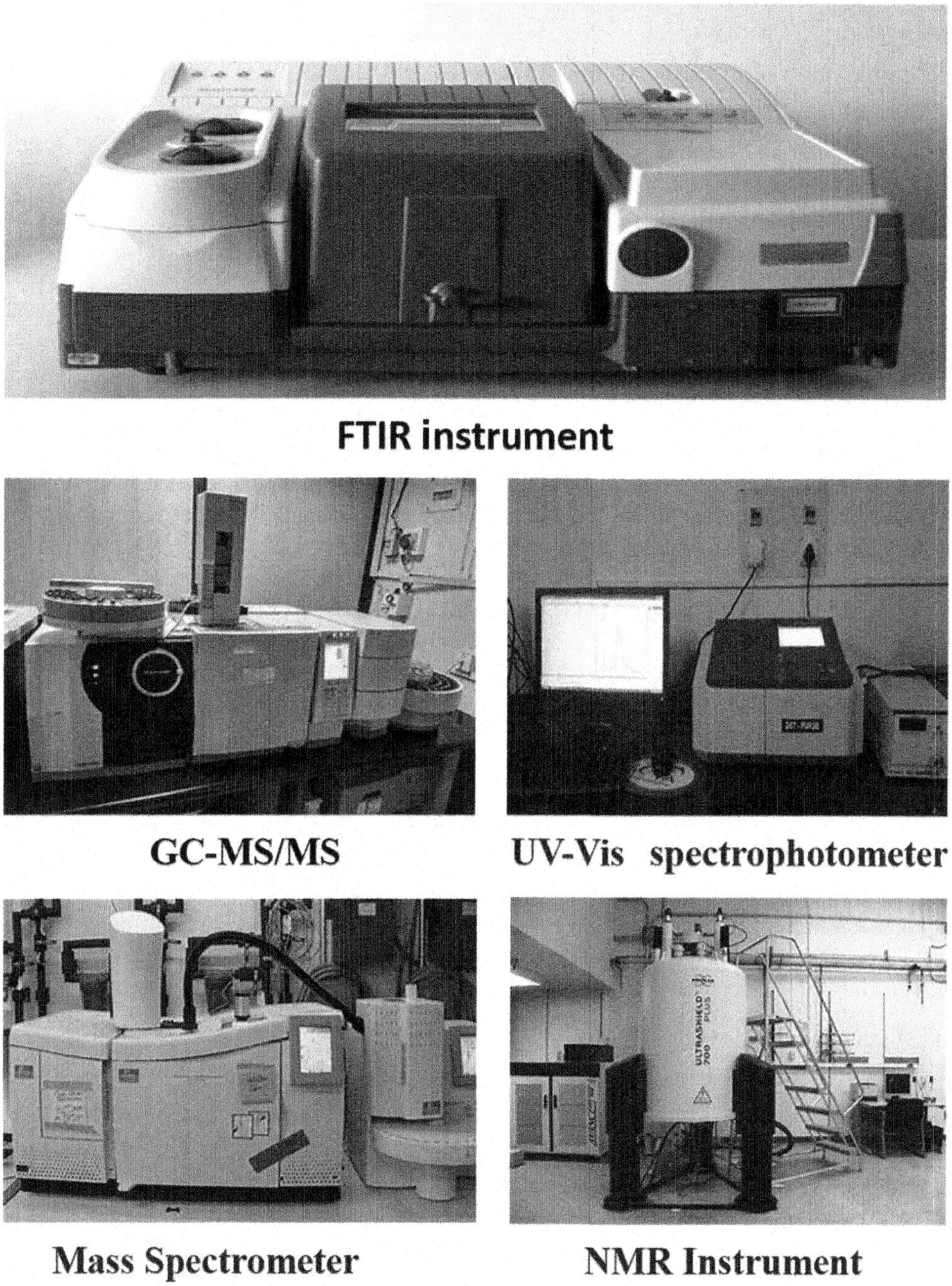

FIGURE 5.3 Photographic image of different types of spectroscopy instruments used for pharmaceuticals analysis.

5.6.1 Ultraviolet-Visible Spectroscopy

UV-visible or UV-spectroscopy spectrophotometry is based on the electromagnetic radiation's absorption in the UV-vis area, with wavelengths between 200 and 400 nm (UV spectroscopy) and 400 and 800 nm (visible spectroscopy). This implies that it makes use of light in the visible and nearby ranges.

TABLE 5.2
Quantitative Analysis of Pharmaceutical Drugs by UV-Visible Spectrophotometric

S. No.	Pharmaceutical drug	Reagent used for analysis	λmax
1	Acetaminophen	m-Cresol	640
2	Lisinopril	Chloranil	525
3	Pregabalin	Ninhydrin	402.6
4	Diclofenac sodium	Tris buffer	284, 305
5	Amoxycillin, ampicillin, carbenicillin	Folin cicalteu phenol	750, 770, 750

Principle: basically, spectroscopy deals with how light and matter interact. The energy content of the atoms or molecules increases as light is absorbed by matter. When ultraviolet rays are absorbed, the electrons are excited from their ground state and moved into a higher energy state. Molecules with n-electrons, which do not form bonds, can absorb energy from UV light to excite these electrons to higher anti-bonding molecular orbitals.

Applications: (i) it ranks among the finest techniques for identifying contaminants in organic compounds; (ii) pharmaceutical compound quantitative analysis; (iii) impurity detection in coloured or organic samples; (iv) spotting the presence of functional groups in the sample. An HPLC detector that uses UV spectrophotometers is possible. The various regions of electromagnetic spectrum are set out and are labelled either according to the wavelength/wave no. range used, or according to the type of the molecular energy levels involved (e.g. UV (electronic) spectra, IR (vibrational) spectra, or RF (NMR) spectra).

5.6.2 Infrared (IR) Spectroscopy

It deals with the infrared area of the electromagnetic spectrum, which is light with a longer wavelength and lower frequency than visible light, known as infrared spectroscopy. It includes a variety of methods, many of which are based on absorption spectroscopy. Identification of functional groups is simple due to their absorption of characteristics.

TABLE 5.3
Various Regions of Electromagnetic Spectrum

Type of radiation	Wavelength	Wave no.	Type of molecular spectrum
RF	> 100 mm	$< 3 \times 10^9$ Hz	NMR (Spin orientation)
Microwave	1–100 mm	10–0.1 cm^{-1}	Rotational
Far IR	50 μm–1 mm	200–10 cm^{-1}	Vibrational fundamental or rotational
Mid IR	2.5 μm–50 μm	4000–667 cm^{-1}	Vibrational fundamental
Near IR	780 nm–2.5 μm	$(13–4) \times 10^3$ cm^{-1}	Vibrational (overtones)
Visible	380 nm–780 nm	$(2.6–1.3) \times 10^4$ cm^{-1}	Electronic (valence orbital)
Near UV	200 nm–380 nm	$(5–2.6) \times 10^4$ cm^{-1}	Electronic (valence orbital)
Vacuum UV	10 nm–200 nm	$(10^2–5) \times 10^4$ cm^{-1}	Electronic (valence orbital)
X-rays	10 pm–10 nm	10^9-10^6 cm^{-1}	Electronic (core orbitals)
Gamma rays	10^{-10} cm	10^{10} cm^{-1}	Mossbauer effect (nuclear transitions) excited states of nuclei

Applications: (i) it includes both research and industry; (ii) it is an efficient method for dynamic measuring and measurement quality control; (iii) drug substance identification; (iv) research on polymers; (v) the proportion of cis-trans isomers in a chemical mixture; (vi) forensic analysis is used in both civil and criminal cases.

5.6.3 Mass Spectroscopy

Mass spectroscopy is a potent analytical technique that may be used to measure known materials, identify unidentified chemicals in a sample, and shed light on the structure and chemical characteristics of various molecules. The mass-to-charge ratio of ions is measured analytically via mass spectroscopy.

Principle: it measures the mass-to-charge ratios of charged molecules or molecular fragments produced by ionising chemical substances. Mass spectrometry principle works by bombarding a natural sample of a volatile molecule (M) with high-energy electrons or electrons of energy greater than the ionisation energy. On electron impact, the molecules are energised sufficiently to eject an electron. The process leads to the formation of a radical cation symbolised more accurately by or M+ in simplification. Fragmentation of the molecular ion reaches a substantial proportion only at higher bombardment energies (70 eV). The resulting positively charged particles accelerate into a bent chamber surrounded by a magnetic field. Varying the strength of the magnetic field, the radius of the curvature changes depending largely on the mass of the ions and the strength of the fields. Therefore, ions are separated on the basis of m/z values. Each positive ion formed directly or by fragmentation of the original molecule has a unique mass/charge ratio. It can be recorded on graph paper to give the mass spectrum of each component. It is also useful for quantitative analysis of mixtures containing closely related compounds.

Applications: (i) to determine a compound's molecular formula and molecular weight; (ii) to prepare natural products, high polymers, and pure isotopes; (iii) to examine free radicals, assess bonding power, and gauge sublimation heat; (iv) to conduct environmental analysis: pollution of the soil and ground water; (v) to conduct clinical research: the use of mass spectroscopy in clinical laboratories has significantly advanced research; (vi) to complete compound structure elucidation: mass spectroscopy plays a significant role in this process.

One study used multiple monitoring reaction mode positive ionisation tandem mass spectrometry (MS/MS) and ultrahigh performance liquid chromatography (UPLC) to analyse seven major pharmaceutical residues in surface water and hospital effluent. These residues contained trimethoprim, ciprofloxacin, ciprofloxacin, ofloxacin, ketoprofen, paracetamol, and sulfamethoxazole. Multimode solid phase extraction was used to purify the water sample. An essential component of electrospray ionisation, tandem mass spectrometry, ionisation suppression/enhancement, was thoroughly assessed using pre-extraction and post-extraction spiking tests. Materials and methods: acetonitrile and methanol were employed as organic solvents, as well as carbamazepine, ciprofloxacin, ofloxacin, sulfamethoxazole, trimethoprim, ketoprofen, paracetamol, formic acid 100% (MS grade), and ammonium hydroxide solution 25%, were also utilised. An adequate quantity of analyte was dissolved in acetonitrile at a concentration of 10mg/L to create an individual stock standard solution. For full dissolution of ciprofloxacin and ofloxacin, 100 L of 10 M NaOH was used. By mixing several stock solutions with an acetonitrile and water mixture that included 0.5% formic acid, the standard solution was produced (mobile phase). A volume of concentrated formic acid in acetonitrile is added to the mobile phase to prepare it. To get rid of any dissolved gas, the mobile phases were filtered and then degassed in an ultrasonic water bath; 200 mL of hospital effluent or surface water were raised to pH 3.0 using 2 M formic acid. A flow of 12 to 15 mL per minute was present, cation operating in mix-mode, the solid phase extraction cartridges were washed with 3 mL of pH 3.0 water and then vacuum dried for 30 minutes. 51 mL of a 90/10 (v/v) combination of MeOH and 2M NH_4OH were used for elution under the best conditions allowed by the International Council on Harmonization (ICH Guidelines Q2 (RI), 2005). Target compounds were separated on a guard column (3.5 m, 2.1 mm i.d. × 10 mm) and a reversed-phase chromatography column (BEH C18 column, 130 A, 1.7 m, 2.1 mm i.d. × 50 mm). With the aid of a gradient of mobile phase, the analyte was eluted. The linear scheduling of 0.5% formic acid in water (A) and 0.5% formic acid in acetonitrile (B) was as follows: it was lowered to the initial mobile phase condition of the equilibrium column after being 100% A for 0–1 min and 70% B for 3–5 min. In 6 minutes, the chromatographic separation was complete. Using an autosampler, a 10 L sample or reference solution was injected into the analytical column equipment. The mobile phase flow rate was maintained at 300L/min. The target compounds were located using the positive electrospray ionisation (+ESI) technique of multiple reaction monitoring (MRM). For the best possible conditioning of the ionisation source, the temperature was adjusted to 350°C, the source temperature to 150°C, the gas flow to 645 L/h to 10 L/h, and the capillary voltage to 3 kV.

5.6.4 Nuclear Magnetic Resonance (NMR) Spectroscopy

Numerous nuclei have spin, and all nuclei are electrically charged, according to the NMR underlying principle. The nuclei align themselves either with or against the field of the external magnet when a magnetic field is applied to them, and this causes the nucleus's spin to enter an excited state. The absorbed radiofrequency energy is released at the same frequency level as the spin returns to its ground state level. The radiofrequency signal that is emitted provides the nucleus in question's NMR spectra. Atomic nuclei in the 4–900 MHz radio frequency range. (i) The sample is placed in a magnetic field and the NMR signal is produced by excitation of the nuclei sample with radio waves into nuclear magnetic resonance, which is detected with sensitive radio receivers. (ii) The intramolecular magnetic field around an atom in a molecule changes the resonance frequency, thus giving access to details of the electronic structure of a molecule and its individual functional groups. (iii) As the fields are unique or highly characteristic to individual compounds, NMR spectroscopy is the definitive method to identify monomolecular organic compounds. (iv) Besides identification, NMR spectroscopy provides detailed information about the structure, dynamics, reaction state, and chemical environment of molecules. (v) The most common types of NMR are proton and carbon-13 NMR spectroscopy, but it is applicable to any kind of sample that contains nuclei possessing spin.

Applications: NMR is used (i) to determine hydrogen bonding; (ii) to determine a compound's aromaticity; (iii) to distinguish cis and trans isomers; (iv) to describe the chemical makeup of organic and inorganic substances; (v) to analyse metabolites with technology.

5.6.5 Fourier-Transform Infrared Spectroscopy (FTIR)

FTIR is a method for obtaining an infrared spectrum of a solid, liquid, or gas's absorption or emission. This technique (FTIR) helped scientist to identify compounds functional groups via their peaks.

Applications: (i) compounds, such as compounded plastics, mixes, fillers, paints, rubber, coatings, resins, and adhesives, can be promptly and accurately identified with this technique; (ii) forensic analysis; (iii) pharmaceutical research; (iv) polymer analysis; (v) fuel additives and lubricant formulation; (vi) studying food; (vii) control and quality assurance.

5.6.6 Phosphorimetry and Fluorimetry

These are used for the analysis of micro samples and have increased steadily in the pharmaceutical industry. These are analytical methods based on the absorption of electromagnetic radiation by the processed sample, which contains or produces molecules that absorb radiation in excited states; when the excited molecules return to the ground state, radiation is emitted, or the excited molecules luminesce. A steady increase in the number of applications for fluorimetry or phosphorimetry was seen in earlier investigations. These approaches for estimating some medications quantitatively are those that have been used in the past and are applicable to drugs that are available in the form of biological fluids.

Advantages: this technology is more sensitive because it can detect concentrations as low as g/ml or even ng/ml.

- Accuracy: phosphorimetry and fluorimetry can readily attain accuracy of up to 100%.
- Precision: more precise than absorption method.
- Scope of use: chemical reactions can also transform non-fluorescent substances into fluorescent ones.

Applications: (i) organic and inorganic substance determination; (ii) measuring thiamine (vitamin B1) and riboflavin (vitamin B2) levels in dietary samples such meat, cereal, fish, seeds, and nuts, among other things; (iii) widely utilised in the realm of nuclear research to identify uranium salt; (iv) quantitative and qualitative research for pharmaceuticals (Masoom Raza Siddiqui, 2017).

5.7 PHARMACEUTICALS' ADVERSE EFFECTS ON THE ENVIRONMENT

Toxicology and pharmacology don't seem to be as crucial to human health problems from drug exposure in the environment as environmental hygiene. For example:

1. Endocrine-active drugs and hormones may prevent human sexual development because they interact with hormone systems and are highly active chemicals.
2. One risk of modern chemotherapy is that some anti-cancer medications, even at very low dosages, may actually induce cancer, and antibiotics may have a role in the evolution of bacteria that are resistant to antibiotics.
3. The toxicology and ecotoxicology fields are still grappling with the problem of medication administered through drinking water. Pregnant women, young children, and elderly persons may be in danger.
4. While diclofenac's effective concentration for chronic fish toxicity was in the range of wastewater concentrations, propranolol and fluoxetine's effective concentrations for zooplankton and benthic species were close to the maximum recorded sewage treatment plant effluent concentrations. By giving the oriental white-backed vultures direct oral exposure to diclofenac residues and treated livestock remains, researchers were able to replicate diclofenac residues and renal illness in vultures in an experimental setting.

There are extremely few documented instances where a pharmaceutical has been demonstrably proved to have a negative impact on an ecosystem, despite the fact that there are many different human medications present in the environment. Sex steroids, in particular oestrogens like 17-ethinylestradiol (EE2), have been discovered in the aquatic environment by the 1990s. They could have an impact on fish reproduction. The tablet is the only product that contains EE2. It is abundant in the aquatic environment, even though most rivers only have sub-ng per L concentrations.

However, fish are so sensitive to this substance that even very low ng L1 concentrations in their water can cause population collapses by shifting the male-to-female sex ratio in favour of females due to males' lack of sexual differentiation. Fortunately, although the margin of safety is not very big, real-world EE2 concentrations are lower than those that significantly impact fish. One of these instances of the effects of pharmaceuticals on wildlife, EE2, may have and probably should have been anticipated, but the other—diclofenac killing off old world vultures—came as a complete shock.

5.8 CHALLENGES

Education and training are lacking: there is a dearth of education and training; many hospitals and healthcare facilities make the error of failing to instruct and train the whole nursing team on correct drug disposal techniques. Nursing staff need to receive professional training on various medications and the proper containers for disposal. This relates to education, training, and a staff of nurses that is overworked (Myranda Eckert, 2020).

Fenfluramine, if found in water, belongs to the anorexia drug class because it quickens testicular development by boosting gonad-stimulating hormone production. Tylosin residues will affect the microbial communities in the soil if they are found there.

Due to budget limitations, only a tiny portion of the pharmaceuticals now in use have been properly investigated; critically required new studies will reveal how more drugs impact the environment.

The current conventional ecotoxicity tests are probably not enough to measure the effects of many medications. Complex endpoints, including changed behaviour, physiology, and biochemistry, seem to be helpful in some circumstances. This endeavour might greatly benefit from the use of a wide range of contemporary molecular biology technology, such as large-scale DNA or protein arrays, proteomics and genomics methods, and genomics approaches, although certain small impacts have been seen following exposure to drugs at levels that resemble environmental circumstances. We still need to explain what these results mean in terms of how the environment functions.

5.9 CONCLUSION

Analytical methods like spectroscopy and chromatography are now recognised as very sensitive and efficient separation techniques. Chromatography is regarded as a recent scientific breakthrough that has had the greatest impact. Although HPLC was once only used by analysts, it is now frequently used by researchers, chemists, biologists, industrial employees, and other labs for quality control and research. The many approaches, UV-visible spectroscopy, mass spectroscopy, infrared spectroscopy, nuclear magnetic resonance, fluorimetry, and phosphorimetry have been used as spectroscopy techniques for the quantitative and qualitative evaluation of pharmaceuticals and their metabolites present in varying amounts in the samples obtained from animals, humans, water, and from soil. Chromatographic and spectroscopic techniques are the main techniques used for the evaluation and quantification of pharmaceutical products and their active or inactive metabolites in the surface water, drinking water, and in the soil sediments. Chromatographic techniques improve chemical and instrumentation productivity by giving more information due to increased productivity reliability, robustness, resolution, speed, and sensitivity. It is possible to cut down on the amount of time needed to refine new techniques.

The most common activities involved in the analytical development of testing techniques are separation and characterisation of impurities as well as degraded products, analytical investigations, studies for identification, and finally fixing of parameter optimisation to specific requirements. The main purposes of analytical method development are for identification, purification, and ultimately quantification of any chemicals or drugs, etc. Therefore, an analyst can greatly benefit from the key aspects in calculating pharmaceutical formulations, bulk medications, and their by-products in the environment. The outcomes of the chromatographic and spectrometry techniques show how accurate, precise, specific, linear, dependable, sensitive, and quick these analytical techniques are. The main goals of developing analytical techniques are to identify, purify, and ultimately qualify any required drug and pharmaceutical pollution in the ecosystem, etc. The creation of analytical techniques aids in precision and accuracy. In order to ensure that quality work is done in the process that supports the creation of medicines and products, validation is a vital approach in the pharmaceutical industry.

REFERENCES

Breton, R. and Boxall, A. Pharmaceuticals and personal care products in the environment: regulatory drivers and research needs. *Qsar & Combinatorial Science*, 22(3), 399–409; 2003.

Bruni, P. Impact of Pharmaceutical pollution on communities and environment in India. This report was researched and prepared for Nordea by Changing markets and ecosystorm. Published in February 2016.

Carmen Lidia Chitescu, Mariana Lupoae, Alina Mihaela Elisei, pharmaceutical residues in the environment – New European Integrated Programs Required, *Revista De Chimie*, 67(5); 2016.

Chitescu, C. L., Lupoae, M. and Elisei, A. M. Pharmaceutical residues in the environment—new European integrated programs required. *Revista De Chimie*, 67(5); 2016.

Coman, V. and Beldean-Galea, S. Chromatographic analysis of some antibiotics in water and sediment samples collected from the Romanian Tisza river watershed. *Studia Universitatis Babeş-Bolyai. Chemia*, 62(4), 129–142; 2017, doi:10.24193/subbchem.2017.4.11.

Daughton, C. G. and Ruhoy, I. S. Environmental footprint of pharmaceuticals: the significance of factors beyond direct excretion to sewers. *Environmental Toxicology and Chemistry*, 28(12), 2495–2521; 2009

Eckert, M. Pharmaceutical Waste Disposal challenges facing pharmaceutical waste management in healthcare, July 22, 2020.

Fick, J., Soderstrom, H., Lindberg, R. H. and Phan, C. Contamination of surface, ground, and drinking water from pharmaceutical production. *Environmental Toxicology and Chemistry*, 28(12), 2522–2527; 2009.

Grabicova, K., Grabic, R., Blaha, M., Kumar, V., Cerveny, D., Fedorova, G. and Randak, T. Presence of pharmaceuticals in Benthic fauna living in a small stream affected by effluent from a municipal sewage treatment plant. *Water Research*, 72, 145–153; 2015.

Holm, J. V., Ruegge, K. and Bjerg, P. and Christensen, T. Occurrence and distribution of pharmaceutical organic compounds in the groundwater downgradient of a landfill (grindsted, Denmark). *Environment Science Technology*, 29, 1415–1420; 1995. doi:10.1021/es00005a039.

ICH guideline Q2(R2) on validation of analytical procedures, International Council for Harmonisation of Technical Requirements for Pharmaceuticals for Human Use, Validation of Analytical Procedures Q2(R2) Draft version Endorsed on 24 March 2022.

Joakim Larsson, D. G. Pollution from drug manufacturing: review and perspectives. *Philosophical Transactions of the Royal Society B*, 369; 2014.

Larsson, D. G. I. Release of active pharmaceutical ingredients from manufacturing sites: need for new management strategies. *Integrated Environmental Assessment and Management*, 6, 184–186; 2010. doi:10.1002/ieam.20.

Mahar, P. and Verma, A. Pharmaceutical process validation: an overview. *International Journal of Pharma and Bio Sciences*, 3, 243–62; 2014.

Myranda Eckert, Pharmaceutical Waste Disposal Challenges Facing Pharmaceutical Waste Management in Healthcare, July 22, 2020.

Ravali, R., Phaneendra, M., Bhanu Jyothi, K., Ramya Santhoshi, L. and Sushma, K. Recent trends in analytical techniques for the development of pharmaceutical drugs. *Journal of Bioanalysis & Biomedicine* s11; 2011. dio:10.4172/1948–593x.s11–002.

Ravisankar, P., Navya, C., Pravallika, D. and Sri, D. A review of step- by-step analytical method validation. *IOSR Journal of Pharmacy*, 5, 7–19; 2015.

Richardson, M. L. and Bowron, J. M. 1985. The fate of pharmaceutical chemicals in the aquatic environment. *Journal of Pharmacy and Pharmacology*, 37(1), 1–12; 1985.

Sherer, J. T. Pharmaceuticals in the environment. *American Journal of Health-System Pharmacy*, 63(2), 174–178; 2006.

Siddiqui, M. R., Alothman, Z. A. and Rahman, N. Analytical techniques in pharmaceutical analysis: a review. *Arabian Journal of Chemistry*, 10, S1409–S1421; 2017.

The analysis of pharmaceutical in environmental samples Oluwole Sam Oloruntola April 2018.

Virginia Coman, Simion Beldean-Galea, Chromatographic analysis of some antibiotics in water and sediment samples collected from the Romanian Tisza River watershed, *studia ubb chemia*, lxii, 4, tom i, doi:10.24193/subbchem.2017.4.11; 129–142. 2017.

6 Advanced Instrumentation Approaches for Quantification of Pharmaceuticals in Liquid and Solid Samples

Israr Masood ul Hasan, Rabia Ashraf, Muhammad Sufhan Tahir, and Farwa

CONTENTS

6.1 INTRODUCTION

During pharmaceutical production, analytical support provides quality assurance and scientific guidelines for the predominant functions of chemicals during production and packaging activities. In the past, with the help of clinical science and pharmacology driven by chemistry, pharmaceutical research has played an important role in the formation of medicines. The contributions of chemistry,

DOI: 10.1201/9781003361091-6

biochemistry, microbiology and pharmacology has set a standard in drug development, where new pharmaceutical products are a result of biologists and chemists exchanging ideas.

The medicine development process begins through the invention of drug substances that have demonstrated treatment to fight, resist, prevent, or cure illnesses. The combination and characterization of these compounds, also known as active pharmaceutical ingredients (APIs), as well as their examination to provide initial safety and therapeutic effectiveness data are all necessary steps in identifying drug candidates for further research (Velagaleti et al., 2003).

In pre-drug discovery, research depends on understanding the root cause of the disease to be treated, how the genes cause mutation or disease, the interface of proteins with the afflicted cells, the alteration brought about by these affected cells and how the potential new drugs influence these cells. Based on these findings, a chemical is developed that relates with the damaged cells and might finally become a medication molecule or active API.

Pharmaceutical research is creating more complicated chemicals and medicine formulations and each new and highly selective one is a game changer. As a result, the analytical approach has a lot of promise. Therefore, pharmaceutical quality control should guarantee that the right analytical methods are used. Because of the importance of pharmaceutical quality control procedures and the need for quick, accurate, and unambiguous analytical techniques, in this chapter we summarize recent instrumentation approaches for quantification of pharmaceuticals in liquid and solid samples.

Analytical techniques play an important role throughout the drug development process, from knowing the physical and chemical permanency of the drug to influencing the dosage form selection and design, assessing the stability of the drug molecules, quantifying impurities and identifying those impurities that are above the established threshold level, and evaluating the toxicity profiles of these impurities to distinguish these from APIs (Siddiqui et al., 2017). In pharmacokinetic investigations, the examination of the drugs and their metabolites, which can be quantitative or qualitative, is extensively used. In view of this discussion, the selected analytical techniques such as spectrophotometry in the ultraviolet/visible (UV/VIS) range, electroanalytical methods, fluorimetry and titrimetry chromatographic techniques (mostly voltammetry)—thin-layer chromatography (TLC), gas chromatography, and, in particular, high-resolution liquid chromatography (HPLC))—Capillary Electrophoresis (CE) helps to separate the molecules in an electric field according to size and charge of molecules. CE and vibrational capillary electrophoresis are used for quantitative research on pharmaceuticals in liquid and solid samples.

6.2 STAGES FOR DRUG DEVELOPMENT

In pharmaceuticals drug development is complex; it is the whole procedure of manufacturing novel drugs/compounds in compliance with regulatory approval (Deore et al., 2019). During new discovery, drug developmental stages (described in Figure 6.1) should be in compliance with high standards in each phase like manufacturing, analysis, and finally application for preclinical and clinical approval. This helps to identify chemical drugs that are useful in treatment and monitoring disease conditions. Often, researchers discover new medicine along with new ideas into a disease process that allows the researchers to design a drug that will control the effect of this disease (Brodniewicz and Grynkiewicz, 2010).

6.3 DISCOVERY AND DEVELOPMENT

6.3.1 Discovery

Generally, scientists discover new drugs through:

- Scientists discovering or designing new products for remediation of a newly identified disease or to stop its effects.

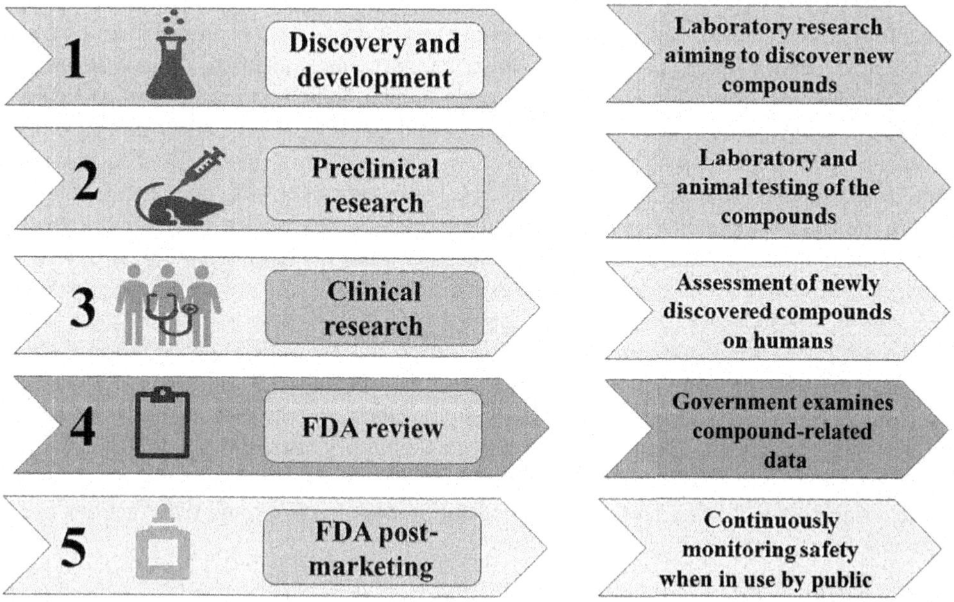

FIGURE 6.1 Drug development stages.

- Assessments of molecular compounds present in drugs show positive effects against identified diseases with minimum side effects.
- Already present treatments have unexpected effects against disease.
- Newly discovered technologies help medicinal products to provide new behaviors of the drug and target specific sites inside the body or to influence on genetic material.

(Deore et al., 2019)

6.3.2 Development

After identification of newly discovered compounds for development, scientists gather information by conducting experiments on:

- How drugs' absorption, distribution, metabolism and excretion occur.
- Mechanism of action of specific drug.
- Best dosage concentration.
- Appropriate drug route in body (such as by mouth or injection).
- How drug toxicity occurs or side effects of the medicine.
- How side effects or toxicity affect different age groups or gender.

(Kaitin, 2010)

6.3.2.1 Preclinical Testing

Preclinical testing of drug development includes newly discovered drugs' efficiency and health safety for treatment process by testing on animals with the perspective of human health. Trials in preclinical testing also have required approved agreement by regulatory authorities. And these regulatory authorities are responsible for giving approval only to drugs whose trials were conducted in a properly safe and effective way and for declaring that these medicines are safe and healthy for usage (Eaton and Gilbert, 2008). Preclinical trials can be conducted in these two ways.

6.3.2.2 Pharmacology

Pharmacology basically deals with these two parameters of drugs, first pharmacokinetics and second pharmacodynamics. These parameters help to determine movements of drugs throughout the body and the specified drug's effect within a suitable animal body. Pharmacodynamics deals with the physiological and biochemical effects of a specified drug. It also is the studying of the mechanism of action within microorganisms, animals, or a combination of organisms. Pharmacokinetic parameters help to determine the safety and efficiency of drugs by knowing the route of administration within the body via injection or oral eating, then their adsorption, distribution within the body, then finally metabolism and excretion from the body (Friedman et al., 2015).

6.3.2.3 Toxicology

Toxicology effects of drugs due to different classification of impurities as described in Figure 6.2 are evaluated by performing in-vitro and in-vivo tests. In-vitro medical tests inspect direct effects on living organism cells or phenotypes. While in-vivo medical experiments conduct qualitative and quantitative analysis of toxicological effects only in a laboratory dish or test tube. Basically, drugs are species specific because toxicity is studied only on appropriate animal species. In-vivo studies help to determine toxicological and pharmacological modes of actions and these actions used for further analysis in clinical studies (Faqi, 2012).

6.3.2.4 Clinical Research

In clinical trials, research of newly discovered treatment methods, therapies, vaccines are directed voluntarily in people and proposed to respond to a specific question with respect to drug safety and efficiency. The specific protocols to follow for drug clinical trials are designed by scientists, researcher manufacturers, or investigators. During clinical study design, researchers design the complete process for investigation in clinical research phases (Table 6.1). Prior to the start of clinical

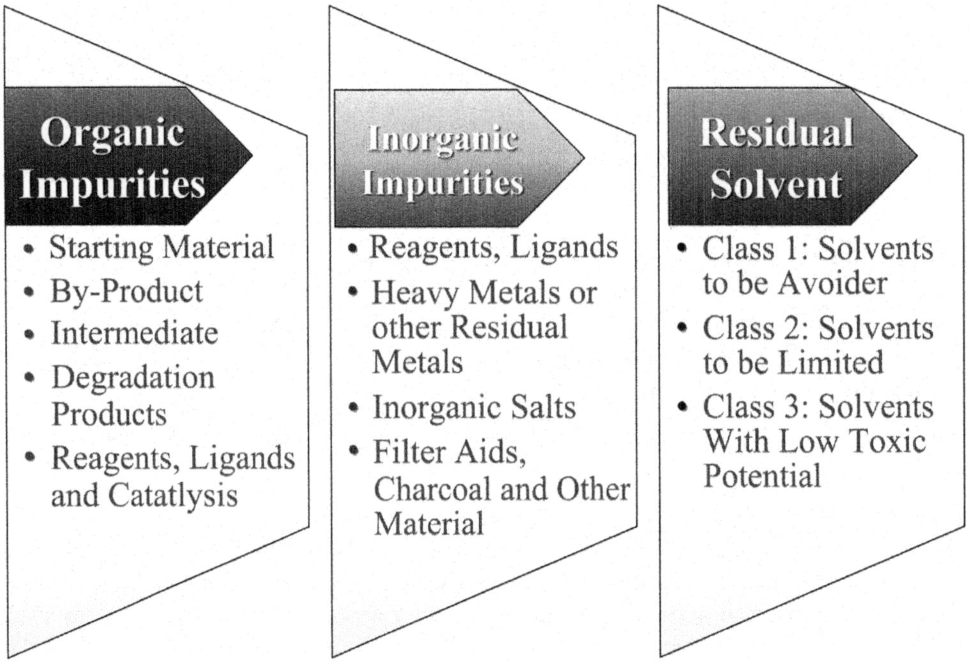
Organic Impurities
• Starting Material
• By-Product
• Intermediate
• Degradation Products
• Reagents, Ligands and Catatlysis
Inorganic Impurities
• Reagents, Ligands
• Heavy Metals or other Residual Metals
• Inorganic Salts
• Filter Aids, Charcoal and Other Material
Residual Solvent
• Class 1: Solvents to be Avoider
• Class 2: Solvents to be Limited
• Class 3: Solvents With Low Toxic Potential

FIGURE 6.2 Classification of impurities.

TABLE 6.1
Phases for Clinical Trials

Phases for clinical trial	Number and type of subject	Investigation
Phase 1	50–200 patients with target disease	• Evaluate safety • Determine safe dosage • Identify side effects
Phase 2	100–400 patients with target disease	• Effectiveness identification • Further evaluation required
Phase 3	1000–5000 patients with target disease	• Confirm effectiveness • Re-monitor side effects • Comparison with other treatment methods • Collection of information
Phase 4	Many thousand or million patients with target disease	• Provide additional information after approval, including benefits and super usage

research for a new drug, researchers gather some information and also have some questions and objectives (Fitzpatrick, 2005). Then, they decide:

- Choice standard for contributors.
- Number of the participants included in research study.
- Specific duration of research study.
- Specific dosage of drug and pathway of administration (injection or oral).
- Risk assessment of specific parameters.
- Information collection and investigation.

(Deore et al., 2019)

6.3.3 Food and Drug Administration Review

Food and Drug Administration (FDA) review requires six to ten months for reviewing and approval of a new drug application (NDA). But in the case of an incomplete NDA, the FDA team refuses NDA approval. FDA approval means that a new medicine for treatment of specific disease has been revealed to be safe and efficient for drug administration and application. Then information up-grade for drug professional labeling (information for how to use drug) and remaining issues resolved before the marketing of the drug. Then the FDA determines whether the drug needs further studies or not, depending upon the FDA reviewing team (Deore et al., 2019).

6.3.4 Post-Market Drug Safety Monitoring

[[Note to comp: Please place the heading “6.3.4 Post-Market Drug Safety Monitoring” after this table. See “AU Original Ms” for styling.]]

After a drug’s FDA approval, a post-market drug safety and monitoring trial is conducted. Post-market drug safety monitoring involves pharmacovigilance. Pharmacovigilance encompasses the activities relating to detection, assessment, and prevention of side effects of a specific drug. It also evaluates the health safety, cost-effectiveness, and efficiency. Post-market safety and drug trials are done by regulatory authorities or by a sponsored company. Therefore, this whole process requires months or even years for completion and bringing into the market. FDA review reports represent prescription and OTC drugs, and safety precautions as well as other adverse drug reactions (Adams and Brantner, 2003).

6.4 ANALYTICAL TECHNIQUES FOR QUANTIFICATION OF PHARMACEUTICALS

6.4.1 High-Performance Liquid Chromatography (HPLC)

High-performance liquid chromatography (HPLC) is the latest form of liquid chromatography that is used to separate a complex mixture of molecules in a chemical system so that individual molecules can be identified. The HPLC technique has great selectivity and achieves appropriate precision at the same time. It is the most commonly used form of chromatographic technique. The choice of the monitoring process is critical in liquid chromatography to ensure that all contents are identified (Siddiqui et al., 2017).

The ultraviolet/visible (UV-Vis) light detector is one of the most commonly used detectors in HPLC, and it may monitor many wavelengths at the same time if a numerous wavelength scanning program is utilized. If light exists in acceptable quantity, the UV detector detects all the UV-absorbing components. A photodiode array (PDA) consists of a lined arrangement of isolated photodiodes on an integrated circuit (IC) chip for spectroscopy. It's attached to a spectrometer's image plane to allow many wavelengths to be perceived at the same time. When using a variable wavelength detector (VWD), a sample must be injected multiple times, each time with a different wavelength, to ensure that all of the peaks are identified. But when using PDA, a wavelength range can be defined and all substances that absorb within that range can be identified in a single study. Peak purity can also be determined with a PDA detector by comparing spectra inside a peak. The PDA detector is being used in the development of an iloperidone detection approach in pharmaceuticals (Manjula Devi and Ravi, 2012).

The fluorescence detector is one of the most sensitive liquid chromatography (LC) detectors. For strong UV absorbing materials, its sensitivity is often 10–1000 times higher than that of the UV detector, giving it an advantage in the measurement of specific fluorescent chemicals in specimens. The estimate of pharmaceuticals is the most important application of fluorescence (Ulu and Tuncel, 2012).

The reversed-phase (RP) method with UV absorbance detection was used by the majority of researchers because it provides high accuracy, study time, repeatability, and sensitivity. HPLC has been used to test several medicines in pharmaceutical formulations. The cost of columns, solvents, and the lack of long-term repeatability due to the proprietary nature of column packing are all constraints of HPLC.

The large range of detection methods and chromatographic columns presented allows for the examination of almost all pharmaceutical molecules. Though, the apparatus essential to conduct HPLC is expensive, this procedure can produce a lot of waste. By using chromatographic columns lined with smaller particles, the problem of waste disposal is reduced, resulting in much faster analyses and, as a result, a reduction in waste formation (Bonfilio et al., 2010).

For iloperidone detection in tablet dosage form, a unique and sensitive RP-HPLC method with PDA detection and UV spectrophotometric method was established. On a LiChrospher® RP-18 HPLC column, chromatographic separation was performed using 0.1 percent trifluoroacetic acid: acetonitrile in a 50:50 v/v (pH 5.02) mobile phase and paracetamol as internal standard. At 275 nm, the effluent was observed. Internal standard and iloperidone produced two sharp peaks at 2.8 and 7.6 minutes, respectively. The UV spectrophotometric technique was used with methanol as the solvent at 229 nm. The HPLC method had a linear range of 1–10 g mL^{-1}, while the UV spectrophotometric approach had a linear range of 2–20 g mL^{-1}. Both methods were shown to be exact, accurate, and sensitive, and can be used for rapid screening of iloperidone in pharmaceutical formulations, according to the International Council for Harmonization of Technical Requirements for Pharmaceuticals for Human Use (ICH) recommendations and statistical analysis (Manjula Devi and Ravi, 2012).

6.4.2 Liquid Chromatography-Mass Spectroscopy (LC-MS)

One of the most important techniques of the last decade of the twentieth century was liquid chromatography combined with mass spectrometry (LC-MS). It has become the technique of choice as analytical support in several stages of pharmaceutical quality control and assurance (Siddiqui, 2017). Since the increased sample amount created by combinational chemistry, the pharmaceutical sector has made significant demand for speedier analytical methods over the last ten years. Through its great sensitivity and specificity, LC-MS has become the most popular technology. Short columns (usually 3–5 cm long) are currently employed with very fast LC gradients of only a few minutes, as opposed to 25 cm long columns with 30–60-minute gradients. Short columns with quick gradients have drawbacks in cases where chromatographic separation is expected to be incomplete.

By using flow injection analysis (FIA), which analyzes samples directly into the MS without any separation, difficulties with ionization suppression are also present. LC-MS and fraction collection have both been presented by various manufacturers. These are identical to earlier systems that collected predetermined fractions based on whether they responded to UV detectors illustrated in Figure 6.3. For MS, the expected mass initiations assortment of an element from the eluent at

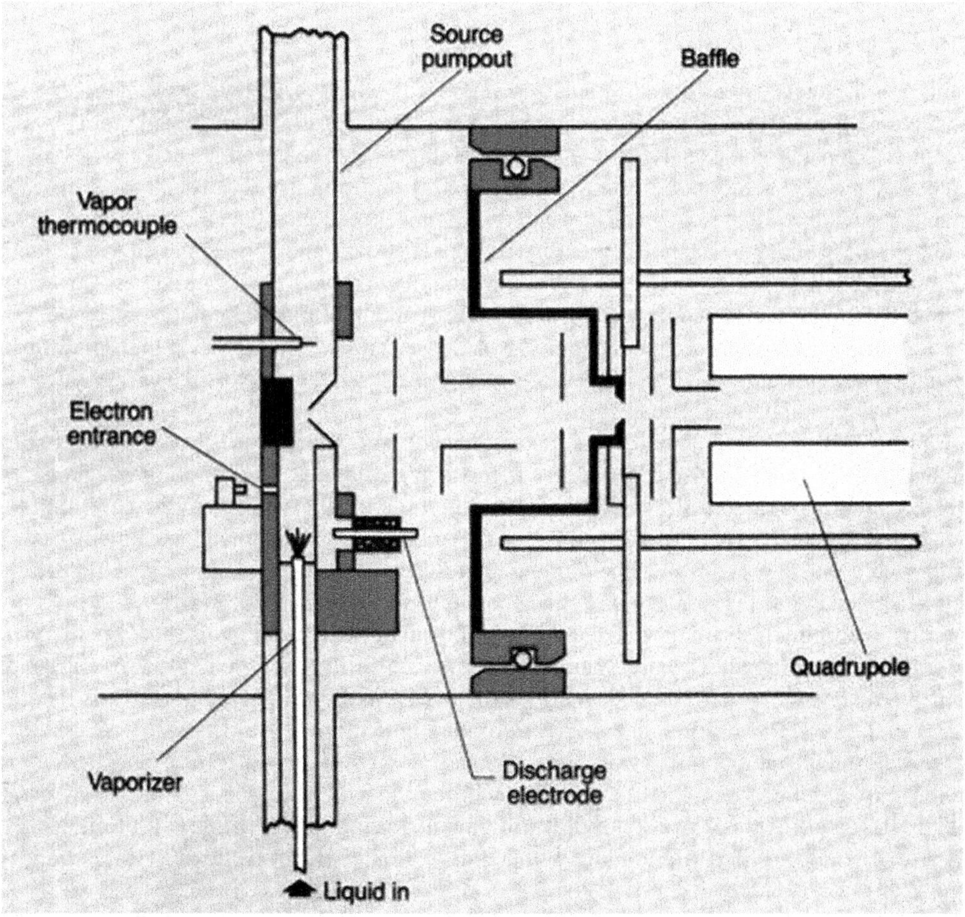

FIGURE 6.3 A thermos-spray LC-MS interfere equipped for three modes of ionization: filament-off ionization, filament ionization, and discharge ionization.

Source: Reproduced with permission of American Chemical Society; Covey et al. (1986).

a fragmented. UV detection in conjunction with MS offers additional structural information and a transparency indication and may reveal components not visible on MS due to failure to ionize or ion conquest (Lim and Lord, 2002).

A multichannel instrument for MS analysis comprise with 96 electrospray tips and the ability to examine 720 samples per hour. Huge amounts of data are produced as an outcome of this amended ability to examine many different samples in a given period of time as a faster procurement of MS equipment. This is rapidly becoming a serious problem, necessitating advances in data management through the use of automated processing and interpretation. Medicinal metabolism research, the study and determination of contaminants and degradation products in pharmaceuticals, and the isolation and characterization of prospective drug compounds from natural or synthetic sources have all been key applications of LC-MS in pharmaceutical investigations because of its sensitivity and specificity (Lim and Lord, 2002).

LC-MS is the preferred process for studying drug metabolism. For structural elucidation, it also offers molecular weight information and fragmentation patterns (Lim and Lord, 2002). LC-MS is used to track impurity profiles during pharmaceutical development and scaling up, as well as to assess the safety of batches utilized in clinical trials (Ermer and Vogel, 2000).

6.4.3 Titrimetric Methods

Titrimetry is a category of quantitative analysis methods in which an analyte's stoichiometric reaction with a reagent of known concentration is evaluated by gradually adding small amounts of the reagent to a sample until the analyte is consumed quantitatively. The reaction's end can be observed visually, using a well-selected indication, or using an instrumental method. The amount of reagent used in the reaction with the analyte, the reagent solution concentration, and the reaction stoichiometry is used to calculate the analyte content in the sample (Kozak and Townshend, 2019).

Titrimetry has been widely used for the examination of pharmaceuticals in past years. Titrimetry has several advantages, including a high processing speed and minimal equipment costs. When conventional separation approaches are compared to advanced techniques such as high-performance liquid chromatography or titrimetry, capillary electrophoresis procedures reveal a lack of selectivity. Their application in pharmaceuticals is given next.

Titrimetry with spectrophotometry was used to develop and optimize procedures for determining stavudine (STV) in high drug and dose forms. In titrimetry technique, STV solution was prepared in an HCl medium with a recognized additional bromate-bromide by followed iodometric back titration to determine the amount of unreacted bromine. According to the authors, the approaches were found to yield satisfactory results when functional to the analysis of STV in tablets and capsules. The titrimetric approach is simple, quick, and economical (Basavaiah et al., 2008).

For the determination of nordiazepam, a spectrophotometric method and two titrimetric methods were developed. The presence of nordiazepam in the precipitated complex is determined by titrimetry, either iodometrically with a standard solution of potassium iodate or with a standard EDTA solution and a xylenol orange indicator (Mostafa and AlGohani, 2010). For the determination of hydroxyzine dihydrochloride (HDH) in pure form and tablets, two titrimetric techniques were established. The methods used acid-base reactions to assess the drug's hydrochloride concentration by titration against standardized NaOH solution, also visually with an indicator phenolphthalein with a glass-calomel electrode device (Rajendraprasad et al., 2010).

For the analysis of doxycycline hyclate (DCH) in large number of medicines and their formulations, one titrimetric and two spectrophotometric methods were described. In titrimetry, DCH was treated in an acid medium with an identified additional bromate-bromide solution; after the reaction between DCH and in-situ bromine the residual bromine was iodometrically titrated again (Jagannathamurthy et al., 2010). Titrimetric procedures are inexpensive, involve minimal instrumentation, and are feasible to use (Bonfilio, 2010).

6.4.4 UV/VIS Spectrophotometric Methods

To measure pharmaceutical active substances, many ultraviolet/visible (UV/VIS) spectrophotometric techniques have been devised. Because most pharmaceuticals have chromophore groups, they can be identified using ultraviolet light. Spectrophotometric approaches are still interesting due to the widespread availability of apparatus, the simplicity to operate, low cost, time efficiency, precision, and accuracy of the methodology (Bonfilio, 2010).

UV spectrometric approaches based on the principles of additivity, absorbance difference, and processing absorption spectra have been established. The absorption of visible and UV radiation (200–400 nm) relates to the excitation of electrons in both atoms and molecules from lower to higher energy levels, according to the basic concept of UV spectroscopy. Because matter's energy levels are quantized, only light with the precise amount of energy required to produce transitions between them will be absorbed (Atole and Rajput, 2018). The following are some examples of spectrophotometric uses in quantitative pharmaceutical analysis.

Pregabalin (PGB) was determined in bulk, pharmaceutical formulations, and human urine samples using a simple UV spectrophotometric approach that was devised and validated (Gujral et al., 2009). Quantitative determination of the lumiracoxib in tablet UV-based spectrophotometry method was devised and validated. The UV method used ethanol as a solvent and measured absorbance at 275 nm. According to the authors, the spectrophotometric approach is a simple, inexpensive, and time-saving method. The spectrophotometric method's detection limit was 0.44 g mL^{-1}, indicating that it is suitably sensitive (Moreira et al., 2008).

Individual determinations of different drugs were made using primary and secondary-order derivative spectrophotometric techniques with methyl alcohol as the solvent at two wavelengths. The scientists found that no interference of matrix constituent was seen and that the advanced method could be effective in drug quality control. Despite its poor sensitivity detection limits of less than 3.23 g mL^{-1} for all drugs, this method worked satisfactorily (Stolarczyk et al., 2009).

6.4.5 Near-Infrared (NIF) Spectroscopy

Spectroscopy in the range of 800–2500 nm is described as near-infrared (NIR) spectroscopy. Light absorption, emission, reflection, and diffuse-reflectance are all examined. NIR spectroscopy includes both electronic and vibrational spectroscopy. When compared to IR and Raman spectroscopy, NIR spectroscopy is particularly distinctive as vibrational spectroscopy because it only deals with overtones and combinations. It is commonly known that electronic transitions in the near-infrared range cause absorption bands in a variety of molecules (Ozaki et al., 2017). The use of near-infrared spectroscopy to interpret the physical and chemical information contained within pharmaceutical items is a unique prospect. Furthermore, NIR spectroscopy enables the separation of molecular-level changes in the material (Awa et al., 2014).

The molecular-level alterations in pharmaceutical tablets were studied using near-infrared spectroscopy. Model tablets containing acetaminophen and microcrystalline cellulose (MCC) were made at various grinding levels in this investigation. Tablet moisture absorbability varied significantly depending on grinding time, and the associated variations in NIR spectra were easily detected. The grinding procedure extensively disintegrates the crystalline and forms a glassy amorphous structure of MCC, which is required to absorb water molecules, according to a comprehensive study employing shifts of amorphous and crystalline peaks at 6970 and 6300 cm^{-1}. This study uses NIR spectroscopy to demonstrate process understanding, which is a goal of process analytical technology (PAT) in the pharmaceutical industry (Awa, 2014).

Near-infrared spectroscopy (NIRS) is a non-destructive, fast method to analyze multi-components in practically any matrix. NIR spectroscopy has achieved widespread acceptance in the pharmaceutical industry in recent years for raw material testing, product quality control, and process monitoring (Siddiqui, 2017). The increasing interest of pharmaceuticals in NIR spectroscopy is most likely via its benefits over other analytical techniques. NIR spectroscopy may samples easily without any pre-treatment, the possibility of separating of the sample measurement position using, fiber optic probes, and it provides physical and chemical parameters in a single spectrum. NIR methods are widely used in major pharmacopeias. In the pharmaceutical study, NIR spectroscopy combined with multivariate data analysis brings up numerous new perspectives, both qualitatively and quantitatively (Siddiqui, 2017).

6.4.6 Electrophoretic Methods

Capillary electrophoresis (CE) is considered as an alternate separation technique in forensic analysis. It can give much faster separations than HPLC and can provide a separation when the analyte in question does not behave well via GC analysis. Capillary electrophoresis produces very efficient, fast separations that can be used on both charged and neutral species. Forensic laboratories prefer capillary electrophoresis due to its excellent separation and resolution power, low mass detection limits, low reagent costs. Capillary electrophoresis, in its most basic form, entails the separation of charged analytes based on differences in their electrophoretic mobilities, which result in distinct migration velocities. These separations are performed out in fused silica capillaries loaded with a background electrolyte and measuring 25–75 m in diameter and 50–100 cm in length (Anastos et al., 2005).

Capillary electrophoresis is a comparatively new analytical technique that involves the separation of charged analytes through a tiny capillary under an electric field. As solutes flow through the detector, they are recognized as peaks, and the area of each peak is proportional to their content, allowing quantitative calculations. CE analysis is relatively more accurate, can be done in a shorter length of time, and only requires a little quantity of injection volume, up to nanoliters. These properties of CE have proven to be advantageous in a variety of pharmaceutical applications. Several studies have been published on the use of this approach in regular drug analysis (Zhang et al., 2009). Several types of capillary electrophoresis have been developed and applied to pharmaceutical purity testing and drug bioanalysis, including capillary zone electrophoresis, isotachophoresis, capillary gel electrophoresis and affinity capillary electrophoresis.

Capillary electrophoresis is an effective separation technique for both large and tiny molecules. This technique provides a high separation efficiency that provides for complex separations, quick method development; it takes a very low amount of sample and consumes less solvent and is automated. Capillary Electrophoresis (CE) instrumentation includes electrodes for sample inserting, a vessel, a high-voltage source of current, a sample detector, and a data output device (Suntornsuk, 2010). Furthermore, various detection methods (mass spectrometry, light-emitting diode, fluorescence, chemiluminescence, and contactless conductivity (C4D) detectors) are employed, providing CE with diversity and sensitivity (Suntornsuk, 2010). The following are a few examples of this technique that can be used in pharmaceutical analysis.

In pharmaceutical formulations for the measurement of mitoxantrone (MTX), a capillary zone electrophoretic (CZE) approach was devised. Separation was obtained using 25 mM ammonium acetate in 50 percent v/v acetonitrile, +30 kV voltage applied, and capillary temperature of 25 °C. The detection wavelength was 242 nm. In less than 7 minutes, MTX and doxorubicin (DOX) both used as an internal standard (ISTD) were entirely separated (Kika et al., 2009).

For the quantification of ethylenediamine, ethanolamine, propylamine, piperazine, and other derivative groups in blood samples and medications, researchers developed a CE approach with UV detection. A 24 cm capillary, UV detection at 214 nm, 20 m mol L^{-1} phosphate consecutively buffer at pH 2, 2 psi s^{-1} injection pressure, and 5 kV voltage were the best conditions for determining antihistamines. The analytical time was less than 10 minutes under these conditions. Furthermore, the approach has a high sensitivity for all of the substances investigated, with detection limits ranging from 4 to 28 pg mL^{-1} (Rambla-Alegre et al., 2010). Capillary electrophoresis was shown to provide substantial benefits in terms of high separation efficiency, low reagent usage (mainly aqueous solutions), and little waste creation. Due to the flow cell's shorter path length, this approach typically has low sensitivity (Bonfilio, 2010).

6.4.7 Fluorimetry

Pharmaceutical companies are always on the lookout for sensitive analytical techniques that use tiny samples. One of the approaches that can achieve high sensitivity without losing precision or specificity is fluorescence spectrometry. In the recent past, there has been a gradual growth in the number of articles on the use of fluorimetry in the quantitative analyses of different drugs in dosage forms and biological fluids (Rahman et al., 2009).

Luminescence spectroscopy is a highly sensitive analytical method that has been widely used to solve problems with low detection limits (Sotomayor et al., 2008). This approach is used to analyze concentrations in complex matrices in two ranges, one in the ng mL^{-1} and, second in some cases, the pg mL^{-1} (Andrade-Eiroa et al., 2010). As a result, light intensity measurements have enabled the specific and complex quantification with a wide range of active medicinal compounds. Fluorescence, phosphorescence, and chemiluminescence are three distinct techniques that can be used in luminescence spectroscopy. Fluorimetry, in pharmaceutical analysis, is most often used luminescence method (Bonfilio, 2010). The following are some examples of fluorimetric procedures used in pharmaceutical analysis.

The determination of individual verapamil hydrochloride, diltiazem hydrochloride, nicardipine hydrochloride, and flunarizine utilized a kinetic spectrofluorimetric technique with water as the diluting solvent. The approach relied on the degradation of the drugs in an acidic media using cerium ammonium sulfate. After excitation at 255 nm, the fluorescence of the generated Ce was analyzed at 365 nm. The proposed method was used to analyze marketable tablets, and the researchers found that it was easy, quick, and economical (Walash et al., 2009).

Based on irinotecan (CPT-11) and topotecan (TPT), a fluorimetric technique for assessing camptothecin (CPT) in anti-cancer drugs was developed. The detection of TPT in samples was done at 368 nm, but the detection of CPT-11 in samples was done at 267 nm, which is a differential wavelength, utilizing the subsequent wavelength derived from the corresponding spectrum. The study concluded that spectrofluorimetry allowed them to make determinations faster, cheaper, and easier. The approach was complex, with detection confines of around 9 ng mL^{-1}, making it an attractive process (Marques et al., 2010).

When compared to spectrophotometric procedures and even chromatographic techniques, spectrofluorimetric techniques have the benefit of being highly sensitive. Furthermore, because few molecules have native luminescence, these approaches are very selective, allowing drug quantification in biological matrices. The ability to analyze only luminous substances, which makes this method more complex and time-consuming, is one of this technique's limitations (Bonfilio, 2010).

6.4.8 Mass Spectrometric Detection

Mass spectrometry (MS) has evolved as a front-runner approach in pharmaceutical analysis over the previous few decades, including both qualitative and quantitative features. MS is just one of several analytical techniques used in drug research and discovery. It is used to characterize and analyze a wide variety of biological and chemical entities. The produced ions are separated, processed, digitized, and identified by ion detectors in mass spectrometers. The detectors amplified the tiny response in such a way that a usable mass spectrum is generated (Goodlett et al., 2006).

Mass spectrometry (MS) is based on creating ions from inorganic or organic molecules using any suitable method, separating them, and detecting them qualitatively and quantitatively based on their mass-to-charge ratio (m/z) and abundance. Thermally, by electric fields, or by striking energetic electrons, ions, or photons, the analyte can be ionized. A mass spectrometer is made up of an ion source, a mass analyzer, and a detector that all work in a high vacuum. Although some material is required for mass spectrometry, the amount of analyte required is in the microgram range, making it a practically non-destructive technology. Mass spectrometry is the preferred method when many other analysis methods fail to give analytical data from nanogram amounts of material due to its extremely low sample consumption. The intensity of a peak, as signals, directly reflects the number of ionic species of that related m/z ratio that have been produced from the analyte within the ion source (Gross, 2006).

6.4.9 Electrochemical Method

Electrochemical approaches have been increasingly popular in the evaluation of pharmaceutical drugs in recent years. In the analysis of drugs and pharmaceuticals, many electrochemical methods such as voltammetry, polarography, amperometry, and potentiometry were used (Siddiqui, 2017). In contrast, voltammetric procedures are by far the utmost extensively utilized electrochemical methodology for analyzing pharmaceuticals in dosage forms (Santos et al., 2009). Electrochemical approaches are characterized by high sensitivity and less limit of detection, as well as the utilization of simple and low-cost equipment. The following sections will describe different electroanalytical methods that have been used in pharmaceutical analysis.

The capsaicin-modified carbon nanotube modified basal-plane pyrolitic graphite electrode has been developed for the determination of benzocaine and lidocaine. The production of the capsaicin-benzocaine adduct, which is initiated electrochemically, results in a linear decrease in the voltammetric signal corresponding to capsaicin, which is proportional to the adding concentration of benzocaine (Kachoosangi et al., 2008). In pharmaceutical formulations, to measure sulfadiazine, researchers used a bismuth-film electrode. Differential-pulse voltammetry was used to examine sulfa drugs in a 0.05 mol L^{-1} Britton-Robinson with pH 4.5 solution. The bismuth-film electrode remained appropriate due to the direct cathodic voltammetric measurement of sulfadiazine. The simplicity, good sensitivity with a limit of detection of 2.1 micro mol L^{-1}, cost effectiveness, and the opportunity of application to pharmaceuticals are all advantages of this technology (Campestrini et al., 2010).

Propranolol (PROP) and atenolol (ATN) in pharmaceutical formulations were measured using square-wave voltammetry and a cathodically produced boron-doped diamond electrode. Propranolol and atenolol electroanalytical measurements were performed in 0.1 mol L^{-1} H_2SO_4 or 0.5 mol L^{-1} $NaNO_3$ and pH 1.0, modified with concentrated nitric acid, correspondingly. In numerous pharmaceutical formulations (tablets), the suggested process was effectively employed in the detection of drugs, according to the authors, with results that were in excellent correlation at a 95 percent confidence level. The procedures described are sensitive, simple, and quick, and they don't require any organic reagents or expensive equipment (Sartori et al., 2010).

High sensitivity, appropriate swiftness, pre-treatment, acceptable discrimination, extensive pertinency, and less instrumentation maintenance costs are all features included in electrochemical techniques (mostly voltammetry) (Bonfilio, 2010).

6.5 CURRENT CHALLENGES IN DETECTION AND ANALYSIS

In the previous sections, different analytical techniques used in pharmaceuticals are described separately. These analytical techniques include a wide variability of products, equipment, and machinery. Each type of instrumentation plays an important role in every step of the medicine's manufacturing process. Instrumentation for the pharmaceutical industry's analytical techniques require the highest level of precision, reliability, and quality for analysis of different drugs. With the highest level of precision and accuracy, analytical instrumentation also faces some sort of challenges, so in this book chapter some analytical instrumental challenges are also describe precisely. Pharmaceuticals drugs have been analyzed using both macro-ATR and micro-ATR imaging. The limitations of each technique are recompences of the other, which is why the two techniques complement each other so well. The macro-ATR approach is a high-speed tool for counterfeit detection, but its usage for determining tablet formulations is restricted. On the other side, the micro-ATR imaging technique is slow and has partial use for detecting counterfeits but it is a more effective tool for characterizing tablet formulations. When it is of interest to both detect and characterize of counterfeit formulations for sourcing purposes, the most effective and inclusive approach involves using both techniques (Mattrey et al., 2017).

The challenges in the potentiometric titration technique should be distinguished noticeably. Specially, titrimetric analysis lacks dynamic range (1–2 orders of magnitude), which often makes it difficult to support dissolution testing of drug formulations and early drug development which comprise the wide range of doses in pharmaceuticals (Samadov et al., 2017).

In addition, analytical interference in drug formulation from other components, such as the preparation excipients, dissolution media, and active pharmaceutical constituents for multi-component dosage formulae, demonstrate another direct UV detection issue. Furthermore, when multi-active ingredients or components are involved in the pharmaceutical for dosage formation, the spectrophotometric technique often cannot be applied, because of cross-interference of active ingredients. Therefore, there is currently a need to develop UV-VIS spectrometers that can be able to perform multi-component analyses for the combinational dosage formation, and develop more including further different method for development counting with investigation of data treatment for routine dissolution testing.

In high-performance liquid chromatography (HPLC), instrument design should quantitively nano-peptides investigated of by-products in one hour. The limitations of technique, however, become apparent when complex matrices have to be analyzed or when trace amounts of by-products have to be quantified. Further studies on the influence of salt and buffer concentrations on temperature reaction, and longer residence times due to lower pressures and diffusion rates have to be considered with respect to fluorescence resolution and yield. In the future, instrumental modification can be projected with both pumping systems and fluorescence detectors to minimize the cost as well as improve the sensitivity. But some of these studies in this regard are currently in progress (Felletti et al., 2019).

Similarly, with HPLC in mass spectrometry (MS) there are many challenges that lie ahead to examine method and technology. However, in a comparatively short period of time huge advancements have made it necessarily to quantitate necessity in quantitating analytes by mass spectrometry. Of particular note, in diagnostic clinical testing new modifications require examining sensitivity and metabolomics. In addition, a great start has also been made by the CDC in standardizing steroid hormone assays. So, mass spectrometry is limited for the identification of hydrocarbons that produce similar ions, and it is unable to express difference between optical and geometrical isomers apart. The shortcomings are compensated for by combining MS with other techniques, such as gas chromatography (GC-MS). Accordingly, based on experience it could take years to achieve such accomplishments. Continuing efforts and funding by different organizations are essential in laboratory testing to attain these goals (Awad and El-Aneed, 2013).

On the basis of theoretical and experimental investigations of the anode workpiece surface and the cathode-tool electrode help to described the affiliation between the characteristics, shape, and the dimensions introduced by the micro features on the anode-workpiece surface and the cathode-tool electrode under machining situations. However, the electrochemical method comprises the narrow or limited temperature range in sample analysis in pharmaceuticals, short or limited shelf life of method, cross-sensitivity for the other gases, greater exposure to the target gas, and a shorter life span. In addition, the limiting conditions of micro-ECM are considered with respect to micro-shaping using non-profiled tool electrodes. So, it is necessary to improve micro-machining capabilities of ECM processes; the application of ultra-short pulse current and ultra-small gap size has been recommended earlier (Ozkan and Uslu, 2016).

In UV-visible spectrometry drugs analysis, stray light can possibly cause serious measurement errors due to wavelength selectors because small amount of light from a wide wavelength range may still be transmitted from the sample. Despite this, stray light sources may come from also the loosely fitted instrument or come from the environment (sun light). In addition, light scattering is often triggered by suspended solids in liquid samples, which may cause the reason for serious measurement errors. The presence of air bubbles in the cuvette or sample will scatter light, resulting in unreliable results. Furthermore, this technique limits the analysis of multiple adsorbents that have similar wavelengths. For example, a sample may have multiple types of the green pigment chlorophyll. So, the different chlorophylls overlap spectra when analyzed simultaneously in the same sample. In this aspect a proper quantitative analysis requires more research for combined analysis for those samples that have similar wavelengths in the same sample (Bakeev, 2010).

The fluorescence method only analyzes fluorescent molecules, not all. In this analysis rapid scanning is not possible to obtain the excitation and emission spectrum of the analyte. Besides, it is not able to analyze reference and sample solutions simultaneously. And the diluted sample solutions of compounds are less stable.

6.5.1 Conclusion and Prospective

The primary goal of pharmaceutical drugs is to help people by preventing or curing diseases or making them immune to potential illnesses. For the medication to accomplish its main goal, it should not contain any impurities or other interference that could be harmful to humans. This book chapter intended to concentrate on the function of different analytical tools in the assessment of pharmaceuticals (both in solid and liquid phase) and giving a comprehensive review of the advance analytical techniques used in pharmaceutical analysis. This book chapter also illustrates the development of the methods starting with the old techniques to the advanced levels of hyphenated methods.

REFERENCES

Adams C, Brantner VV (2003): New drug development: Estimating entry from human clinical trials. FTC Bureau of Economics Working Paper.

Anastos N, Barnett NW, Lewis SW (2005): Capillary electrophoresis for forensic drug analysis: a review. *Talanta* 67, 269–279 1–24.

Andrade-Eiroa A, de-Armas G, Estela J-M, Cerda V (2010): Critical approach to synchronous spectrofluorimetry II. *TrAC Trends in Analytical Chemistry* 29, 902–927.

Atole DM, Rajput HH (2018): Ultraviolet spectroscopy and its pharmaceutical applications: a brief review. *Asian Journal of Pharmaceutical and Clinical Research* 11, 59–66.

Awa K, Shinzawa H, Ozaki Y (2014): An effect of cellulose crystallinity on the moisture absorbability of a pharmaceutical tablet studied by near-infrared spectroscopy. *Applied Spectroscopy* 68, 625–632.

Awad H, El-Aneed A (2013): Enantioselectivity of mass spectrometry: challenges and promises. *Mass Spectrometry Reviews* 32, 466–483.

Bakeev KA (2010): *Process analytical technology: spectroscopic tools and implementation strategies for the chemical and pharmaceutical industries.* John Wiley & Sons.

Basavaiah K, Ramakrishna V, Somashekar C, Kumar URA (2008): Sensitive and rapid titrimetric and spectrophotometric methods for the determination of stavudine in pharmaceuticals using bromate-bromide and three dyes. *Anais da Academia Brasileira de Ciências* 80, 253–262.

Bonfilio R, De Araujo MB, Salgado HRN (2010): Recent applications of analytical techniques for quantitative pharmaceutical analysis: a review. *WSEAS Transactions on Biology and Biomedicine* 7, 316–338.

Brodniewicz T, Grynkiewicz G (2010): Preclinical drug development. *Acta Poloniae Pharmaceutica* 67, 578–585.

Campestrini I, de Braga OC, Vieira IC, Spinelli A (2010): Application of bismuth-film electrode for cathodic electroanalytical determination of sulfadiazine. *Electrochimica Acta* 55, 4970–4975.

Covey TR, Lee ED, Bruins AP, Henion JD (1986): Liquid chromatography/mass spectrometry. *Analytical Chemistry* 58, 1451A–1461A.

Deore AB, Dhumane JR, Wagh R, Sonawane R (2019): The stages of drug discovery and development process. *Asian Journal of Pharmaceutical Research and Development* 7, 62–67.
Eaton DL, Gilbert SG (2008): Principles of toxicology. In CD Klaassen (ed.) *Casarett & Doull's Toxicology. The Basic Science of Poisons*, McGraw Hill. 11–34.
Ermer J, Vogel M (2000): Applications of hyphenated LC-MS techniques in pharmaceutical analysis. *Biomedical Chromatography* 14, 373–383.
Faqi AS (2012): *A comprehensive guide to toxicology in preclinical drug development*. Academic Press.
Felletti S, Ismail OH, De LuCa C, Costa V, Gasparrini F, Pasti L, Marchetti N, Cavazzini A, Catani M (2019): Recent achievements and future challenges in supercritical fluid chromatography for the enantioselective separation of chiral pharmaceuticals. *Chromatographia* 82, 65–75.
Fitzpatrick S (2005): *The clinical trial protocol*. Institute of Clinical Research (ICR).
Friedman LM, Furberg CD, DeMets DL, Reboussin DM, Granger CB (2015): *Fundamentals of clinical trials*. Springer.
Goodlett DR, Gale DC, Guile SS, Crowther JB (2006): Mass spectrometry in pharmaceutical analysis. In *Encyclopedia of Analytical Chemistry: Applications, Theory and Instrumentation*.
Gross JH (2006): *Mass spectrometry: a textbook*. Springer Science & Business Media.
Gujral RS, Haque SM, Shanker P (2009): Development and validation of pregabalin in bulk, pharmaceutical formulations and in human urine samples by UV spectrophotometry. *International Journal of Biomedical Science: IJBS* 5, 175.
Jagannathamurthy P, Basavaiah KR, Ranganath M, Nagaraju D, Kanakapura R, Vinay B (2010): Titrimetric and spectrophotometric determination of doxycycline hyclate using bromate-bromide, methyl orange and indigo carmine. *Chemical Industry and Chemical Engineering Quarterly/CICEQ* 16, 139–148.
Kachoosangi RT, Wildgoose GG, Compton RG (2008): Using capsaicin modified multiwalled carbon nanotube based electrodes and p-chloranil modified carbon paste electrodes for the determination of amines: application to benzocaine and lidocaine. *Electroanalysis: An International Journal Devoted to Fundamental and Practical Aspects of Electroanalysis* 20, 2495–2500.
Kaitin KI (2010): Deconstructing the drug development process: the new face of innovation. *Clinical Pharmacology & Therapeutics* 87, 356–361.
Kika FS, Zacharis CK, Theodoridis GA, Voulgaropoulos AN (2009): Validated assay for the determination of mitoxantrone in pharmaceuticals using capillary zone electrophoresis. *Analytical Letters* 42, 842–855.
Kozak J, Townshend A (2019): *Titrimetry: overview*. Elsevier Inc.1–10.
Lim C-K, Lord G (2002): Current developments in LC-MS for pharmaceutical analysis. *Biological and Pharmaceutical Bulletin* 25, 547–557.
Manjula Devi A, Ravi T (2012): Validation of UV spectrophotometric and HPLC methods for quantitative determination of iloperidone in pharmaceutical dosage form. *International Journal of PharmTech Research* 4, 576–581.

Marques FFDC, da Cunha AL, Aucélio RQ (2010): Selective spectrofluorimetric method and uncertainty calculation for the determination of camptothecin in the presence of irinotecan and topotecan. *Analytical Letters* 43, 520–531.
Mattrey FT, Makarov AA, Regalado EL, Bernardoni F, Figus M, Hicks MB, Zheng J, Wang L, Schafer W, Antonucci V (2017): Current challenges and future prospects in chromatographic method development for pharmaceutical research. *TrAC Trends in Analytical Chemistry* 95, 36–46.
Moreira T, Pierre M, Fraga C, Sousa V (2008): Development and validation of HPLC and UV spectrophotometric methods for the determination of lumiracoxib in tablets. *Revista de Ciências Farmacêuticas Básica e Aplicada* 29.
Mostafa N, AlGohani E (2010): Spectrophotometric and titrimetric methods for the determination of nordiazepam in pure and pharmaceutical dosage form. *Journal of Saudi Chemical Society* 14, 9–13.
Ozaki Y, Genkawa T, Futami Y (2017): Near-infrared spectroscopy. *Encyclopedia of Spectroscopy and Spectrometry*, 40–49.
Ozkan SA, Uslu B (2016): From mercury to nanosensors: past, present and the future perspective of electrochemistry in pharmaceutical and biomedical analysis. *Journal of Pharmaceutical and Biomedical Analysis* 130, 126–140.
Rahman N, Siddiqui S, Azmi SNH (2009): Spectrofluorimetric method for the determination of doxepin hydrochloride in commercial dosage forms. *AAPS Pharmscitech* 10, 1381–1387.
Rajendraprasad N, Basavaiah K, Vinay BK (2010): Acid-base titrimetric assay of hydroxyzine dihydrochloride in pharmaceutical samples. *Chemical Industry and Chemical Engineering Quarterly* 16, 127–132.

Rambla-Alegre M, Peris-Vicente J, Esteve-Romero J, Capella-Peiró M-E, Bose D (2010): Capillary electrophoresis determination of antihistamines in serum and pharmaceuticals. *Analytica Chimica Acta* 666, 102–109.

Samadov B, Sych I, Shpychak T, Kiz O (2017): Quantitative determination by potentiometric titration method of active pharmaceutical ingredients in complex dosage form.

Santos AL, Takeuchi RM, Stradiotto NR (2009): Electrochemical, spectrophotometric and liquid-chromatographic approaches for analysis of tropical disease drugs. *Current Pharmaceutical Analysis* 5, 69–88.

Sartori ER, Medeiros RA, Rocha-Filho RC, Fatibello-Filho O (2010): Square-wave voltammetric determination of propranolol and atenolol in pharmaceuticals using a boron-doped diamond electrode. *Talanta* 81, 1418–1424.

Siddiqui MR, Alothman ZA, Rahman N (2017): Analytical techniques in pharmaceutical analysis: a review. *Arabian Journal of Chemistry* 10, S1409–S1421.

Sotomayor MD, Dias ILT, Lanza MR, Moreira AB, Kubota LT (2008): Application and advances in the luminescence spectroscopy in pharmaceutical analyses. *Química Nova* 31, 1755–1774.

Stolarczyk M, Apola A, Krzek J, Sajdak A (2009): Validation of derivative spectrophotometry method for determination of active ingredients from neuroleptics in pharmaceutical preparations. *Acta Poloniae Pharmaceutica. Drug Research* 66.

Suntornsuk L (2010): Recent advances of capillary electrophoresis in pharmaceutical analysis. *Analytical and Bioanalytical Chemistry* 398, 29–52.

Ulu ST, Tuncel M (2012): Determination of bupropion using liquid chromatography with fluorescence detection in pharmaceutical preparations, human plasma and human urine. *Journal of Chromatographic Science* 50, 433–439.

Velagaleti R, Burns PK, Gill M (2003): Analytical support for drug manufacturing in the United States-from active pharmaceutical ingredient synthesis to drug product shelf life. *Therapeutic Innovation & Regulatory Science* 37, 407.

Walash M, El-Brashy A, El-ENany N, Kamel M (2009): Spectrofluorimetric and spectrophotometric determination of rosiglitazone maleate in pharmaceutical preparations and biological fluids. *Pharmaceutical Chemistry Journal* 43, 697–709.

Zhang Z-X, Zhang X-W, Zhang S-S (2009): Heart-cut capillary electrophoresis for drug analysis in mouse blood with electrochemical detection. *Analytical Biochemistry* 387, 171–177.

7 Electrochemical Methods for the Detection of Pharmaceutical Residues in Environmental Samples

Shreanshi Agrahari, Ankit Kumar Singh, Ravindra Kumar Gautam, and Ida Tiwari

CONTENTS

7.1 INTRODUCTION

Pharmaceuticals are chemicals with biological effects that have been discovered in the environment even at low quantities from ng L^{-1} to μg L^{-1} (Phillips et al. 2004). These compounds have a distinct role in the environment and human health since their presence, even at low concentration, may cause irreparable damage (Kronacher and Hogreve 1936). Moreover, it has been discovered that the products from pharmaceutical industries such as medicines are posing a new sort of environmental pollution and potential health hazards to people (Wintgens et al. 2008). Pharmaceutical consumption is predicted to rise in the future as a result of increasing living standards as well as due to increase in consumption of medicines with rise in age.

The pharmaceutical products most commonly found in water are anti-inflammatories (e.g., diclofenac), antiepileptics (e.g., carbamazepine), beta-blockers (e.g., propranolol), antibiotics (e.g.,

DOI: 10.1201/9781003361091-7

sulfamethoxazole), antidepressants (e.g., venlafaxine), antimicrobials (e.g., triclosan), and other medicines (Sauberan and Bradley 2018; Phillips et al. 2004; Cháfer-Pericás et al. 2010). Pharmaceuticals are found in medical, industrial, agricultural, and domestic wastes, and accumulate in the environment (Figure 7.1). Pharmaceuticals accumulate over time, posing long-term risks. The presence of these pharmaceuticals in surface and drinking water has a variety of detrimental consequences for people and ecosystems (Cizmas et al. 2015). Pharmaceuticals in drinking water may affect newborns, children, elderly persons and cause renal or liver disease, as well as cancer (Aschengrau et al. 2011). It is critical to determine the effect of these pharmaceuticals' residues on human health after long-term exposure and appropriately measure the presence of pharmaceuticals residues in the environment. As a result, pharmaceuticals residues must be monitored and removed from the environmental matrices.

Pharmaceuticals and their metabolites must be removed from the environment to prevent toxicity and other health risks. Since pharmaceuticals and their metabolites are tested in the $\mu g\ L^{-1}$ to $ng\ L^{-1}$ concentration range, high sensitivity and sample preparation are two important factors to be considered during evaluation (Ribeiro et al. 2013). Sorption and biodegradation are the most common methods for removing pharmaceuticals from diverse environmental matrices (Golet et al. 2002). Furthermore, liquid and gas chromatography are used to detect new toxins based on their polarisation, volatility, and thermal characteristics. Identifying thousands of compounds using standard gas

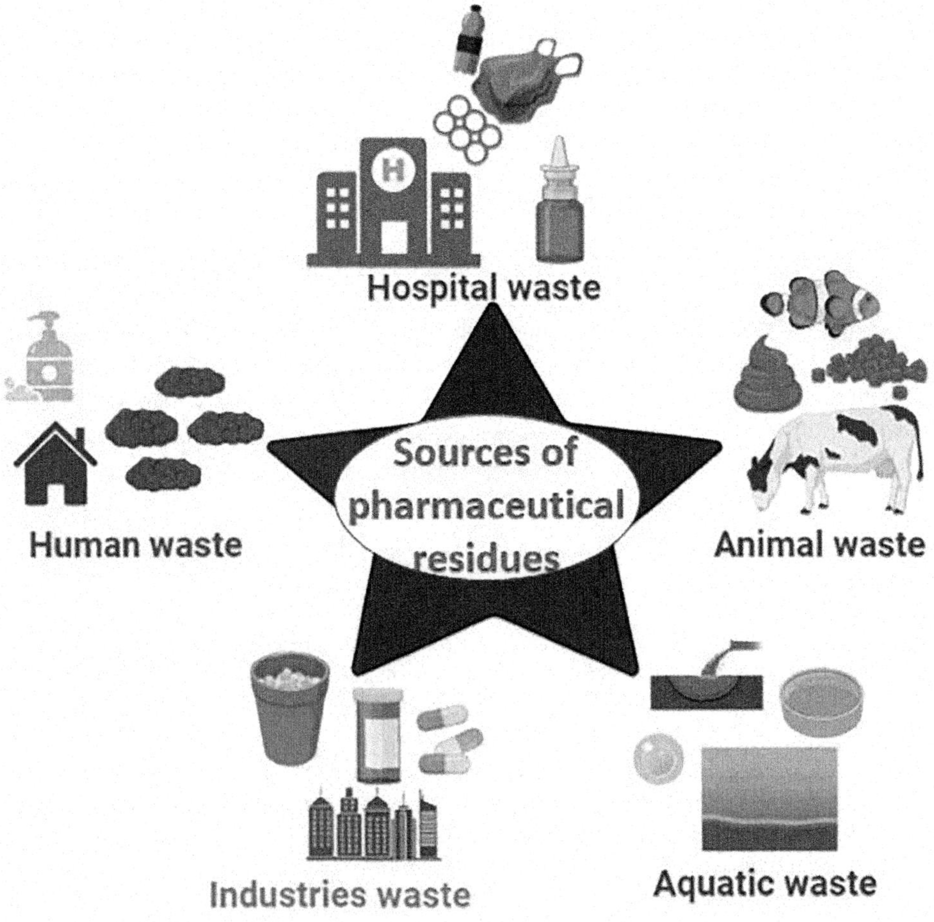

FIGURE 7.1 Various sources of pharmaceutical residues in the environment.

chromatography–mass spectrometry is a difficult process (Rezaei Kahkha et al. 2018; Quintanilla et al. 2018). Significant advances in liquid chromatography–mass spectrophotometry have made it an essential pollutant assessment technology (Gros et al. 2006). Recently developed and extensively used devices include GC-MS, LC-MS (Riva et al. 2018). Carbon dioxide is the most extensively utilised supercritical fluid chromatography mobile phase since it is non-toxic, nonexplosive, and easy to convert to liquid. It facilitates greater column flow rates with minimal pressure dips, ensuing in faster study and less organic solvent consumption.

Moreover, current techniques for assessing pharmaceutical residues need sample collection, treatment, and lab technology. These methods and expenses limit sample numbers. Medical residues are measured using expensive chromatography such as HPLC, GC-MS, etc. (De Saeger and Van Peteghem 1999; Gros et al. 2006). ELISA immunological tests are used in laboratories. High-portability, field-deployable assays would enable large-scale monitoring of drug residues and exposure dangers. Field-deployable, cost-effective equipment can analyse pharmaceutical residues in the environment, water, air, or food chain. Affordable technology may improve rural sustainability, healthcare, and food and water safety. Electrochemical methods such as voltammetry, amperometry, and electrochemical impedance spectroscopy may be miniaturised and used in the field. Recent improvements in electrochemical technologies have allowed the identification of new pollutants via employing electrodes and microelectrode arrays. Nanotechnology allows designers to regulate the size, surface qualities, and composition of useful nanoparticles. Recent advances in nano-patterning and printing methods, such as nano-drop, 2D and 3D printing, allow scalable, controlled production of electrochemical sensors (Baron et al. 2007).

The main objective of this book chapter is to explain how pharmaceutical products act as contaminants and how several analysis and detection techniques are used for the determination of pharmaceutical residues in environmental samples. Advantages of the electrochemical techniques over conventional detection techniques are also discussed. Further, challenges and future aspects in the pharmaceutical residues' detection is also discussed.

7.2 POTENTIAL EFFECTS OF PHARMACEUTICAL RESIDUES IN ENVIRONMENTAL MATRICES

Over the last 15 years, pharmaceuticals have received the attention of environmental scientists because they act as potential bioactive chemicals in the environment (Kümmerer 2009). They are considered as emerging pollutants in the environment (Figure 7.2) because they still remain unregulated, although the directives and legal frameworks are not yet set up (Kolpin et al. 2002). Pharmaceutical residues are defined from these compounds and their metabolites which are biologically active, which are constantly brought together in the aquatic environment and are detected at trace amounts. In fact, it seems that most urban waste is probably contaminated with medicinal compounds because of their pseudo-persistent nature (Jones and Voulvoulis 2001). This affects the environment and may constitute a potential risk for the ecosystems and human and animal welfare in the long term (Klavarioti et al. 2009).

Pharmaceuticals in surface waters are hazardous to aquatic organisms, whilst their existence in drinking water may enhance the occurrence of definite illnesses, such as cancer, asthma, and lung cancer (Phillips et al. 2004). Antibiotics may cause bacteria, especially harmful germs, to become more drug resistant. The incidence of oestrogens in drinking water has been linked to decreased male fertility and an increased risk of breast and testicular cancer. Anticancer medicines, which are often found in drinking water, may cross the blood-placenta barrier, causing teratogenic and embryotoxic effects, and are especially risky to pregnant women owing to their cytotoxic nature (Zwiener 2007).

Filtration, pH correction, framework segregation, and sample concentration are necessary before testing. Each emerging contaminant analysis involves pre-treatment and instrumentation. Sample pre-treatment analysis techniques include multiple calibrated, integrated operations for greater

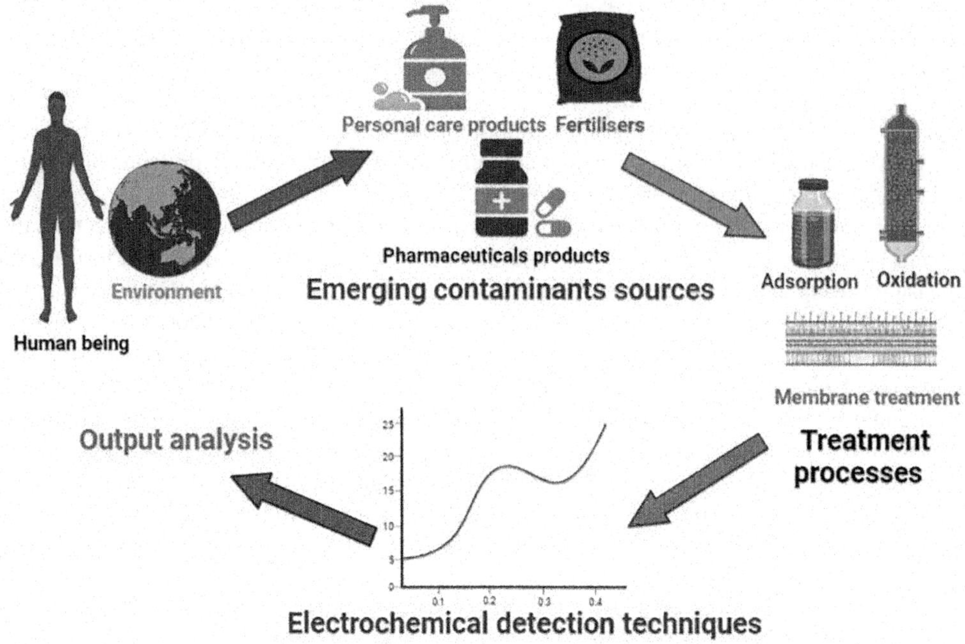

FIGURE 7.2 Steps of the emergence of pharmaceutical residues from several sources with its involved detection process.

performance. The material is processed to collect, separate, or concentrate analytes for instrumental tests. Sample preparation creates a concentration that can be analysed using instruments. Derivatisation is used to modify properties of polar reagents so that they may be analysed using a specific technique. Standard reagents to be derivatised contain large molecules with thermos-labile classes like hydroxyl, carboxyl, or amino groups or small multifunctional unit compounds (Alvarez and Jones-Lepp 2011).

Various pharmaceutical residues are present in the environment including acidic, neutral, and iodinated pharmaceutical compounds. Acidic pharmaceutical residues include compounds containing carboxylic and phenolic hydroxy groups (Ternes et al. 1999). Lipid regulators and their metabolites were tested with phenolic antiseptics. Both beta-blockers and L2-sympathomimetics have secondary aminoethanol structures and hydroxy groups. Due to the large number of functional groups, their polarity is high, hence efficient derivatisation is required for GC analysis (Ternes et al. 1999).

Neutral medicines are molecules from several medicinal families that have no acidic functional groups. Neutral and weak basic medicines include antiphlogistics, lipid regulators, antiepileptic agents, psychiatric treatments, and vasodilators. Hence, neutral medicines can be analysed by GC/MS without derivatisation (Ternes et al. 1999). Iodinated X-ray contrast media are frequently used compounds in medicine. They show high polarity and resistance against metabolism by the organisms and environmental deterioration. Oestrogens are steroid molecules having phenolic or aliphatic hydroxy groups. Because of their identical physical and chemical characteristics, natural oestrogens (e.g., 17β-estradiol, estrone) and synthetic contraceptives may be tested concurrently (Ternes 2001).

Bioaccumulation is not precisely defined; it can mean uptake of substances from the environment or accumulation over time. Bioaccumulation is evaluated by comparing the chemical's concentration in biota (plants, animals) to its concentration in the surrounding medium (e.g., soil, sediment, or water) (Taei et al. 2016). Further, the stability of pharmaceutical residues depends on temperature, concentration, pH, and some external factors (Singh and Tiwari 2020).

7.3 ANALYSIS AND DETECTION OF PHARMACEUTICALS IN ENVIRONMENTAL SAMPLES

The separation procedures of gas and liquid chromatography, particularly the latter, are the most widely utilised for the examination of pharmaceutical residues (Figure 7.3). The creation of reversed-phase liquids chromatography has enhanced the separation of organic compounds, which permits polar compounds to be used as solvents. Separation by chromatography C18 is the most often utilised phase. In the last several years, ultra-high performance liquid chromatography (UHPLC) was developed, allowing for a better resolution and bigger peak size (Jennings and Poole 2012; Careri 2013). The development of atmospheric pressure chemical ionisation (APCI) and electrospray ionisation (ESI) aids in coupling LC with MS, HRMS, and MS/MS using triple quadrupoles. As previously stated, the choice of methodologies is mostly determined by the financial resources of laboratories. As a result, GC/MS is employed to identify the most volatile substances, with derivatisation being required, making GC/MS a lengthier and time-taking procedure. Since the most significant part of mass spectrometry analysis and identification is the ionisation of molecules in the sample mixture, additives are sometimes used to boost ionisation efficiency. This also explains the benefit of this technique in the study of drugs, owing to the ability to switch between negative and positive modes (in the analysis of neutral and basic analytes). It is important to note that the detected fragments for the same molecule may vary depending on investigational circumstances and the equipment utilised. It is also crucial to think about intervention and the medium. Although the APCI and the use of labelled standards may help to eliminate interferences, these standards are not always accessible for all compounds of interest. Finally, it's worth noting that even the most powerful and sophisticated equipment isn't enough. Technical expertise is also necessary for successful functioning, as well as correct column selection and troubleshooting abilities (Rimmer 2015; Scognamiglio et al. 2016).

The most extensively used HPLC/MS/MS technique is EPA-1694 (USEPA 2007). For the analysis of various pharmaceutical residues in environment, this approach employs solid phase extraction (SPE) and HPLC/MS/MS. SPE samples are first acid-treated (pH 2), then alkaline. This is not a

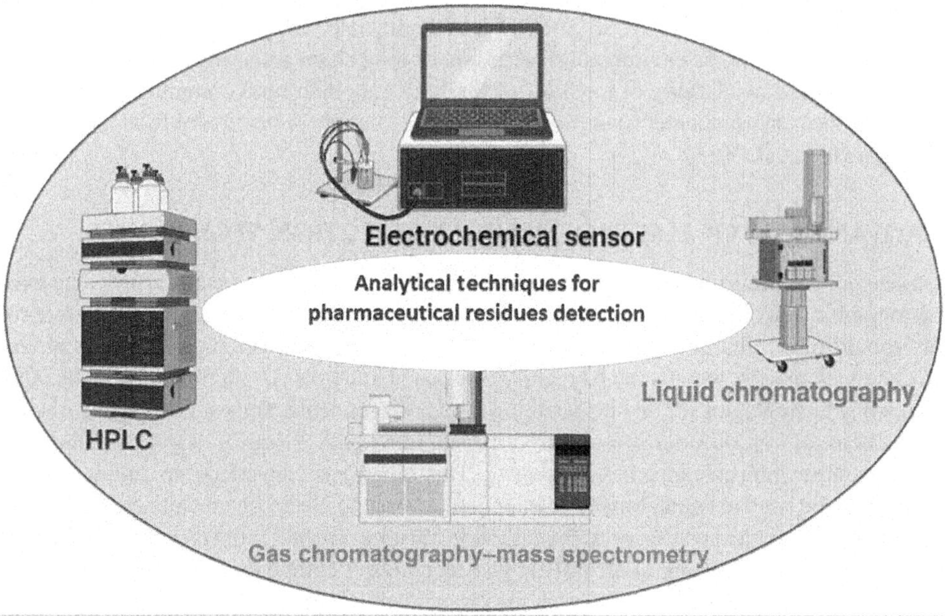

FIGURE 7.3 Several analytical techniques used for detection of pharmaceutical residues in the environment.

regulated technique, but it is a great starting point for constructing a method adapted to the needs and analytes-of-interest, which explains the production of multiple variants of the standard method (Cahill et al. 2004). The US Geological survey (USGS) approach for the detection of 14 pharmaceuticals in the environment by HPLC/MS is another effective method. The filtrate of the sample is separated using SPE with a styrene-divinylbenzene cartridge in this procedure. The product is reduced with nitrogen before being reconstituted with the HPLC main eluent. For detection, identification, and quantification, a reverse-phase octadecylsilane HPLC column is used, which is connected to an electrospray ionisation interface and quadrupole mass spectrometer. The success of GC/MS procedures is dependent on column selection and analytical conditions, including derivatisation, which allows for the identification and quantification of the target analyte (Cahill et al. 2004).

As previously indicated, the most often employed analytical approach for the analysis of pharmaceutical pollutants in waterways is LC-MS/MS, which is followed by gas GC/MS (Wooding et al. 2017). This trend is seen in a recently published study. The approach used relies on the volatility, polarity, and amount of analytes in the sample. These approaches are often reliable and exact, but they need costly equipment and a lengthy analysis period (Sghaier et al. 2017). Some pharmaceuticals residues, such as erythromycin, remain in water, whereas others, such as diclofenac, are "pseudo-persistent." A number of residues are not removed easily from several treatment facilities, indicating that current removal procedures may be ineffective. Advanced oxidation technologies using photocatalysis, ozonation, Fenton and photo-Fenton oxidation, UV/H_2O_2 oxidation, ionising radiation, non-thermal plasma, and sonolysis have all been developed to remove them (Magureanu et al. 2015). Recently, several promising techniques based on degradation using oxidoreductase enzymes found in fungal species have been identified. Because of its simplicity of use, cheap cost, and suitability for a broad variety of chemicals, including organic, inorganic, and ionic species, electrochemical techniques for treatment have received a lot of interest. Their sole stipulation is that big particles be removed from water prior to application during standard treatment (Naghdi et al. 2018).

Current contaminants of emerging concern (CEC) measurement methods include lengthy sample assembly, and handling of the instruments, all of which necessitate many stages and high operating costs, limiting the number of samples that can be evaluated (Campos-Mañas et al. 2017). Chromatographic techniques (HPLC, GC-MS) are used to test CECs; however they are costly and need specialised staff. Immunological tests, such as ELISA are also used, although they are not intended for usage outside of labs (James et al. 2016). Field-deployable, cost-effective equipment that can quickly evaluate CECs in the environment, water, air, or food chain are needed for frequent contamination analysis. The availability of low-cost technology may help rural communities with limited resources enhance environmental friendliness, healthcare, and the protection of food and water supplies (Magureanu et al. 2015).

7.4 ADVANTAGES OF ELECTROCHEMICAL DETECTION TECHNIQUES

A good sensor for measuring pharmaceutical residues should be equivalent to or better than existing chromatographic devices in terms of detection performance. Portable, low-cost, and simple-to-use sensors should offer reliable real-time data with little sample preparation (Agrahari et al. 2022). Electrochemical sensors, which may be readily automated and downsized, provide a viable alternative to existing technologies for on-site detection of pharmaceutical residues. Embedding a sensor in an electrochemical transducer creates electrochemical sensors. A chemical or physical change happens when a target molecule attaches to a sensor. This binding is converted into an understandable signal that may be detected qualitatively or quantitatively using electrochemical methods. Materials and chemical or biological processes at the interface must be carefully considered when developing electrochemical sensors (Meng et al. 2019). The recognition components selected and generated at the electrode surface are compatible with the target analyte. Changes in colour, current, voltage, resistance, and impedance are employed as signals. Work function reveals how much thermodynamic work is necessary to remove an electron from a material's Fermi level, whereas permittivity demonstrates how calm EMR may travel through it (Xu et al. 2019). Further, electrochemical biosensors have a number of benefits, including ease of manufacture, cheap cost, sensitivity, mobility, low power consumption, and adaptability for in situ water monitoring analysis (Singh et al. 2022). Electrochemical biosensors are divided into two types. The first one is a biocatalytic device which includes biosensors based on enzymes, cells, or tissues that produce electroactive species upon interaction with the target analyte. The second one is an affinity device that includes biosensors based on DNA or antibodies or receptors and recognises the selective binding between the target analyte and biological sensing elements (Ronkainen et al. 2010).

In recent years, the use of electrochemical methods in the determination of medicines and pharmaceuticals in the environmental matrices has grown dramatically. The main benefit of this approach is that it is a more advanced instrumentation technique and simple to learn. Low detection limits, simple instrumentation, in situ analysis, low cost of analysis, and no requirement for trained personnel are some of the benefits of electrochemical detection techniques over the other analytical methods presently employed for the determination of pharmaceuticals in the environment (Figure 7.4) (Singh et al. 2021). Pharmaceutical compounds such as imipramine, trimipramine, and desipramine were determined using CV, chronocoulometry, EIS, and ASDPV, in which an electrode made of amberlite XAD-2- and TiO_2 nanoparticle modified with glassy carbon paste electrode was used in the electrochemical analysis of pharmaceuticals (Sanghavi and Srivastava 2013). Electrochemical sensors

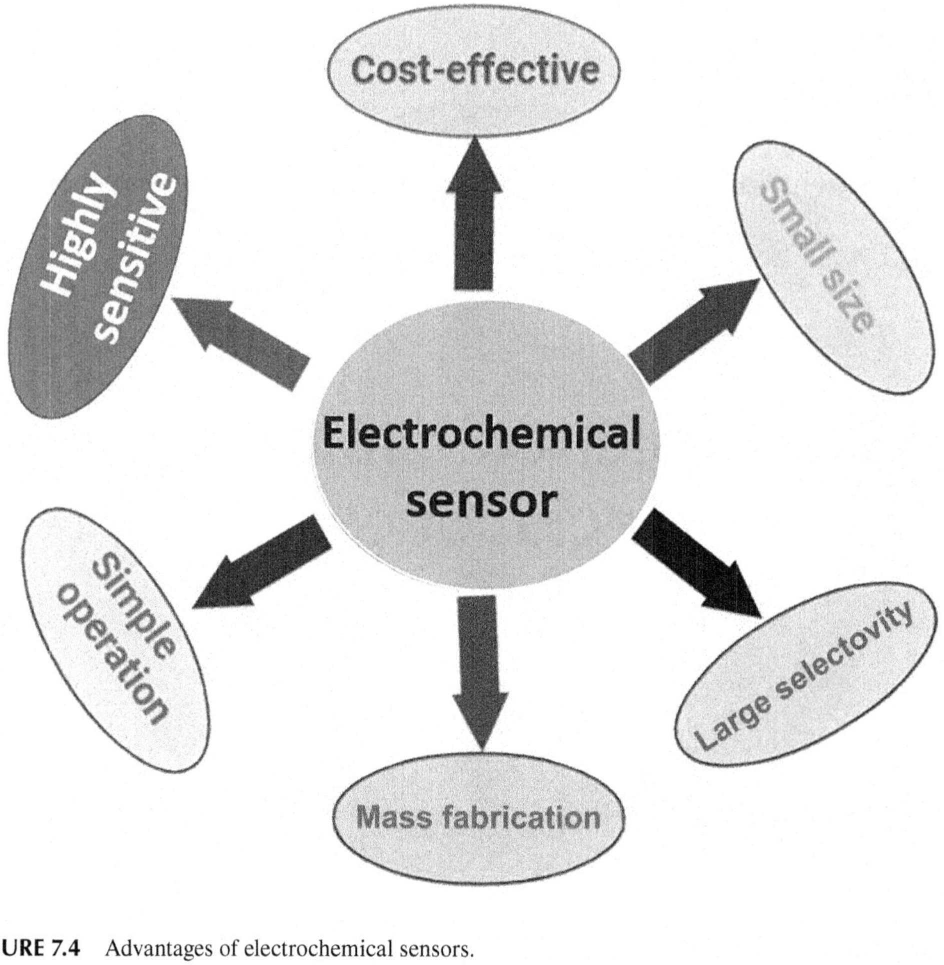

FIGURE 7.4 Advantages of electrochemical sensors.

for detecting pharmaceutical residues need precise binding sites and recognition capabilities. Transducers must be combined in a portable sensing interface. For sensitivity and selectivity, specialised recognition sites are needed. Electrochemical procedures have a cheap operating cost, are simple to use, and have a high sensitivity. A wide range of one-use electrodes and self-powered devices are available, and they provide great platforms for simple, low-cost, in situ analysis of pharmaceutical residues (Agrahari et al. 2022).

7.5 VARIOUS ELECTROCHEMICAL TECHNIQUES FOR THE DETECTION OF PHARMACEUTICAL RESIDUES

Electrochemical techniques are very useful for providing quantitative analytical information. There are many ways for the electrochemical analysis of the analyte that depends analyte oxidation or reduction reaction, and the output signal can be measured in the form of current or potential. Changes in solution properties owing to electron production/consumption are monitored compared to a stable reference electrode. The procedure is focused on the working electrode surface, not the solution concentration. A broad variety (Figure 7.5) of electrical techniques such as potentiometry, amperometry, voltammetry, conductometry, and impedance are used in measurement of electrochemical properties (Bettazzi et al. 2017). Table 7.1 represents the use of various electrochemical detection techniques for the investigation of various pharmaceutical contaminants present in environmental matrices.

Ibuprofen is a common pharmaceutical contaminant, and dozens of electrochemical detection methods have been made for it. Nagaraj et al. (2014) constructed an alumina nanochannel-based

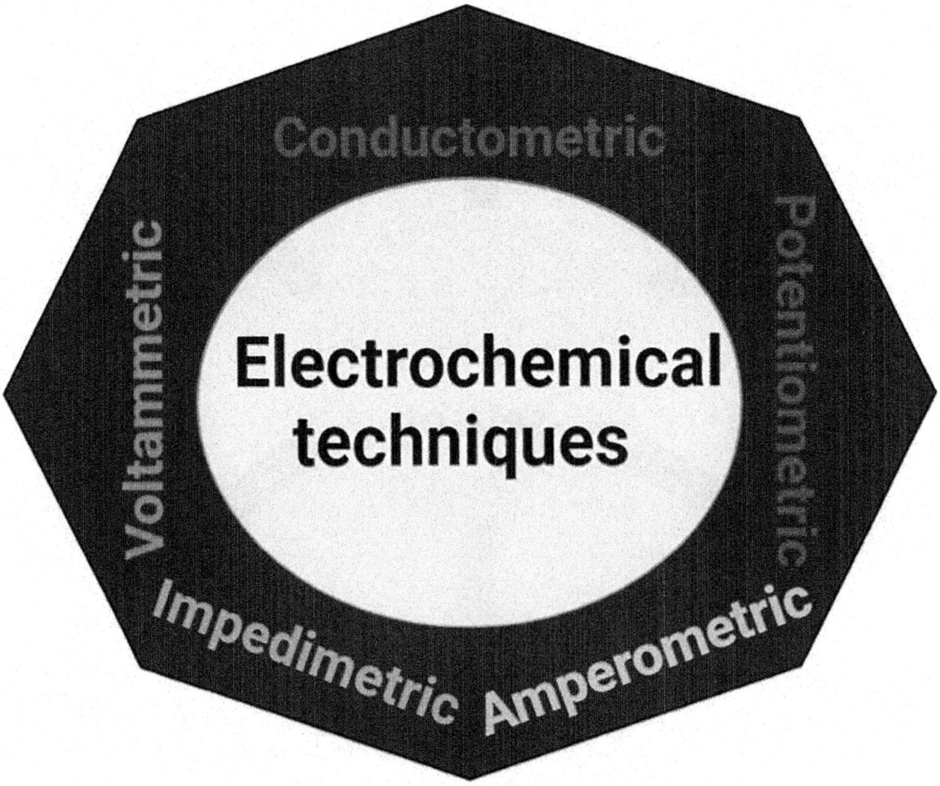

FIGURE 7.5 Various types of electrochemical techniques for the detection of pharmaceutical residues in the environment.

TABLE 7.1
Summary for the Detection of Various Pharmaceutical Products Based on Several Electrochemical Techniques

Analyte	Type of nanosensor	Techniques	LOD	Linear ranges	Application	References
Acetaminophen	AuNPs/MWCNT/GCE	DPV	0.03 μM	1.0–150.0 μmol L^{-1}	Tablets, blood serum	(Madrakian et al. 2014)
Caffeine	$MnFe_2O_4$/CNT-N/GCE	SWV	0.83 μM	1.0×10^{-6} to 1.1×10^{-3} mol dm^{-3}	Bulk	(Fernandes et al. 2015)
Levodopa	MWCNT-PAH/GCE	DPV, CV	0.84 μmol L^{-1}	2.0 to 27 μmol L^{-1}	Tablets, blood serum	(Takeda et al. 2016)
Carbidopa	MWCNT-PAH/GCE	DPV, CV	0.65 μmol L^{-1}	2.0 to 23 μmol L^{-1}	Tablets, blood serum	(Takeda et al. 2016)
Methadopa	MWCNT-PGE	DPV	87 nM	0.1–15 μM	Human serum, urine	(Alipour et al. 2015)
Morphine	MWCNTs/SnO_2—Zn_2SnO_4/CPE	CV, DPV, chronoamperometry	9 nM	0.1–310 μmol L^{-1}	Urine and pharmaceutical samples	(Taei et al. 2016)
Codeine	MWCNTs/SnO_2—Zn_2SnO_4/CPE	CV, DPV, chronoamperometry	9 nM	0.1–600.0 mol L-1,	Urine and pharmaceutical samples	(Taei et al. 2016)
Paracetamol	CuNPs/fullerene-C_{60}/MWCNTs/CPE	CV, EIS, SWV, chronocoulometry	73 pM	4.0×10^{-9} to 4.0×10^{-7} M	Human blood serum, plasma and urine	(Brahman et al. 2016)
Methadone	fMWCNT/MGCE	DPV	0.28 μM	0.5–100.0 μM	Tablets	(Amiri-Aref et al. 2013)
Tadalafil	TiO_2-MWCNTPE	SWV	0.11 μM	0.27–15.2 μM	Tablets, blood serum	(Demir et al. 2014)
Nifedipine	MWCNTs/β-CD	CV, DPV	2.5×10^{-8} M	7.0×10^{-8} to 1.5×10^{-5} M	Drug dosage, blood serum	(Kor and Zarei 2013)
Ganciclovir	Fe_3O_4/cMWCNTs/GCE	SWV	20 nM	80–53,000 nM	Human blood serum and urine	(Paimard et al. 2016)

immunosensor to promote ibuprofen-antibody interaction. The sensor detected ibuprofen in water from several sources, signifying its approach for real-time on-site monitoring (Nagaraj et al. 2014). Antibiotic erythromycin may be detected using a particular antibody (Jacobs et al. 2013). Venlafaxine was electrochemically detected using a nanocellulose-based sensor. Synthesised nano-composite Fe_3O_4@cellulose nanocrystals/Cu was utilised with graphite to make a screen-printed carbon electrode (SPCE) sensor for detecting venlafaxine by CV, chronoamperometry, and DPV. The sensor's LOD was 0.01 M and its linear dynamic range was 0.05 to 600 M (Khalilzadeh et al. 2020). Recent sensing technologies detect drugs using MOFs. Tobramycin (TOB) was detected utilising a Fe-PEI-functionalised MOF aptasensor. The sensor's LOD was 56 pM and its linear range was 100–500 nM. The stability of the sensor was found to be 15 days, with its degradation rate of 6.8% (Zhang et al. 2021).

7.5.1 Conductometry

Conductometric transducers detect changes in an ionic strength of solution, which affects current flow and electrical conductivity. Despite a few benefits such as low-cost thin-film applications, direct real-time monitoring, no requirement for a reference electrode, and the ability to miniaturise, this technology produces less sensitive responses than other electrochemical techniques (Dziąbowska et al. 2018). A conductometric biosensor based on inhibition analysis was initially developed for organophosphorus pesticide detection. A sensitive element was acetyl- and butyrylcholinesterase. Organophosphates inhibit acylcholinesterase by phosphorylating the serin group. The sensitivity of the sensor towards different pesticides (such as diisopropyl fluorophosphate, paraoxon-ethyl, paraoxon-methyl, trichlorfon) was investigated. The minimal detection limits for inhibitor concentrations were 5×10^{-11} M for diisopropyl fluorophosphate, 10^{-8} M for paraoxon-ethyl, 5×10^{-7} M for paraoxon-methyl, and 5×10^{-7} M for trichlorfon (Jaffrezic-Renault and Dzyadevych 2008).

7.5.2 Potentiometry

Potentiometric sensors assess the potential difference between two electrodes when one is referenced to the other. For the detection of chloramphenicol, a selective carbon paste nano-composite sensor was developed using molecular imprinted polymer as a sensing element. Carbon paste was made using molecular imprinted polymer, MWCNT, nanosilica, graphite powder, and room temperature ionic liquid. The nano-composite carbon paste sensor has wide detection range between 1.0×10^{-6} to 1.0×10^{-2} mol L^{-1} with a detection limit of 1.0×10^{-6} mol L^{-1} (Ganjali et al. 2012). Environmentally friendly ion selective electrodes were developed for detecting diclofenac in pharmaceutical effluent. The sensor has a wide detection range from 0.1 μM–10 mM of diclofenac and a detection limit of 0.11 μM. The electrode demonstrated high diclofenac potentiometric selectivity across many interfering ions and organic compounds (Elbalkiny et al. 2019).

7.5.3 Amperometry

Under a constant voltage applied to a working electrode, amperometric transducers detect the direct current from a redox reaction. Before and after engagement with a target molecule, the activity of a recognition element fluctuates. The finished product must be electroactive and undergo a redox reaction. The current is a rate of electron transfer that is proportional to the concentration of analyte. For ultrasensitive detection of kanamycin, a label-free amperometric immunosensor based on graphene sheet-nafion/thionine/Pt nanoparticles is presented. The suggested immunosensor has a low detection limit (5.74 pg/mL), broad linear range (0.01 to 12.0 ng/mL), excellent stability, and good selectivity for kanamycin detection (Wei et al. 2012). This work coupled a biomimetic sensor to a flow injection device to detect paracetamol. Under optimal circumstances, a large linear response range ($1.0 \times 10^{-5} - 5.0 \times 10^{-2}$ mol/L) was produced with a LOD of 1.0 μmol/L (Oliveira et al. 2010).

7.5.4 Voltammetry

Voltammetric transducers are the most thorough and are often employed in biosensing analyses by research organisations. The current-potential connection is measured by the sensor. The potential is detection in no current situations (Caygill et al. 2010). The potential for redox peaks is unique to the species under investigation, and the existing peak size is proportional to the species. In one test, multiple substances with distinct characteristic potentials may be detected. There are several voltammetry techniques for the detection of pharmaceutical residues such as CV, DPV, stripping voltammetry, AC voltammetry, polarography, LSV, and other voltammetric techniques; however, CV, DPV, and LSV are the most widely employed. The distinction is in the manner of possible application. CV is most frequently used in the starting of the experiment for the electrochemical analysis of a medicinal molecule. Organic chemistry, biochemistry, and environmental electrochemistry are just a few of the fields where it is useful. The utility of CV lies in its capacity to quickly observe redox activity over a broad voltage range and to provide preliminary information regarding the electrode/pharmaceutical chemical interface. The active principle does not need to be pre-treated in any way before it may be identified. The main benefits of DPV are its ease of use and the availability of low-cost equipment. Because of its low detection limit of roughly 10^{-8} M, DPV is often used for therapeutic dosage analysis. The simplest is LSV, in which the applied potential grows linearly in time at the working electrode. According to the Faraday law, the flowing current has two parts: first is the Faradaic current due to flow of electrons from the active substance; second is the capacitive current which is generated by the formation of two electric layers between the solution. This technique is useful for the determination of redox reactions but is unsuitable for quantitative analysis because of limited sensitivity. During linear scanning, the DPV concept is used to apply a periodic constant potential pulse. The before and after current difference were measured after the pulse was applied, yielding a single peak-graph. With detection limits of 10–100 μg/L, this approach is very sensitive (Ziąbowska et al. 2018).

7.5.5 Impedimetry

Impedimetric transducers are a kind of transducer that is widely used. EIS is a technique for determining the structure and function of electrodes, particularly those that have been modified with biological material. Depending on whether a redox probe is present in the solution, it may be categorised as Faradaic or non-Faradaic. Because no reagents are required, the second option is favoured POC devices. The resistance and capacitance of a double layer alter when one electrode surface is immobilised, generating an impedance shift. On the sensor surface, the biorecognition process and label-free interactions may therefore be observed. Nyquist and Bode plots are two of the most common outcomes. SAM or conductive polymer base layer methods are often used to make EIS sensors. In comparison to potentiometric or amperometric approaches, detection limits are lower. The biggest disadvantage is that the electrolytes might provide false positive readings. It may be reduced by inhibiting the binding sites of electrode surfaces which are non-specific, such as using bovine serum albumin (BSA) protein. Impedance changes are immediately shown by the immunoreaction between antigen and antibody (Ziąbowska et al. 2018).

7.6 ECENT TRENDS IN ELECTROCHEMICAL SENSING OF PHARMACEUTICALS IN THE ENVIRONMENTAL SAMPLES

Clinical trials evaluate novel medications and medical treatments to enhance quality of life. All pharmaceuticals, treatments, and medical instruments must complete all clinical trial stages to fulfil regulatory standards and enter the market. Researchers review clinical outcomes to establish doses, effectiveness, and safety, dependability, and benefits of new medical technologies. All pharmaceutical medications, vaccinations, cosmetic goods, and many medical advances must be authorised via clinical research and trials before routine practice or accessing the market. Clinical trials are necessary to assess the safety and effectiveness of pharmaceutical substances, medicines, and therapies. Analytical methods have a particular function in clinical trials for determining the concentrations of pharmaceutical substances in biological matrices (Figure 7.6) and monitoring the conditions of patients (or volunteers) throughout different clinical stages.

SWV (square wave voltammetry) uses voltammetry or potentiometry to quantitatively measure ions in clinical studies. Adsorptive, anodic, cathodic, and potentiometric methods include preconcentration, deposition, and stripping. There are two steps for the identification of analyte. First is the accumulation step through which analyte gets accumulated at the working electrode surface through electrolysis. Second step is of detection, which is exactly opposite to the first step, where the analyte concentration decreases and is stripped from the electrode. This approach is commonly utilised in clinical research since it is sensitive and has a low detection limit (Ghoneim and Beltagi 2003). Jain et al. (2010) employed HMDE to assess Cefixime in biological matrices. In another study, differential pulse cathodic adsorptive stripping voltammetry (DPCAdSV) was compared with square-wave cathodic adsorptive stripping voltammetry (SWCAdSV) for analysing antibiotic concentrations (Parnsubsakul et al. 2017). This approach is sensitive, affordable, and repeatable, similar to many spectroscopic methods for determining medicinal substances, especially metal ions. Stripping voltammetry has limits for complicated metal detection and mercury-coated electrode waste.

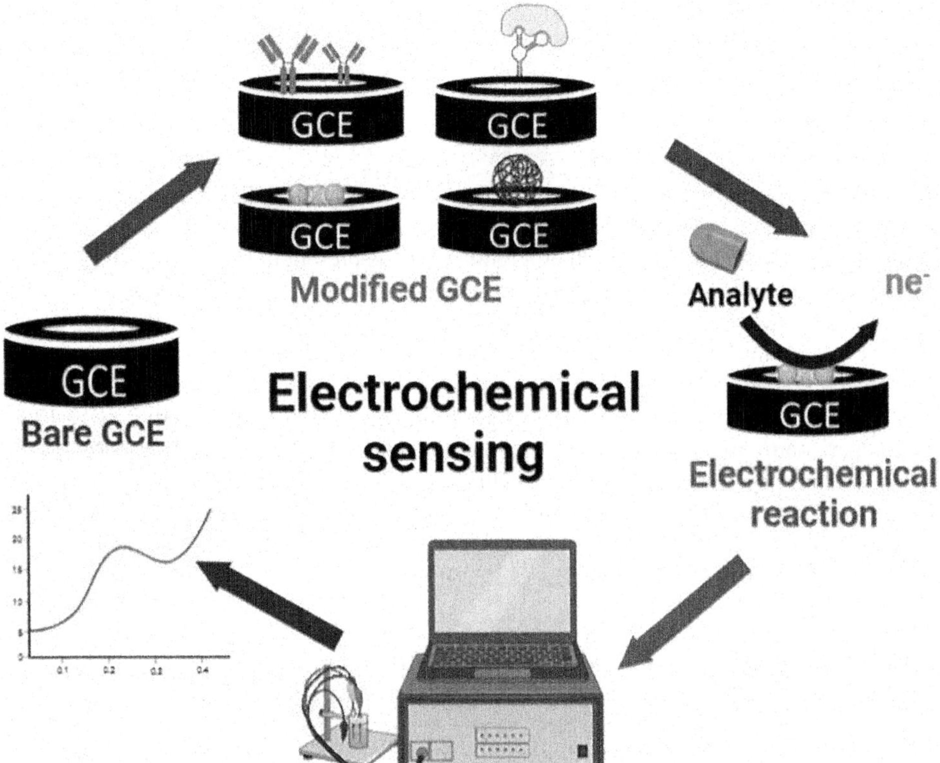

FIGURE 7.6 Stages involved in the electrochemical analysis of pharmaceutical residues.

7.7 CHALLENGES IN ELECTROCHEMICAL SENSING OF PHARMACEUTICAL RESIDUES

Pharmaceuticals are often discovered in the environment, and food supplies offer serious environmental and human health risks. Many medications operate as endocrine disruptors, causing reproductive impacts in animals as well as toxicity at low doses (USEPA 2007). Sensitive testing procedures are needed to detect and monitor environmental consequences. Because of the extensive use and dangerous consequences of pharmaceuticals, the EPA has set standards to analyse and manage environmental concerns about specific chemicals. The development of indirect electro sensing techniques that should be able to enhance the selectivity offered by bioreceptors or biomimetic receptors coupled with nano-composite materials is an open problem in terms of repeatability and automation of the modification operations. Despite this, electrochemical sensors have yet to be used in large-scale environmental and food contamination research. The development of new technologies and materials, such as low-cost disposable paper sensors or smart nanomaterials, may make automated manufacture and, as a result, everyday usage of these devices easier (Zwiener 2007). Despite the established potential of indirect detection methodologies, additional work is needed to apply and technologically transfer them as analytical instruments to address environmental and food-related analytical issues. Despite this, owing to the vast quantity of organics and the low concentration of medicines presently found in waste streams, the detection processes are far from being usable with actual samples. When dealing with waste, preconcentration procedures such as flow injection analysis might be used. Things are complicated due to a lack of standardised regulations concerning the maximum permissible limits of various drug classes, a lack of standardised sampling and pre-treatment methods from country to country, a lack of electrode materials, and the need for improved detection sensitivity and selectivity (Agrahari et al. 2022).

Several routes may be taken to improve electrochemical technologies for field detection of pharmaceutical residues: (i) employing sensibly planned 2D and 3D nanomaterials in combination with natural particles as sensing components; (ii) using scalable manufacturing processes to build electrodes and increase repeatability at a huge scale; (iii) utilising large multi-array devices for detecting many CECs concurrently, maybe incorporating self-reference electrodes to increase selectivity and reduce interferences; (iv) using electrochemical tests with pattern recognition and artificial intelligence, with wireless data processing and remote monitoring; (v) utilising advances in nano-impact electrochemistry, which might permit ultrasensitive analysis of analytes. Electrochemical sensors might be interfaced with a sample unit to form a portable device with separation, identification, and detection capabilities. In brief, electrochemical devices may be cost-effective, usable platforms for detecting environmental pollutants (Hassan and Khan 2021).

7.8 CONCLUSIONS AND FUTURE ASPECTS

The presence of pharmaceutical residues in the environment have shown various negative impacts. Hence, water, soil, plants, and food should be certainly monitored on a regular basis. The maximum allowed and tolerable concentrations must now be regulated, and once set, they must be reviewed on a regular basis. As a result, effective, economical, and low-cost approaches are required for this work. Effective strategies for the degradation or removal of these compounds from natural and residual waters must be made. Analytical chemistry must develop eco-friendly, sensitive techniques that need minimum sample volumes, respond rapidly and precisely, and have low detection limits. Moreover, electrochemical procedures are low-cost, easy-to-use, and sensitive. Reusable mono-use electrodes enable in situ pharmacological analysis in dispersed laboratories. Recent breakthroughs in sophisticated materials research, microfabrication, and additive manufacturing processes including inkjet and 3D printing have improved sensor selectivity, stability, and shelf life. Despite development, various upgrades are required to get this technology from the lab to the field. First, most research still uses conventional solutions that ignore matrix effect and sample composition. Furthermore, interferences, organic compounds, and electrode passivation must also be identified. Second, selectivity and sensitivity must be shown in situ, with or without sample pre-treatment. Electrochemical approaches can remove pharmaceutical contaminants from the environment effectively. Electrochemical detection approaches provide benefits over analytical methods presently utilised for drug removal from the environment, which includes low detection limits, simple instrumentation, in situ analysis, and minimal analysis and electrochemical equipment costs.

AUTHOR CONTRIBUTIONS

SA: Methodology, conceptualisation, visualisation, writing—original draft; AKS: investigation, writing—review and editing; RKG: writing—review and editing, data curation, formal analysis; IT: supervision, validation.

CONFLICTS OF INTEREST

The authors declare that they have no known competing financial interests, personal relationships that could have appeared to influence the work reported in this paper.

ACKNOWLEDGEMENTS

The financial support provided by Scheme for Promotion of Academic and Research Collaboration (SPARC-6019) and IOE incentive grant for faculty (Scheme Number-6031) are much appreciated. SA (Chem./2019–2020/RET-2/Sept-19-term/1/975) and AKS (Chem./2018–19/RET/Sept.18-term/1/4809) are thankful to UGC for funding their research fellowships. The illustrations are created with BioRender.com and hence, the service provider is highly acknowledged.

REFERENCES

Agrahari, Shreanshi, Ravindra Kumar Gautam, Ankit Kumar Singh, and Ida Tiwari. 2022. "Nanoscale Materials-Based Hybrid Frameworks Modified Electrochemical Biosensors for Early Cancer Diagnostics: An Overview of Current Trends and Challenges." *Microchemical Journal* 172 (PB): 106980. https://doi.org/10.1016/j.microc.2021.106980.

Alipour, Esmaeel, Mir Reza Majidi, and Omid Hoseindokht. 2015. "Development of Simple Electrochemical Sensor for Selective Determination of Methadone in Biological Samples Using Multi-Walled Carbon Nanotubes Modified Pencil Graphite Electrode." *Journal of the Chinese Chemical Society* 62 (5): 461–68. https://doi.org/10.1002/jccs.201400391.

Alvarez, David A., and Tammy L. Jones-Lepp. 2011. "And Analysis 11 Sampling of Emerging Pollutants." In *Water Quality Concepts, Sampling and Analysis*, 199–226. Taylor and Francis Group, LLC.

Amiri-Aref, Mohaddeseh, Jahan Bakhsh Raoof, and Reza Ojani. 2013. "Electrocatalytic Oxidation and Selective Determination of an Opioid Analgesic Methadone in the Presence of Acetaminophen at a Glassy Carbon Electrode Modified with Functionalized Multi-Walled Carbon Nanotubes: Application for Human Urine, Saliva and Pharm." *Colloids and Surfaces B: Biointerfaces* 109: 287–93. https://doi.org/10.1016/j.colsurfb.2013.03.055.

Aschengrau, Ann, Janice M. Weinberg, Patricia A. Janulewicz, Megan E. Romano, Lisa G. Gallagher, Michael R. Winter, Brett R. Martin, et al. 2011. "Affinity for Risky Behaviors Following Prenatal and Early Childhood Exposure to Tetrachloroethylene (PCE)-Contaminated Drinking Water: A Retrospective Cohort Study." *Environmental Health : A Global Access Science Source* 10: 102. https://doi.org/10.1186/1476-069X-10-102.

Baron, Ronan, Bilha Willner, and Itamar Willner. 2007. "Biomolecule-Nanoparticle Hybrids as Functional Units for Nanobiotechnology." *Chemical Communications* 4: 323–32. https://doi.org/10.1039/b610721b.

Bettazzi, Francesca, Giovanna Marrazza, Maria Minunni, Ilaria Palchetti, and Simona Scarano. 2017. "Biosensors and Related Bioanalytical Tools." *Comprehensive Analytical Chemistry* 77: 1–33. https://doi.org/10.1016/bs.coac.2017.05.003.

Brahman, Pradeep Kumar, Lakkavarapu Suresh, Veeramacheneni Lokesh, and Syed Nizamuddin. 2016. "Fabrication of Highly Sensitive and Selective Nanocomposite Film Based on CuNPs/Fullerene-C60/MWCNTs: An Electrochemical Nanosensor for Trace Recognition of Paracetamol." *Analytica Chimica Acta* 917: 107–16. https://doi.org/10.1016/j.aca.2016.02.044.

Cahill, Jeffery D., Edward T. Furlong, Mark R. Burkhardt, Dana Kolpin, and Larry G. Anderson. 2004. "Determination of Pharmaceutical Compounds in Surface- and Ground-Water Samples by Solid-Phase Extraction and High-Performance Liquid Chromatography-Electrospray Ionization Mass Spectrometry." *Journal of Chromatography A* 1041 (1–2): 171–80. https://doi.org/10.1016/j.chroma.2004.04.005.

Campos-Mañas, Marina Celia, Patricia Plaza-Bolaños, José Antonio Sánchez-Pérez, Sixto Malato, and Ana Agüera. 2017. "Fast Determination of Pesticides and Other Contaminants of Emerging Concern in Treated Wastewater Using Direct Injection Coupled to Highly Sensitive Ultra-High Performance Liquid Chromatography-Tandem Mass Spectrometry." *Journal of Chromatography A* 1507: 84–94. https://doi.org/10.1016/j.chroma.2017.05.053.

Careri, Maria. 2013. "Colin Poole (Ed.): Gas Chromatography." *Analytical and Bioanalytical Chemistry* 405 (18): 5855–56. https://doi.org/10.1007/s00216-013-7007-x.

Caygill, Rebecca L., G. Eric Blair, and Paul A. Millner. 2010. "A Review on Viral Biosensors to Detect Human Pathogens." *Analytica Chimica Acta* 681 (1–2): 8–15. https://doi.org/10.1016/j.aca.2010.09.038.

Cháfer-Pericás, Consuelo, Ángel Maquieira, and Rosa Puchades. 2010. "Fast Screening Methods to Detect Antibiotic Residues in Food Samples." *TrAC—Trends in Analytical Chemistry* 29 (9): 1038–49. https://doi.org/10.1016/j.trac.2010.06.004.

Cizmas, Leslie, Virender K. Sharma, Cole M. Gray, and Thomas J. McDonald. 2015. "Pharmaceuticals and Personal Care Products in Waters: Occurrence, Toxicity, and Risk." *Environmental Chemistry Letters* 13 (4): 381–94. https://doi.org/10.1007/s10311-015-0524-4.

Demir, Ersin, Recai Inam, Sibel A. Ozkan, and Bengi Uslu. 2014. "Electrochemical Behavior of Tadalafil on TiO2 Nanoparticles—MWCNT Composite Paste Electrode and Its Determination in Pharmaceutical Dosage Forms and Human Serum Samples Using Adsorptive Stripping Square Wave Voltammetry." *Journal of Solid State Electrochemistry* 18 (10): 2709–20. https://doi.org/10.1007/s10008-014-2529-5.

Dziąbowska, Karolina, Elżbieta Czaczyk, and Dawid Nidzworski. 2018. "Application of Electrochemical Methods in Biosensing Technologies." In *Biosensing Technologies for the Detection of Pathogens: A Prospective Way for Rapid Analysis*, 151–71. Intech Open Science. https://doi.org/10.5772/intechopen.72175.

Elbalkiny, Heba T., Ali M. Yehia, Safa'a M. Riad, and Yasser S. Elsaharty. 2019. "Potentiometric Diclofenac Detection in Wastewater Using Functionalized Nanoparticles." *Microchemical Journal* 145 (October 2018): 90–95. https://doi.org/10.1016/j.microc.2018.10.017.

Fernandes, Diana M., Nádia Silva, Clara Pereira, Cosme Moura, Júlia M. C. S. Magalhães, Belén Bachiller-Baeza, Inmaculada Rodríguez-Ramos, Antonio Guerrero-Ruiz, Cristina Delerue-Matos, and Cristina Freire. 2015. "MnFe2O4@CNT-N as Novel Electrochemical Nanosensor for Determination of Caffeine, Acetaminophen and Ascorbic Acid." *Sensors and Actuators, B: Chemical* 218: 128–36. https://doi.org/10.1016/j.snb.2015.05.003.

Ganjali, M. R., T. Alizade, and P. Norouzi. 2012. "Chloramphenicol Biomimetic Molecular Imprinted Polymer Used as a Sensing Element in Nano-Composite Carbon Paste Potentiometric Sensor." *International Journal of Electrochemical Science* 7 (5): 4800–10.

Ghoneim, M. M., and A. M. Beltagi. 2003. "Adsorptive Stripping Voltammetric Determination of the Anti-Inflammatory Drug Celecoxib in Pharmaceutical Formulation and Human Serum." *Talanta* 60 (5): 911–21. https://doi.org/10.1016/S0039-9140(03)00151-6.

Golet, Eva M., Alfredo C. Alder, and Walter Giger. 2002. "Environmental Exposure and Risk Assessment of Fluoroquinolone Antibacterial Agents in Wastewater and River Water of the Glatt Valley Watershed, Switzerland." *Environmental Science and Technology* 36 (17): 3645–51. https://doi.org/10.1021/es0256212.

Gros, Meritxell, Mira Petrović, and Damià Barceló. 2006. "Multi-Residue Analytical Methods Using LC-Tandem MS for the Determination of Pharmaceuticals in Environmental and Wastewater Samples: A Review." *Analytical and Bioanalytical Chemistry* 386 (4): 941–52. https://doi.org/10.1007/s00216-006-0586-z.

Hassan, Mohamed H., Reem Khan, and Silvana Andreescu. 2021. "Advances in Electrochemical Detection Methods for Measuring Contaminants of Emerging Concerns." *Electrochemical Science Advances*. doi.org/10.1002/elsa.202100184.

Jacobs, Michael, Vinay J. Nagaraj, Tim Mertz, Anjan Panneer Selvam, Thi Ngo, and Shalini Prasad. 2013. "An Electrochemical Sensor for the Detection of Antibiotic Contaminants in Water." *Analytical Methods* 5 (17): 4325–29. https://doi.org/10.1039/c3ay40994e.

Jaffrezic-Renault, Nicole, and Sergei V. Dzyadevych. 2008. "Conductometric Microbiosensors for Environmental Monitoring." *Sensors* 8 (4): 2569–88. https://doi.org/10.3390/s8042569.

Jain, Rajeev, Vinod K. Gupta, N. Jadon, and K. Radhapyari. 2010. "Voltammetric Determination of Cefixime in Pharmaceuticals and Biological Fluids." *Analytical Biochemistry* 407 (1): 79–88. https://doi.org/10.1016/j.ab.2010.07.027.

James, C. Andrew, Justin P. Miller-Schulze, Shawn Ultican, Alex D. Gipe, and Joel E. Baker. 2016. "Evaluating Contaminants of Emerging Concern as Tracers of Wastewater from Septic Systems." *Water Research* 101: 241–51. https://doi.org/10.1016/j.watres.2016.05.046.

Jennings, Walter G., and Colin F. Poole. 2012. "Milestones in the Development of Gas Chromatography." *Gas Chromatography*, 1–17. https://doi.org/10.1016/B978-0-12-385540-4.00001-8.

Jones, O. A. H, N. Voulvoulis, and J. N. Lester. 2001. "Human Pharmaceuticals in the Aquatic Environment a Review." *Critical Reviews in Environmental Science and Technology* 35 (February 2013): 37–41.

Kahkha, Rezaei, Mohammad Reza, Ali Reza Oveisi, Massoud Kaykhaii, and Batool Rezaei Kahkha. 2018. "Determination of Carbamazepine in Urine and Water Samples Using Amino-Functionalized Metal—Organic Framework as Sorbent." *Chemistry Central Journal* 12 (1): 1–12. https://doi.org/10.1186/s13065-018-0446-x.

Khalilzadeh, Mohammad A., Somayeh Tajik, Hadi Beitollahi, and Richard A. Venditti. 2020. "Green Synthesis of Magnetic Nanocomposite with Iron Oxide Deposited on Cellulose Nanocrystals with Copper (Fe3O4@CNC/Cu): Investigation of Catalytic Activity for the Development of a Venlafaxine Electrochemical Sensor." *Industrial and Engineering Chemistry Research* 59 (10): 4219–28. https://doi.org/10.1021/acs.iecr.9b06214.

Klavarioti, Maria, Dionissios Mantzavinos, and Despo Kassinos. 2009. "Removal of Residual Pharmaceuticals from Aqueous Systems by Advanced Oxidation Processes." *Environment International* 35 (2): 402–17. https://doi.org/10.1016/j.envint.2008.07.009.

Kolpin, Dana, Edward Furlong, Steven Zaugg, and Herbert Buxton. 2002. "DigitalCommons @ University of Nebraska—Lincoln Pharmaceuticals, Hormones, and Other Organic Wastewater Contaminants in U.S. Streams, 1999–2000: A National Reconnaissance." *Environmental Science and Technology* 36: 1202–11.

Kor, Kamalodin, and Kobra Zarei. 2013. "β-Cyclodextrin Incorporated Carbon Nanotube Paste Electrode as Electrochemical Sensor for Nifedipine." *Electroanalysis* 25 (6): 1497–504. https://doi.org/10.1002/elan.201200652.

Kronacher, C., and F. Hogreve. 1936. "Röntgenologische Skelettstudien an Dahlemer Binder-Drillingen Und -Zwillingen." *Zeitschrift Für Züchtung. Reihe B, Tierzüchtung Und Züchtungsbiologie Einschließlich Tierernährung* 36 (3): 281–94. https://doi.org/10.1111/j.1439-0388.1936.tb00094.x.

Kümmerer, Klaus. 2009. "The Presence of Pharmaceuticals in the Environment Due to Human Use—Present Knowledge and Future Challenges." *Journal of Environmental Management* 90 (8): 2354–66. https://doi.org/10.1016/j.jenvman.2009.01.023.

Madrakian, Tayyebeh, Esmaeel Haghshenas, and Abbas Afkhami. 2014. "Simultaneous Determination of Tyrosine, Acetaminophen and Ascorbic Acid using Gold Nanoparticles/Multiwalled Carbon Nanotube/Glassy Carbon Electrode by Differential Pulse Voltammetric Method." *Sensors and Actuators, B: Chemical* 193: 451–60. https://doi.org/10.1016/j.snb.2013.11.117.

Magureanu, Monica, Nicolae Bogdan Mandache, and Vasile I. Parvulescu. 2015. "Degradation of Pharmaceutical Compounds in Water by Non-Thermal Plasma Treatment." *Water Research* 81: 124–36. https://doi.org/10.1016/j.watres.2015.05.037.

Meng, Zheng, Robert M. Stolz, Lukasz Mendecki, and Katherine A. Mirica. 2019. "Electrically-Transduced Chemical Sensors Based on Two-Dimensional Nanomaterials." Review-article. *Chemical Reviews* 119 (1): 478–598. https://doi.org/10.1021/acs.chemrev.8b00311.

Nagaraj, Vinay J., Michael Jacobs, Krishna Mohan Vattipalli, Venkata Praveen Annam, and Shalini Prasad. 2014. "Nanochannel-Based Electrochemical Sensor for the Detection of Pharmaceutical Contaminants in Water." *Environmental Sciences: Processes and Impacts* 16 (1): 135–40. https://doi.org/10.1039/c3em00406f.

Naghdi, Mitra, Mehrdad Taheran, Satinder Kaur Brar, Azadeh Kermanshahi-pour, Mausam Verma, and R. Y. Surampalli. 2018. "Removal of Pharmaceutical Compounds in Water and Wastewater Using Fungal Oxidoreductase Enzymes." *Environmental Pollution* 234: 190–213. https://doi.org/10.1016/j.envpol.2017.11.060.

Oliveira, Mariana C. Q., Marcos R. V. Lanza, Auro A. Tanaka, and Maria D. P. T. Sotomayor. 2010. "Flow Injection Analysis of Paracetamol Using a Biomimetic Sensor as a Sensitive and Selective Amperometric Detector." *Analytical Methods* 2 (5): 507–12. https://doi.org/10.1039/b9ay00283a.

Paimard, Giti, Mohammad Bagher Gholivand, and Mojtaba Shamsipur. 2016. "Determination of Ganciclovir as an Antiviral Drug and Its Interaction with DNA at Fe3O4/Carboxylated Multi-Walled Carbon Nanotubes Modified Glassy Carbon Electrode." *Measurement: Journal of the International Measurement Confederation* 77: 269–77. https://doi.org/10.1016/j.measurement.2015.09.019.

Parnsubsakul, Attasith, Rika Endara Safitri, Patsamon Rijiravanich, and Werasak Surareungchai. 2017. "Electrochemical Assay of Proteolytically Active Prostate Specific Antigen Based on Anodic Stripping Voltammetry of Silver Enhanced Gold Nanoparticle Labels." *Journal of Electroanalytical Chemistry* 785: 125–30. https://doi.org/10.1016/j.jelechem.2016.12.010.

Phillips, Ian, Mark Casewell, Tony Cox, Brad De Groot, Christian Friis, Ron Jones, Charles Nightingale, Rodney Preston, and John Waddell. 2004. "Does the Use of Antibiotics in Food Animals Pose a Risk to Human Health? A Critical Review of Published Data." *Journal of Antimicrobial Chemotherapy* 53 (1): 28–52. https://doi.org/10.1093/jac/dkg483.

Quintanilla, Paloma, Mª Carmen Beltrán, Bernardo Peris, Martín Rodríguez, and Mª Pilar Molina. 2018. "Antibiotic Residues in Milk and Cheeses after the Off-Label Use of Macrolides in Dairy Goats." *Small Ruminant Research* 167: 55–60. https://doi.org/10.1016/j.smallrumres.2018.08.008.

Ribeiro, Ana Rita, Carlos Magalhães Afonso, Paula M. L. Castro, and Maria Elizabeth Tiritan. 2013. "Enantioselective HPLC Analysis and Biodegradation of Atenolol, Metoprolol and Fluoxetine." *Environmental Chemistry Letters* 11 (1): 83–90. https://doi.org/10.1007/s10311-012-0383-1.

Rimmer, Catherine A. 2015. "Salvatore Fanali, Paul Haddad, Colin Poole, Peter Schoenmakers, and David Lloyd (Eds.): Liquid Chromatography: Applications." *Analytical and Bioanalytical Chemistry* 407 (1): 7–8. https://doi.org/10.1007/s00216-014-8239-0.

Riva, Francesco, Sara Castiglioni, Elena Fattore, Angela Manenti, Enrico Davoli, and Ettore Zuccato. 2018. "Monitoring Emerging Contaminants in the Drinking Water of Milan and Assessment of the Human Risk." *International Journal of Hygiene and Environmental Health* 221 (3): 451–57. https://doi.org/10.1016/j.ijheh.2018.01.008.

Ronkainen, Niina J., H. Brian Halsall, and William R. Heineman. 2010. "Electrochemical Biosensors." *Chemical Society Reviews* 39 (5): 1747–63. https://doi.org/10.1039/b714449k.

Saeger, Sarah De, and Carlos Van Peteghem. 1999. "Flow-through Membrane-Based Enzyme Immunoassay for Rapid Detection of Ochratoxin A in Wheat." *Journal of Food Protection* 62 (1): 65–69. https://doi.org/10.4315/0362-028X-62.1.65.

Sanghavi, Bankim J., and Ashwini K. Srivastava. 2013. "Adsorptive Stripping Voltammetric Determination of Imipramine, Trimipramine and Desipramine Employing Titanium Dioxide Nanoparticles and an Amberlite XAD-2 Modified Glassy Carbon Paste Electrode." *Analyst* 138 (5): 1395–404. https://doi.org/10.1039/c2an36330e.

Sauberan, Jason B., and John S. Bradley. 2018. "Antimicrobial Agents." *Principles and Practice of Pediatric Infectious Diseases*, 1499–531.e3. https://doi.org/10.1016/B978-0-323-40181-4.00292-9.

Scognamiglio, Viviana, Amina Antonacci, Luisa Patrolecco, Maya D. Lambreva, Simona C. Litescu, Sandip A. Ghuge, and Giuseppina Rea. 2016. "Analytical Tools Monitoring Endocrine Disrupting Chemicals." *TrAC—Trends in Analytical Chemistry* 80: 555–67. https://doi.org/10.1016/j.trac.2016.04.014.

Sghaier, Rafika Ben, Sopheak Net, Ibtissem Ghorbel-Abid, Salma Bessadok, Maïwen Le Coz, Dalila Ben Hassan-Chehimi, Malika Trabelsi-Ayadi, Michele Tackx, and Baghdad Ouddane. 2017. "Simultaneous Detection of 13 Endocrine Disrupting Chemicals in Water by a Combination of SPE-BSTFA Derivatization and GC-MS in Transboundary Rivers (France-Belgium)." *Water, Air, and Soil Pollution* 228 (1). https://doi.org/10.1007/s11270-016-3195-2.

Singh, Ankit Kumar, Ravindra Kumar Gautam, Shreanshi Agrahari, Jyoti Prajapati, and Ida Tiwari. 2022. "Graphene Oxide Supported Fe_3O_4-MnO_2 Nanocomposites for Adsorption and Photocatalytic Degradation of Dyestuff: Ultrasound Effect, Surfactants Role and Real Sample Analysis." *International Journal of Environmental Analytical Chemistry* 1–27. https://doi.org/10.1080/03067319.2022.2091930.

Singh, Ankit Kumar, Nandita Jaiswal, Ravindra Kumar Gautam, and Ida Tiwari. 2021. "Development of G-C3N4/Cu-DTO MOF Nanocomposite Based Electrochemical Sensor towards Sensitive Determination of an Endocrine Disruptor BPSIP." *Journal of Electroanalytical Chemistry* 887: 115170. https://doi.org/10.1016/J.JELECHEM.2021.115170.

Singh, Ankit Kumar, and Ida Tiwari. 2020. *Nanomaterial Synthesis and Mechanism for Enzyme Immobilization: Part II.* https://doi.org/10.1007/978-981-13-9333-4_8.

Taei, M., F. Hasanpour, V. Hajhashemi, M. Movahedi, and H. Baghlani. 2016. "Simultaneous Detection of Morphine and Codeine in Urine Samples of Heroin Addicts Using Multi-Walled Carbon Nanotubes Modified SnO_2-Zn_2 SnO_4 Nanocomposites Paste Electrode." *Applied Surface Science* 363: 490–98. https://doi.org/10.1016/j.apsusc.2015.12.074.

Takeda, Humberto Hissashi, Tiago Almeida Silva, Bruno Campos Janegitz, Fernando Campanhã Vicentini, Luiz Henrique Capparelli Mattoso, and Orlando Fatibello-Filho. 2016. "Electrochemical Sensing of Levodopa or Carbidopa Using a Glassy Carbon Electrode Modified with Carbon Nanotubes within a Poly(Allylamine Hydrochloride) Film." *Analytical Methods* 8 (6): 1274–80. https://doi.org/10.1039/c5ay03041b.

Ternes, Thomas A. 2001. "Analytical Methods for the Determination of Pharmaceuticals in Aqueous Environmental Samples." *TrAC—Trends in Analytical Chemistry* 20 (8): 419–34. https://doi.org/10.1016/S0165-9936(01)00078-4.

Ternes, Thomas A., M. Stumpf, J. Mueller, K. Haberer, R. D. Wilken, and M. Servos. 1999. "Behavior and Occurrence of Estrogens in Municipal Sewage Treatment Plants—I. Investigations in Germany, Canada and Brazil." *Science of the Total Environment* 225 (1–2): 81–90. https://doi.org/10.1016/S0048-9697(98)00334-9.

USEPA. 2007. "Method 1694 : Pharmaceuticals and Personal Care Products in Water, Soil, Sediment, and Biosolids by HPLC/MS/MS." *EPA Method*, 77.

Wei, Qin, Yanfang Zhao, Bin Du, Dan Wu, He Li, and Minghui Yang. 2012. "Ultrasensitive Detection of Kanamycin in Animal Derived Foods by Label-Free Electrochemical Immunosensor." *Food Chemistry* 134 (3): 1601–6. https://doi.org/10.1016/j.foodchem.2012.02.126.

Wintgens, T., F. Salehi, R. Hochstrat, and T. Melin. 2008. "Emerging Contaminants and Treatment Options in Water Recycling for Indirect Potable Use." *Water Science and Technology* 57 (1): 99–107. https://doi.org/10.2166/wst.2008.799.

Wooding, Madelien, Egmont R. Rohwer, and Yvette Naudé. 2017. "Determination of Endocrine Disrupting Chemicals and Antiretroviral Compounds in Surface Water: A Disposable Sorptive Sampler with Comprehensive Gas Chromatography—Time-of-Flight Mass Spectrometry and Large Volume Injection with Ultra-High Performance Li." *Journal of Chromatography A* 1496: 122–32. https://doi.org/10.1016/j.chroma.2017.03.057.

Xu, Leimeng, Jianhai Li, Tao Fang, Yongli Zhao, Shichen Yuan, Yuhui Dong, and Jizhong Song. 2019. "Synthesis of Stable and Phase-Adjustable CsPbBr3@Cs4PbBr6 Nanocrystals: Via Novel Anion-Cation Reactions." *Nanoscale Advances* 1 (3): 980–88. https://doi.org/10.1039/c8na00291f.

Zhang, Youxiong, Bing Li, Xianhu Wei, Qihui Gu, Moutong Chen, Jumei Zhang, Shuping Mo, et al. 2021. "Amplified Electrochemical Antibiotic Aptasensing Based on Electrochemically Deposited AuNPs Coordinated with PEI-Functionalized Fe-Based Metal-Organic Framework." *Microchimica Acta* 188 (8). https://doi.org/10.1007/s00604-021-04912-z.

Zwiener, C. 2007. "Occurrence and Analysis of Pharmaceuticals and Their Transformation Products in Drinking Water Treatment." *Analytical and Bioanalytical Chemistry* 387 (4): 1159–62. https://doi.org/10.1007/s00216-006-0818-2.

8 Exposure and Health Impact of Pharmaceutical Residue Ingestion via Dietary Sources and Drinking Water

Haitham G. Abo-Al-Ela, Francesca Falco, and Caterina Faggio

CONTENTS

8.1 INTRODUCTION

The occurrence of pharmaceutical residue in the environment can be linked to several effectors, such as high consumption of medicine(s) in human and/or veterinary sectors, massive production of pharmaceuticals, and/or inappropriate discard of unused or expired medicines. The presence of pharmaceutical compounds (PCs) in the environment has grown to be a global problem during the last few decades. Pharmaceuticals belong to a large class of xenobiotics (Faggio et al., 2018) that are used extensively in medications, making them of special importance because many of these compounds may wind up in the environment where they are persistent or pseudo-persistent (Burgos-Aceves et al., 2018; Faggio et al., 2016; Huerta et al., 2013a).

A wide range of pharmaceuticals has continually polluted the environment, which is dangerous for both humans and animals because of direct contact with the skin, mucous membranes, or lungs or from indirect contact with food and water (Kuch and Ballschmiter, 2001; Webb et al., 2003). This raises significant worries about pharmaceutical residues in the environment, which have negative effects on, for instance, human health, including an increased risk of developing cancer (Wee and Aris, 2017), aquatic creature reproductive toxicity (Fent, 2008), and the development of antibiotic

DOI: 10.1201/9781003361091-8

resistance in aquatic microorganisms as well as in terrestrial invertebrates (García et al., 2020; González-Alcaraz et al., 2020). Additionally, pharmaceutical residues impact terrestrial, freshwater, and marine ecosystems, which disrupts organism physiology and behavior and creates an environment that may directly or indirectly pose health concerns to humans and animals. There is growing evidence that pharmaceutical residues are present in various environmental components, such as soils, sediments, and water bodies (Cabello, 2006; Sui et al., 2015; Verlicchi and Zambello, 2015). In general, earlier studies have found pharmaceutical residues in a variety of environments and at low concentrations (from ng/L to g/L). Ecosystem alteration significantly drives the transmission, emergence, and distribution of many infectious diseases (Arnold et al., 2014; Myers et al., 2013), leading to an increased prevalence of opportunistic infections and risks of uncommon or low-incidence diseases. A large variety of pharmaceuticals have been detected in wastewater treatment plants (WWTPs) influent and effluents, seawater and freshwater worldwide (Table 8.1).

TABLE 8.1
The Main Pharmaceutical Compounds Found Globally

		SWT (main conc., ng/L)		Seawater (mean conc., ng/L)	Freshwater (mean conc., ng/L)	Reference
	Pharmaceuticals	Influence	Effluence			
Switzerland	Carbamazepine	950	–	–	35–60	(Bendz et al., 2005; Öllers et al., 2001; Tixier et al., 2003)
	Clofibric acid	60	–	–	5–10	
	Diclofenac	990	–	–	n.d.–10	
	Ibuprofen	1300	–	–	5–15	
	Ketoprofen	1800	–	–	n.d.	
	Naproxen	2600	–	–	n.d.–10	
UK	Paracetamol	–	–	503	–	(Ashton et al., 2004; Letsinger et al., 2019; Paxéus, 2004)
	Propranolol	–	80		29	
	Ibuprofen	–	410	3481	826	
	Extropropoxyphene	–	–	–	58	
	Mefenamic acid	–	–	–	62	
	Dextropropoxyphene	–	–	–	58	
	Diclofenac	–	290	302	<20	
	Acetyl sulfamethoxazole	–	–	–	<50	
Germany	Ibuprofen	–	–	–	n.d.–200	(Daughton, 2001; Heberer et al., 2002)
	Clofibric acid	–	415–510	–	n.d.–7300	
	Diclofenac	3200	210–1110	–	n.d.–380	
	Gemfibrozil	–	n.d.	–	n.d.–340	
	Ketoprofen	–	n.d.	–	n.d.–30	
	Mefenamic acid	–	n.d.	–	–	
	Naproxen	–	n.d.–120	–	–	
	Primidone	–	n.d.–880	–	n.d.–690	
	Propyphenazone	–	n.d.–740	–	n.d.–1465	
	Salicylic acid	54	n.d.–65	–	n.d.–1225	
	Clofibric acid derivate	–	–	–	n.d.–2900	
	N-(phenylsulfonyl)-sarcosine	–	–	–	n.d.–470	

	Pharmaceuticals	SWT (main conc., ng/L)		Seawater (mean conc., ng/L)	Freshwater (mean conc., ng/L)	Reference
		Influence	Effluence			
South Korea	Diclofenac	–	210–1110		n.d.–380	(Choi et al., 2008; Kim et al., 2009)
	Gemfibrozil	–	n.d.	–	n.d.–340	
	Ketoprofen	–	n.d.	–	n.d.–30	
	Mefenamic acid	–	n.d.	–	–	
	Naproxen	–	n.d.–120	–	–	
	Primidone	–	n.d.–880	–	n.d.–690	
Spain	Propyphenazone	–	n.d.–740	–	n.d.–1465	(Carmona et al., 2014; Paíga et al., 2015)
	Salicylic acid	–	n.d.–65	–	n.d.–1225	
	Clofibric acid derivate	–	–	–	n.d.–2900	
	N-(phenylsulfonyl)-sarcosine	–	–	–	n.d.–470	
	Ibuprofen	–	–	–	up to 414	
	Atenolol	–	–	–	up to 690	
	Carbamazepine	up to 451	up to 195	–	up to 595	
	Acetaminophen	up to 56.94	up to 9	–	–	
Northern Taiwan	Cimetidine	up to 17.651	up to 7763	–	–	(Jiang et al., 2014)
	Diltiazem	up to 19	up to 13	–	–	
	Noproxen	–	–	30	–	
	Chlorfibric acid	12.8	–	–	17	
	Chloramphenicol	11	–	–	3	
South Africa	Diclofenac	90	–	17	49	(K'Oreje et al., 2016)
	Ethylparaben	72	–	–	16	
	Flufenamic acid	57	–	–	21	
	Gemfibrozil	336	–	–	77	
	Ibuprofen	21	–	40.5	830	
Norway	Clofibric acid	–	101.7	55.1	–	(Ngubane et al., 2019; Weigel et al., 2004)
	Diclofenac	–	131	53.6	–	
	Ibuprofen	–	1600	n.d.–57.1	–	
	Acetaminophen	–	–	2.6–16.7	–	
	Ketoprofen	–	128	n.d.–23.3	–	
	Diclofenac	930	40	–	463	
	Ibuprofen	8400	1990	170	6993	
	Indomethacin	980	n.d.	–	77	
	Paracetamol	110887	100	–	41129	
	Naproxen	–	–	160	–	
	Ibuprofen	0.6	0.68	0.3	–	
	Hydroxy-ibuprofen	1.32	1.13	0.75	–	
	Carboxy-ibuprofen	1.63	1.27	4.15	–	

Note: n.d., not detected; SWT, sewage water treatment.

Source: Han et al. (2010).

It is strongly believed that the levels of pharmaceuticals will continually increase because of human activities and the increasing demand for securing food by expanding animal production and the aquaculture industry (Fonseca et al., 2021). With efforts to preserve and solve the problems of limited water resources, the reuse of treated wastewater and sewage sludge are highly encouraged; however, such practices with improper treatment or detection of pharmaceutical residues help disseminate more residues in the environment and increase the problem of antimicrobial resistance. Convincing evidence shows that wastewater reuse and biosolid application in agricultural activities disrupt the soil resistome and microbiota, help accumulate pharmaceuticals in plants, and increase the development rate of antibiotic resistance (Sorinolu et al., 2021).

8.2 OCCURRENCE OF PHARMACEUTICALS IN THE DIFFERENT SYSTEMS (TERRESTRIAL, FRESHWATER, AND MARINE ECOSYSTEMS)

Several pollutants are of growing concern, including those found in personal care and pharmaceutical products, and are presently being assessed for potential new regulations of their use under the EU Water Framework Directive or by the US EPA because of their extensive overuse (Zhang et al., 2021a). These products are composed of diverse active ingredients with complex molecules that have unique structures and activities in addition to a wide range of biological and physical features. Pharmaceutical active molecules are among the chemicals that are becoming pollutants, and they continue to be present in the environment, including groundwater, surface water, soil, and sediments (Beretta et al., 2014; Kumar et al., 2016; Sharma and Kaushik, 2017). Certain PCs are partially metabolized by human and animal bodies before being released into the environment. Due to their bioactivity, they pose a major danger to both aquatic and terrestrial organisms.

8.2.1 Terrestrial Ecosystem

Modern human or veterinary medications are increasingly being used, which has resulted in an increase in anthropogenic pharmaceuticals frequently introduced into terrestrial environments through direct discharge of untreated or treated wastewater from municipal sources (Zhang et al., 2021a), industrial (WWTPs) sectors, hospitals, landfill leachate, sewage sludge as soil amendments, land application of municipal wastewater onto permitted lands (Pan and Chu, 2016), sewer leakage/overflow, animal farm waste (Gottschall et al., 2013; McEachran et al., 2016), surface runoff, and unmetabolized products by humans from household sewers (Zhang et al., 2021a). The pharmaceutical compounds enter wastewater, which reaches different environments (Ternes et al., 2004). Drugs of humans and animals have been shown in concentrations at ng/g in solid samples. The main pharmaceuticals introduced to the soil environment were nonsteroidal anti-inflammatory drugs (NSAIDs), antibiotics, psychostimulants, antihypertensive drugs, cardiovascular medications (β-blockers/diuretics), hormones such as estrogens, and antiepileptic drugs (Li, 2014).

The accumulation of pharmaceuticals in aquatic and terrestrial ecosystems may have negative consequences on the environment and living beings. It might limit plant development and be toxic or cause serious allergies in individuals (Miller et al., 2016). For instance, nonylphenol ethoxylates (NPEs), carrier molecules in an antibacterial iodine complex, are used to treat mastitis (inflammation of mammary tissue) in cattle. Nonylphenol (NP) is a rapidly formed breakdown product of the NPE carrier material, which has a long shelf life with moderate to high toxicity to aquatic and soil species. It is anticipated that NPE concentrations in soil and grasses will exceed the trigger threshold of 100 μg/kg (Koschorreck et al., 2002; VICH, 2005). The high levels of NP in plants are not poisonous, while it is mildly toxic in earthworms and very toxic in aquatic species (Hansen et al., 2002). The environmental threshold trigger for the veterinary antiparasitic benzimidazole and its metabolite sulfon/sulfoxide is estimated to be below 100 g/kg (VICH, 2005), in which the antiparasitic itself is promptly degraded into persistently bound metabolites. The ratio of direct input from pastured animals and runoff from manured lands is significantly higher than 1 μg/kg (VICH, 2005). Such metabolites are very hazardous to fish and Daphnia in aquatic environments (Koschorreck et al., 2002).

The presence of antibiotic residues in animal manure that are used in agriculture can contaminate soil and water (Elmund et al., 1971; Thiele-Bruhn, 2003). Among veterinary antibiotics, tetracyclines are commonly added to animal food and used for the treatment of chicken, calf, and pig diseases (Ahmed, 2017; Xu et al., 2020). The predicted environmental concentrations of tetracyclines are more than 100 μg/kg in soil (VICH, 2005). Because of their high consumption, tetracyclines are frequently detected in manures (Yang et al., 2021; Zhang et al., 2015a). Among 17 antibiotics analyzed in different animal manures, oxytetracycline showed a maximum concentration of 416.8 μg/g in chicken manure (Yang et al., 2021; Zhang et al., 2015a). Xu et al. (2020) found that oxytetracycline was 541.020 μg/g in pig manure. According to Solliec et al. (2016), the concentrations of tetracyclines in swine dung slurry varied from 53 to 137 g/L, while their degradation products, including 4-epitetracycline, anhydrotetracycline, and 4-epianhydrotetracycline, were in the range of 118 to 663 g/L.

Once pharmaceuticals have entered the environment, biodegradation, sorption, and photolysis can take place as natural attenuation processes to breakdown these toxins. The spontaneous degradation of pharmaceuticals by microbial communities in soil and water limits their environmental persistence (Biel-Maeso et al., 2019). According to Biel-Maeso et al. (2019), several compounds (e.g., sulfamethoxazole, sulfamethizole, sulfamethopyridazone, carbamazepine, diclofenac, ibuprofen, and hydrochlorothiazide) were sensitive to microbial breakdown under aerobic conditions, with half-lives ranging from 1 to 18 days. Doretto et al. (2014) detected sorption of sulfonamides in soils at a depth of up to 20 cm, and sorption of trimethoprim and sulfonamide was detected in agricultural soil at three distinct depths (i.e., 0–20, 20–80, and 80–100 cm) (Zhang et al., 2014). PC adsorption behaviors vary not only among compounds but are also challenging to predict because the sorption process is frequently governed by interactions with complex pH-dependent speciation or certain functional groups, including electrostatic interactions between ionized compounds and the particle surface (Biel-Maeso et al., 2019; Kibbey et al., 2007; Zhang et al., 2014). For example, the physicochemical qualities of antibiotics, such as hydrophobicity, polarizability, polarity, and spatial configuration, which are mostly dictated by their molecular structures, have a substantial impact on how they adsorb in soil (Biel-Maeso et al., 2019; Pan and Chu, 2016; Wu et al., 2013). However, little is known about how the physicochemical characteristics of the various soils may affect the sorption of PCs. Temperature, oxygen content, and interactions between environmental variables and medications affect the decomposition rate of biological materials (Zhi et al., 2019).

8.2.2 Aquatic Environment

Pharmaceutical contaminants are easily accessible and have the potential to accumulate in aquatic systems. According to previous studies, several PCs are present in aquatic ecosystems around the world in the range of ng/L to mg/L (Chen et al., 2008; Huerta et al., 2013b; Wang and Gardinali, 2012). They have an impact on aquatic species on a variety of biological levels throughout their life cycle due to their chemical characteristics that make them bioactive compounds, mostly functioning at very low concentrations (Mezzelani et al., 2016). Prevention and treatment of water pollution depend on the awareness of people of the potential hazards of such compounds (Zicarelli et al., 2022).

8.2.2.1 Freshwater Ecosystem

The occurrence of pharmaceuticals in freshwater represents one of the main concerns that ecotoxicology is facing now. Pharmaceutical substances and complicated mixes can enter surface water in rivers and streams (Buerge et al., 2009) via different sources, including onsite wastewater treatment facilities, septic tanks, leachate, runoff from farms, terrestrial sources, and recreational activities

(Bernot et al., 2013). In a comprehensive environmental assessment of human and veterinary pharmaceuticals, different chemicals have been found in water samples around the world; among all different pharmaceutical substances analyzed, 38 were found in surface water, groundwater, or tap/drinking water (Aus Der Beek et al., 2016). Based on these studies, antibiotics, analgesics, and estrogens are among the therapeutic classes of drugs that are most often examined (Aus Der Beek et al., 2016). Previous studies by Mezzelani et al. (2018), Cahill et al. (2004), and Heberer and Stan (1997) reported that the concentration of PCs in surface water ranged from a few ng/L to hundreds of µg/L, and their concentrations in influent and effluents from WWTPs exceeded 1 µg/L (Koutsouba et al., 2003). In addition to their high biological activity, they may pose a serious risk not only to humans and environmental health (Gavrilescu et al., 2015) but also to nontarget species, leading to different toxic effects (Fent, 2008; Parolini, 2020).

Fekadu et al. (2019) and Parolini (2020) suggested that NSAIDs are one of the most common pharmacological classes to be seen in freshwater globally. Indeed, NSAIDs, including ibuprofen, diclofenac, acetylsalicylic acid, and paracetamol, were among the 16 compounds found in the groundwater and surface and drinking water of five of the United Nations countries (Aus Der Beek et al., 2016). These drugs may have direct or indirect effects on living beings. According to Aliko et al. (2021), fluoxetine and ibuprofen negatively affected tadpole development and behavior, which exhibited delayed metamorphosis time and reduced body weight, in addition to significant increases in unresponsiveness to different stimuli. Moreover, Bartoskova et al. (2013) suggested that ibuprofen induces some antioxidant and biotransformation enzymes to function more actively in zebrafish. In a study by Turani et al. (2019), a species of true frog (*Pelophylax shqipericus*) was subjected to ibuprofen at a sublethal dosage (5 g/L) for 48 hours, which showed collapse and rupture of the cell membrane and abnormalities in cellular and nuclear vacuolization.

Although researchers claim that antibiotics do not harm human health, they are another class of xenobiotics that are more widely distributed in aquatic environments (Sanseverino et al., 2018). Despite the low concentrations of antibiotics in aquatic environments, these concentrations are sufficient to develop resistant microbial strains and exert high selective pressure on environmental microbes (Maul et al., 2006; Stanton et al., 2020). The level of antibiotics in rivers is determined by the likelihood of biodegradation or antibiotic degradation. According to Krzeminski et al. (2019), lipophilic chemicals with relatively high octanol/water partition coefficients (K_{OW}s) largely tend to sorb onto solid phases, while chemicals that have relatively low K_{OW}s and low molecular weights (less than 1000 g/mol), such as antibiotics, are easily dissolved in water bodies. Some antibiotics, such as aminoglycosides, β-lactams, lincosamides, macrolides, amphenicols, nitrofurans, quinolones, phosphonates, and fluoroquinolones, have been found to be relatively persistent in the environment. For example, tetracyclines form a relatively persistent complex with suspended matter, which is frequently detected in aqueous matrices at extremely low concentrations (<10 ng/L) (Carvalho and Santos, 2016; Wernersson et al., 2015). Instead, the environmental fluoroquinolone antibiotic concentrations in Polish rivers can reach 2.7 µg/L (Wagil et al., 2014). In the effluents of wastewater treatment facilities in the Czech Republic, norfloxacin, levofloxacin, and ciprofloxacin recorded the highest concentrations at 0.25 g/L, 0.069 g/L, and 0.22 g/L, respectively, according to Golovko et al. (2014).

8.2.2.2 Marine Ecosystem

Years ago, relatively little research was performed on the understanding of PC discharge into coastal-marine ecosystems and their potential deleterious impact on the marine ecosystem because it was believed that pharmaceutical products may be highly diluted in marine waters to safe limits. The first report of pharmaceutical contamination in the environment was reported by Heberer et al. (1998), which is followed by several other studies that have shown that hundreds of drugs are found in aquatic systems (Heberer et al., 1998; Rossknecht et al., 2001; Ternes and Hirsch, 2000), in which municipal sewage treatment facilities are considered as the main sources (Ojemaye and Petrik, 2019). This might be because most pharmaceutical compounds are designed to be stable, robust, polar, and nonvolatile, which allows them to pass through wastewater treatment plants and be found in marine water. Wastewater treatment plants are not specifically designed to breakdown most pharmaceutical compounds (Ojemaye and Petrik, 2019). Furthermore, pharmaceuticals can also enter the marine environment through septic tanks, landfills, urban wastewater, agricultural practices, and industrial effluent (Boxall et al., 2012; Lambropoulou and Nollet, 2014; Rodil et al., 2012; Verlicchi et al., 2012). Pharmaceutical compounds, such as antibiotics and NSAIDs, are frequently detected in marine habitats (Parolini, 2020). A wide variety of quantities of these chemicals are continuously released into surface and marine waters (Evgenidou et al., 2015; Gros et al., 2010; Verlicchi et al., 2012).

8.3 OCCURRENCE OF PHARMACEUTICAL RESIDUE INGESTION VIA DIETARY SOURCES

The most frequent types of pharmaceuticals used to treat and prevent infections in animals raised for food are antibacterial agents, while less common PCs include antifungals, antiadrenergic agonists, corticosteroids, diuretics, dyes, NSAIDs, sedatives, β-blockers, and thyreostatics (Botsoglou and Fletouris, 2001). Among PCs, residual antibiotics seem to have a major impact on health due to their ability to alter the microbiome of humans and animals, encouraging the establishment and selection of bacterial resistance (McDermott et al., 2002), as well as their potential toxicity and allergy (Lillehaug et al., 2003; Liu et al., 2017). All these consequences have exponentially increased, which is a health concern about pharmaceutical residues in the environment (Ben et al., 2019) or in food (Schmerold and Ungemach, 2004). Additionally, there is also the potential risk of applying selection pressure on the microbiome of the environment, which might eventually result in antibiotic resistance reservoirs (Hernando-Amado et al., 2019).

Generally, dietary sources are considered the main methods for unintentional human exposure to pollutants (Cabello, 2006). Furthermore, antibiotic resistance can be transmitted locally across areas through several ecosystems, including farms, hospitals, wastewater treatment facilities, and natural settings (Hernando-Amado et al., 2019). Holzbauer and Chiller (2006) described several factors that contribute to antibiotic resistance that were developed in aquatic environments and have the ability to spread through horizontal gene transfer to bacteria in the terrestrial environment (Kruse and Sørum, 1994).

8.4 OCCURRENCE OF PHARMACEUTICAL RESIDUE INGESTION VIA DRINKING WATER

Pharmaceuticals are widely polluting drinking water worldwide, which raises large concerns about potential negative impacts on human and animal health and ecosystems. Many residual pharmaceuticals have been detected in source water in Japan, and their levels were negligible (below 30 ng/L) in finished water after treatments with conventional and/or advanced methods for processing drinking water, except for six pharmaceuticals and one metabolite (Simazaki et al., 2015). In particular, Simazaki et al. (2015) detected iopamidol in the range of 1000–10,000 ng/L in the source water. Iopamidol is used in angiography that allows visualization of blood vessels by X-ray. Of note, iopamidol showed a maximum concentration of 2400 ng/L in finished water (Simazaki et al., 2015).

Approximately 16 pharmaceutical residues were detected, ranging from 42 to 1762 ng/L, in estuarine waters. Of these, losartan and irbesartan (angiotensin II receptor blockers) demonstrated high ecological risk, while antidepressants, anxiolytics, antiepileptics, and beta-blockers demonstrated moderate ecological risk. Of note, their accumulated mixture showed over 380-fold risk increases (Fonseca et al., 2021). From 51 compounds used in the analysis of pharmaceutical residues in source water, tap water, and finished drinking water, the most frequently detected pharmaceuticals were sulfamethoxazole, trimethoprim, phenytoin, atenolol, carbamazepine, gemfibrozil, naproxen, estrone, and meprobamate (Benotti et al., 2009). Approximately 284 PCs were detected in the surface waters of European countries according to studies and governmental reports published between 1998 and 2016 (Zhou et al., 2019). In developing countries such as Malaysia, sulfamethoxazole, amoxicillin, chloramphenicol, ciprofloxacin, triclosan, nitrofurazone, caffeine, dexamethasone, and diclofenac were detected in drinking water (Mohd Nasir et al., 2019; Wee et al., 2022).

Some pharmaceuticals act as indicators of human activities. Caffeine and its derivatives are considered efficient markers of anthropogenic pollution in urban environments (Buerge et al., 2008; Ferreira et al., 2005). Among the studied pharmaceuticals (i.e., caffeine, dexamethasone, diclofenac, triclosan, ciprofloxacin, amoxicillin, chloramphenicol, nitrofurazone, and sulfamethoxazole), caffeine recorded the highest concentration of 0.38 ng/L, and diclofenac recorded the lowest concentration of 0.14 ng/L in drinking water in Putrajaya (Malaysia) (Praveena et al., 2019; Praveena et al., 2021). Caffeine is a main component of many beverages (e.g., tea, coffee, and other soft drinks, for example, energy drinks) and a variety of foods (e.g., chocolate, chewing gum, and coffee-containing foods, for example, protein bars). Caffeine could be disseminated into water sources via unconsumed beverages or foods or via urine that could contain 0.5–10% unmetabolized caffeine. In the Madrid region (Spain), caffeine and cotinine (a nicotine metabolite found in tobacco) are detected in all surface water samples with variable concentrations of the other studied pharmaceuticals, providing supporting evidence for considering these stimulants as interesting indicators of human activities (anthropogenic contamination) (Valcárcel et al., 2011). Additionally, caffeine and cotinine were found not only in surface water but also in tap water (Valcárcel et al., 2011), which can reach directly to humans.

Antimicrobials, particularly antibiotics, are the most frequently used and highly occurring pharmaceuticals in dietary sources, ranging from less than 1 ng/L to more than 1000 μg/L (Anh et al., 2021). The main sources of antibiotics in surface water are livestock and aquaculture production activities, urban sewage, and wastewater treatment plants, according to studies from 2007 to 2020 (Anh et al., 2021; Rana et al., 2019). For example, sulfonamides are widely noticed in several types of water (e.g., surface, ground, and wastewater) (Fonseca et al., 2021; Zhou et al., 2022). The major classes detected in the environment were β-lactams, tetracyclines, fluoroquinolones, sulfonamides, and macrolides, in addition to other classes. Sulfamethoxazole was one of the most frequently detected pharmaceuticals in source water (12 ng/L) in the USA between 2006 and 2007 (Benotti et al., 2009). In addition, the antibiotic amoxicillin demonstrated high ecological risk in environmental risk assessment (Fonseca et al., 2021). The concentrations of ciprofloxacin were the highest (0.667 ng/L) among the tested pharmaceuticals in drinking water in Malaysia (Mohd Nasir et al., 2019). The occurrence of antibiotic residues was mostly seen in large pharmaceutical producers and markets, such as China and the USA, European countries, and large antibiotic consumer countries, such as East and Southeast Asia, and many developing countries. However, many studies have reported pharmaceutical residues in many countries around the globe. We must admit that the problem of residues is a global issue. The major concern about antimicrobial residues is the rapid increases in the incidence of antibiotic-resistant bacteria and the occurrence of antibiotic resistance genes, particularly in aquatic environments. More details about the antibiotic residues in water, food, and the environment and their potential health hazards will be discussed in the next specific sections.

Other classes of pharmaceuticals, such as antidepressant medications like fluoxetine, paroxetine, and norfluoxetine, have been detected in surface water in countries such as the USA, Canada, and Portugal, and venlafaxine, bupropion, citalopram, fluoxetine, and sertraline are the predominant antidepressants observed (Brooks et al., 2005; Chu and Metcalfe, 2007; Fonseca et al., 2021; Schultz et al., 2010).

Many of the studies on the occurrence of pharmaceutical residues via dietary sources showed that hazard quotient (HQ) values indicated low potential health hazards at ages between 13 and 75 years; however, we have experienced and should consider the accumulative and chronic deleterious effects of pharmaceutical residues. Long-term exposure to chemicals with low or very low concentrations is often difficult to track, especially if they disappear from nature.

8.5 HEALTH IMPACT OF PHARMACEUTICAL RESIDUE INGESTION

The impact of pharmaceutical residues could be direct via exposure (i.e., ingestion, contact with skin and mucous membranes, and inhalation), leading to the development of diseases and health problems; or indirect, which could affect Earth's natural systems (terrestrial, freshwater, and marine ecosystems) by accumulating in objects (living or nonliving) or in humans or animals, resulting in health risks (e.g., increased prevalence of opportunistic infections and risks of uncommon or low-incidence diseases) (Figure 8.1).

Several pharmaceuticals are used by humans that reach the environment. The next sections will discuss the most frequently detected molecules and their potential impact on health.

8.5.1 Antibiotics and Antibacterials

There are several sources of antibiotics and antibacterial residues in the environment, including sources of drinking water and food. In addition to the great concern about the rapid development of antibiotic resistance that represents a major crisis, antibiotics and antibacterial residues, like any chemicals, could cause long-term deleterious impacts on health.

Antibiotics are used not only for treatment and disease prevention but also for animal production. They are used as growth promotors and to improve feed efficiency. Veterinary antibiotics represent approximately 75% of the total antibiotics produced in America (Fauci and Marston, 2014). The global need and increasing demand for animals as a source of food prompted animal breeders to

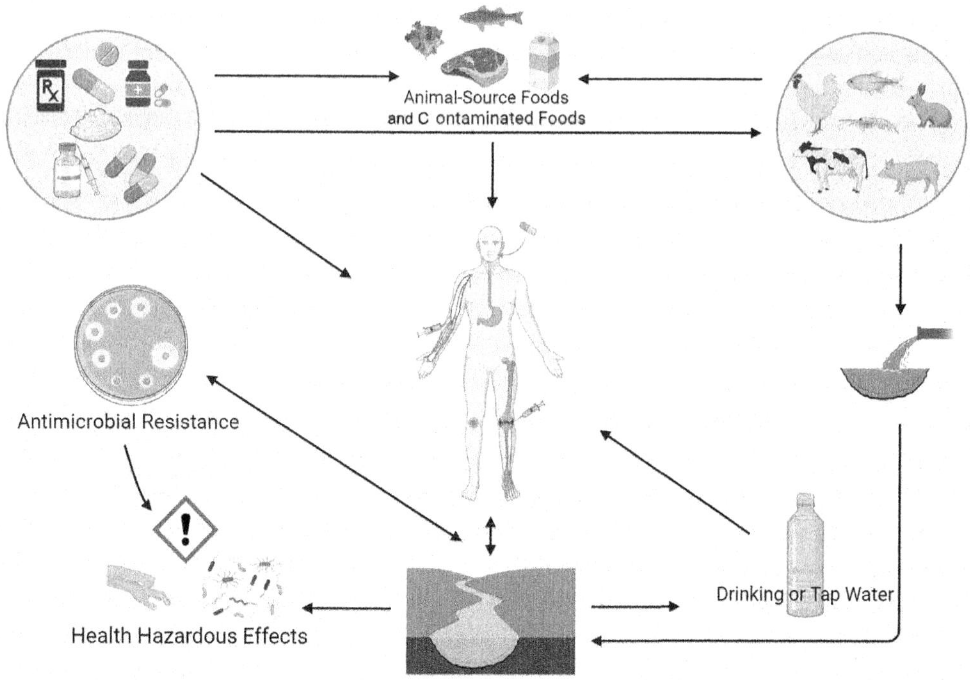

FIGURE 8.1 Potential dietary sources of pharmaceutical residues and their health risks.

Note: Massive and inappropriate prescription of medications in human and veterinary medicine contribute to dissemination of residues in the environment. In addition, animal-source foods and contaminated food act as potential sources that can affect health. Wastes (e.g., manure, biosolids and wastewater) and water from animal rearing and aquaculture contaminate the environment and thus drinking and tap water. These factors help promote antimicrobial resistance. Ultimately, these residues result in hazardous health effects.

search for methods to increase animal production, for example, the use of pharmaceuticals that promote growth and decrease the rearing period. In many countries, especially developing countries, there is unregulated and uncontrolled access to veterinary drugs, particularly antibiotics. This uncontrolled access is accompanied by deficient knowledge about correct dosing and duration, proper use, and holding (withdrawal) periods, which increase the problem of residues.

The global use of antimicrobials in the animal industry (i.e., chicken, cattle, and pigs) is expected to rise from 93,309 tons in 2017 to 104,079 tons in 2030, representing an increase of 11.5% (Tiseo et al., 2020). According to published studies from 1999 until 2021, tetracyclines were the most antimicrobial residues found in food of animal origin, followed by penicillin G, chloramphenicol, and amoxicillin; milk was the most contaminated food, followed by seafood and poultry (Treiber and Beranek-Knauer, 2021). Of note, seafood showed elevated levels of antibiotic residues among the other foodstuffs (Treiber and Beranek-Knauer, 2021; Wang et al., 2021). Some of these residues exceeded the maximum allowed residue limits, particularly milk, eggs, poultry, and beef, which can result in a potentially serious impact on health (Treiber and Beranek-Knauer, 2021). The residues of antibiotics occur not only from their use (either in appropriate or inappropriate ways) but also possibly from their difficult metabolism, such as tetracyclines (Scaria et al., 2021).

Chloramphenicol is frequently used in animal medication, although it has been banned in the treatment of food-producing animals (Treiber and Beranek-Knauer, 2021). Chloramphenicol has a damaging effect on the liver and bone marrow. Although chloramphenicol is banned, which means it should not be detected in food, the presence of chloramphenicol at levels greater than those thought to cause serious health problems (i.e., 0.3 μg/kg) (EFSA CONTAM Panel, 2014; Doğan et al., 2020; Tajik et al., 2010; Wang et al., 2021) could allow its escape to premature and newborn babies via placenta or mother's milk from mothers eating contaminated food or water, which in turn may cause the generation of aplastic anemia and/or gut microbiota dysbiosis or other problems, such as genotoxic effects (Principi and Esposito, 2016). In addition, chloramphenicol residues showed a relevant health hazard in 3–6-year-old children in Guangzhou, China (Wang et al., 2021).

It has been noted that frequent use of antibiotics is markedly linked to obesity in adults and children (Del Fiol et al., 2018; Principi and Esposito, 2016). Antibiotics at therapeutic or very low doses result in dysbiosis or imbalances in the intestinal microbiome (Del Fiol et al., 2018; Principi and Esposito, 2016; Qian et al., 2021). This means that beneficial bacteria, such as butyrate-producing bacteria and *Bacteroides* strains, will decrease in favor of increases in pathobionts (opportunistic bacteria that may cause diseases under certain circumstances), leading to decreased diversity of gut microbiota. Subchronic exposure to oxytetracycline, florfenicol, and doxycycline caused hepatic metabolic disorder in zebrafish, which disrupted the expression of glucose and lipid metabolism-related genes (Qian et al., 2021). The antibiotic tylosin, widely used as a growth promoter for animals, alters the gut microbiota and results in obesity (elevated body weight and visceral and relative fat mass) and insulin resistance in mice, and the changes in the gut microbiota are long-lasting and sufficient to disrupt host metabolic homeostasis, leading to further metabolic disorders in later periods of life (Chen et al., 2022). These findings were confirmed when the fecal microbiota of tylosin-exposed mice was transported to germ-free mice. In addition, the primary bile acids (i.e., chenodeoxycholic acid, cholic acid, α-MCA, and β-MCA) were increased, while the secondary bile acids (i.e., ω-MCA and ursodeoxycholic acid) were decreased, which was accompanied by a decrease in plasma fibroblast growth factor 19 and its expression in the intestine (Chen et al., 2022). Plasma fibroblast growth factor 19 is secreted by the intestinal epithelium and binds to a receptor in the liver to regulate lipid and glucose metabolism (Ge et al., 2014; Vrieze et al., 2014). The majority of changes in the gut microbiota from antibiotics include *Bacteroides*, *Prevotella*, and *Ruminococcus*, which are significantly linked to chronic disease risk (Wilkins et al., 2019). These findings indicate the long-lasting effects of antibiotic residues on health.

Other antibiotics, such as sulfamethazine, oxytetracycline, and furazolidone, have shown carcinogenic effects and are used in animal production (Bacanlı and Başaran, 2019). It is noteworthy that these antibiotics are the same as those used in human treatment. Tetracyclines widely contaminate and represent a concern for the environment (Scaria et al., 2021). Tetracyclines at low concentrations induce developmental toxicity and oxidative stress in aquatic organisms, such as the animal model zebrafish (Zhang et al., 2015b). Chronic exposure to steatogenic drugs induces liver injury. Tetracycline, as a steatogenic compound, showed a hepatotoxic effect, which caused decreased mitochondrial functions and lipid metabolism (Szalowska et al., 2014). In addition, subchronic exposure to dietary sulfamethazine altered gut microbial communities and induced oxidative stress in adults and the offspring of marine medaka (*Oryzias melastigma*) (He et al., 2022; Zhang et al., 2021b).

Oxytetracycline and fluoroquinolones (i.e., ciprofloxacin, ofloxacin, enrofloxacin, enoxacin, and sarafloxacin) residues were also reported in farmed fish and foods of animal origin (Liu et al., 2018; Vishnuraj et al., 2016). In Asian shrimp farming, oxytetracycline has been a popular antibiotic for several decades; however, quinolones and the combined formulations of sulfadiazine and trimethoprim have become frequently used in the last decade (Thuy et al., 2011). We should consider the synergistic toxic effect of antibiotics if they are combined in dietary sources and how this could increase the toxicity of residues. To illustrate, the combination of enrofloxacin with ciprofloxacin, florfenicol, or sulfadimidine showed a more pronounced toxic effect than enrofloxacin alone (Luan et al., 2022). Furthermore, if we investigate more, we will observe and detect the use of other antibiotics and antifungals, such as rifampicin and griseofulvin, in aquaculture (Thuy et al., 2011), indicating uncontrolled use of pharmaceuticals in animal production, especially in developing countries.

In humans, misuse of antibiotics contributes to the dissemination of antibiotics in the environment. The prescription of antibiotics increased during 1995–2011 in the United Kingdom, and most of these prescriptions could be avoided. For example, approximately 69% of patients with sore throat received antibiotics in 2011; however, it is noteworthy that approximately 90% of sore throats were cured without antibiotics (Spinks et al., 2021), which represents a significant deviation from best practice (Hawker et al., 2014). In Japan and many other countries, the overuse of antibiotics is attributed to the inaccurate distinguishing between viral and bacterial upper respiratory tract infections (Hawker et al., 2014; Higashi and Fukuhara, 2009). Recently, some countries, such as Japan, have taken steps to reduce the prescription of antibiotics through the release of a health care policy that provides a financial reward for medical facilities not prescribing antibiotics for early-stages of gastrointestinal and respiratory infections (Okubo et al., 2022). However, according to a recent study, a contrary behavior was observed, in which cephem antibacterial agents were the most inappropriately prescribed pharmaceuticals for patients with a common cold with no symptoms of bacterial infections in Japan (Nakano et al., 2022). Of interest, the majority of the prescribed medications for common cold symptoms were inappropriate or not indicated (Nakano et al., 2022), which to an extent increases the release of pharmaceuticals and their metabolites in the environment. Despite the efforts to minimize the use of antibiotics worldwide and only direct their prescription when indicated, studies are continually increasing, highlighting the inappropriate and massive use of antibiotics. It is possible that the laws and regulations are not sufficient, and we must educate people about the deleterious impacts of misuse of antibiotics and the beneficial effects when appropriate medications are used in general. This could facilitate the work of the physician and may decrease the potential negative feelings and resistance of patients if they are not prescribed antibiotics in their first visits to general practitioners.

8.5.1.1 Antimicrobial Residues and Resistance

Antimicrobial resistance, particularly antibiotic resistance, is a global issue that threatens humanity and health sectors worldwide. According to the World Health Organization, it is considered one of the biggest problems and threats to health and food and economic security. Antibiotic resistance is naturally occurring, but it is safe when kept at the lowest natural levels. Our practices and misuse of these medications accelerate the process and rapidly lead to the occurrence of antibiotic-resistant bacteria and thereby could lead to uncurable infectious diseases.

Resistance genes are developed anywhere, without a clear understanding of the nature of the underlying mechanism. However, we, to a great extent, know that resistance could be developed when microorganisms are exposed to low concentrations or unsuitable antimicrobials that allow time for the microbes to explore these molecules as potential toxins that should be detoxified. The majority of antimicrobials work against growing microbes that target certain stages of growth. Thus, when administered to a patient, several targeted microbes or even the host's microbiota will be exposed to this antimicrobial, and if not killed in the next doses, they may escape, develop resistance factors and transfer these factors to other microbes. This elucidates that misuse or overuse of antimicrobials is a major contributor to the problem. For more information about the environmental factors that help develop antibiotic resistance and their global health problem, you are recommended to read works by Bengtsson-Palme et al. (2018), Zhang et al. (2022), Davies and Davies (2010), Bush et al. (2011), and Cars et al. (2021).

To provide an example of such development of bacterial resistance in nature, as we have mentioned before, antibiotics are massively used in shrimp farming, which in turn results in antibiotic residues in water and mud. Resistant bacteria were isolated in locations of shrimp farming in mangrove areas in Vietnam, where the bacteria showed resistance to norfloxacin, oxolinic acid, sulfamethoxazole, and trimethoprim (Le et al., 2005). It has been strongly thought that aquatic environments are ideal for the evolution of antibiotic resistance and the horizontal transfer of mobile genetic elements, including antibiotic resistance genes (Biyela et al., 2004; Marti et al., 2014). A possible way that bacterial resistance was developed and transferred among bacteria in these locations is that plasmid-mediated quinolone resistance genes—encoding proteins that protect DNA gyrase from quinolones—are horizontally transferred among bacteria (Robicsek et al., 2006; Suzuki and Hoa, 2012). The study by Biyela et al. (2004) showed a strong correlation between the incidence of antibiotic resistance in bacteria isolated from the Mhlathuze River in South Africa and clinical isolates from diarrheic patients. This provides evidence that the aquatic environment could act as a reservoir for resistant bacteria and shows how the massive use of antibiotics easily develops resistant microbes that could be transferred to humans and animals with the potential to cause health problems.

8.5.1.2 CNS Medications

Antidepressants such as fluoxetine (a selective serotonin reuptake inhibitor) and central nervous system stimulants such as methamphetamine were detected in the environment, such as surface water (Weinberger and Klaper, 2014). This, in turn, showed an accumulation in fish tissues such as brain, muscle, and liver (Brooks et al., 2005; Chu and Metcalfe, 2007; Schultz et al., 2010). *Pimephales promelas* (fathead minnow) is considered an excellent model of the chemical impacts of studies on brain function. The *Pimephales promelas* exposed to environmentally relevant concentrations of fluoxetine showed significant dose-dependent behavioral changes (i.e., reproduction, predator avoidance, and feeding). At the lowest concentration (i.e., 1 μg/L), males had a significant impact on mating behavior, with specific changes in nest building and defending; in both sexes, changes were observed in predator avoidance behaviors. At the highest concentration (i.e., 100 μg/L), aggressive behavior of males limited egg production because of female deaths, in addition to the detected isolation and repetitive behaviors in males. These changes were not accompanied by changes in plasma testosterone and 17β-estradiol (Weinberger and Klaper, 2014). However, fluoxetine and fluvoxamine strongly inhibited the activity of C17,20-lyase and CYP11*β* enzymes that are involved in the synthesis of androgens (Fernandes et al., 2011). Such antidepressants are suggested to be neuroendocrine disruptors (Mennigen et al., 2011). This might be because high concentrations can induce endocrine changes or various environmental and/or rearing factors that could affect the response of living organisms. Although CNS-related medications such as fluoxetine showed a lower concentration of residues, which is not lethal, their bioaccumulation in the brains of fish produces marked behavioral changes. This raises a concern about the effect of these pharmaceutical residues on the nervous system of humans and animals.

On another level, the transformation of certain pharmaceuticals to toxins can contaminate the environment with other chemicals. During wastewater treatment, the ozonation process can convert methamphetamine and ephedrine and certain antidepressants, such as fluoxetine, to nitromethane, which is transformed to chloropicrin (nitrochloroform) by the chlorination process (Shi and McCurry, 2020). Nitrochloroform is a potent toxin and was used in World War I. This indicates that certain pharmaceuticals can be converted to more toxic chemicals, and we must take care when detecting potential chemicals in the environment or reuse and disinfect wastes.

8.5.1.3 Other Pharmaceuticals

Many other pharmaceuticals, such as antihistamines, antineoplastic and bronchodilators, anti-asthma drugs, immunosuppressants, gastrointestinal drugs, and local anesthetics, have been reported in dietary sources such as surface and drinking water (Kosonen and Kronberg, 2009; Kristofco and Brooks, 2017; Styszko et al., 2021), and they are used in human and animal medications, which allow their accumulation in food, including animal tissues and water. Research is not sufficient on the long-term effect of such residues or low doses on health. The minimum therapeutic dose (MTD) is used as an indicator of health risks because of the lack of available toxicological data. For example, hospital workers exposed to low doses of antineoplastic drugs showed increased DNA damage and elevated levels of oxidative stress biomarkers (Mrdjanović et al., 2021). Other indicators, such as acceptable daily intake (ADI) and tolerable daily intake (TDI), are also used to help assess health risks. These limitations, with little availability of information, challenge our understanding of health risks from residues.

Animal-derived foods are large sources of pharmaceutical residues, which allow exposure to many pharmaceuticals with high potential health hazards. The ease of obtaining medicines, particularly in developing countries, facilitates their improper and massive use in humans and animals. Such cases potentiate the problem and expose global health to hazards.

In addition, the combined residues of different compounds in dietary sources can result in a more deleterious impact on health than each compound alone (Luan et al., 2022). For example, quinolone antibiotics affect the central nervous system, and this effect is augmented by NSAIDs, leading to seizures (Kim et al., 2009).

8.6 SUMMARY AND CONCLUSIONS

There are several sources of pharmaceutical residues. For example, the animal husbandry and aquaculture industries are rapidly growing, resulting in more release of pharmaceuticals into the environment and food sources. In addition, inappropriate human use of pharmaceuticals throughout life greatly contributes to high levels of residues in the environment. This constantly urges researchers to develop methods to remove or deactivate such molecules from food, water, and the environment. Several pharmaceuticals continuously reach the environment, and a considerable amount of them accumulates or may degrade other molecules that may also be harmful to the environment or living organisms. We must admit that we cannot, at least at the current time, prevent the presence of pharmaceutical residues in our environment, but we should work to keep them at their lowest levels and to minimize their deleterious impacts. This could be accomplished by following the best practice when prescribing or using pharmaceuticals for health problems and in veterinary medicine, optimizing the use of pharmaceuticals in our life, enacting new laws and regulations that limit the uncontrolled use of medications in animal production, finding or searching for safer or easily biodegradable alternatives for frequently used pharmaceuticals, and developing more efficient methods to detect and deactivate these residues.

REFERENCES

Ahmed, M.J. 2017. Adsorption of quinolone, tetracycline, and penicillin antibiotics from aqueous solution using activated carbons: Review. *Environ Toxicol Pharm* 50:1–10.

Aliko, V., Korriku, R.S., Pagano, M. and Faggio, C. 2021. Double-edged sword: Fluoxetine and ibuprofen as development jeopardizers and apoptosis' inducers in common toad, Bufo bufo, tadpoles. *Sci Total Environ* 776:145945.

Anh, H.Q., Le, T.P.Q., Da Le, N., Lu, X.X., Duong, T.T., Garnier, J., et al. 2021. Antibiotics in surface water of East and Southeast Asian countries: A focused review on contamination status, pollution sources, potential risks, and future perspectives. *Sci Total Environ* 764:142865.

Arnold, K.E., Brown, A.R., Ankley, G.T. and Sumpter, J.P. 2014. Medicating the environment: Assessing risks of pharmaceuticals to wildlife and ecosystems. *Phil Trans R Soc B* 369:20130569.

Ashton, D., Hilton, M. and Thomas, K.V. 2004. Investigating the environmental transport of human pharmaceuticals to streams in the United Kingdom. *Sci Total Environ* 333:167–184.

Aus Der Beek, T., Weber, F.-A., Bergmann, A., Hickmann, S., Ebert, I., Hein, A., et al. 2016. Pharmaceuticals in the environment—global occurrences and perspectives. *Environ Toxicol Chem* 35:823–835.

Bacanlı, M. and Başaran, N. 2019. Importance of antibiotic residues in animal food. *Food Chem Toxicol* 125:462–466.

Bartoskova, M., Dobsikova, R., Stancova, V., Zivna, D., Blahova, J., Marsalek, P., et al. 2013. Evaluation of ibuprofen toxicity for zebrafish (*Danio rerio*) targeting on selected biomarkers of oxidative stress. *Neuro Endocrinol Lett* 34 (Suppl 2):102–108.

Ben, Y., Fu, C., Hu, M., Liu, L., Wong, M.H. and Zheng, C. 2019. Human health risk assessment of antibiotic resistance associated with antibiotic residues in the environment: A review. *Environ Res* 169:483–493.

Bendz, D., Paxéus, N.A., Ginn, T.R. and Loge, F.J. 2005. Occurrence and fate of pharmaceutically active compounds in the environment, a case study: Höje River in Sweden. *J Hazard Mater* 122:195–204.

Bengtsson-Palme, J., Kristiansson, E. and Larsson, D.G.J. 2018. Environmental factors influencing the development and spread of antibiotic resistance. *FEMS Microbiol Rev* 42.

Benotti, M.J., Trenholm, R.A., Vanderford, B.J., Holady, J.C., Stanford, B.D. and Snyder, S.A. 2009. Pharmaceuticals and endocrine disrupting compounds in U.S. Drinking water. *Environ Sci Technol* 43:597–603.

Beretta, M., Britto, V., Tavares, T.M., da Silva, S.M.T. and Pletsch, A.L. 2014. Occurrence of pharmaceutical and personal care products (PPCPs) in marine sediments in the Todos os Santos Bay and the north coast of Salvador, Bahia, Brazil. *J Soils Sediments* 14:1278–1286.

Bernot, M.J., Smith, L. and Frey, J. 2013. Human and veterinary pharmaceutical abundance and transport in a rural central Indiana stream influenced by confined animal feeding operations (CAFOs). *Sci Total Environ* 445–446:219–230.

Biel-Maeso, M., González-González, C., Lara-Martín, P.A. and Corada-Fernández, C. 2019. Sorption and degradation of contaminants of emerging concern in soils under aerobic and anaerobic conditions. *Sci Total Environ* 666:662–671.

Biyela, P.T., Lin, J. and Bezuidenhout, C.C. 2004. The role of aquatic ecosystems as reservoirs of antibiotic resistant bacteria and antibiotic resistance genes. *Water Sci Technol* 50:45–50.

Botsoglou, N.A. and Fletouris, D.J. 2001. *Drug Residues in Foods: Pharmacology, Food Safety, and Analysis.* New York, NY: Dekker.

Boxall, A.B.A., Rudd, M.A., Brooks, B.W., Caldwell, D.J., Choi, K., Hickmann, S., et al. 2012. Pharmaceuticals and personal care products in the environment: What are the big questions? *Environ Health Perspect* 120:1221–1229.

Brooks, B.W., Chambliss, C.K., Stanley, J.K., Ramirez, A., Banks, K.E., Johnson, R.D., et al. 2005. Determination of select antidepressants in fish from an effluent-dominated stream. *Environ Toxicol Chem* 24:464–469.

Buerge, I.J., Buser, H.-R., Kahle, M., Müller, M.D. and Poiger, T. 2009. Ubiquitous occurrence of the artificial sweetener acesulfame in the aquatic environment: An ideal chemical marker of domestic wastewater in groundwater. *Environ Sci Technol* 43:4381–4385.

Buerge, I.J., Kahle, M., Buser, H.-R., Müller, M.D. and Poiger, T. 2008. Nicotine derivatives in wastewater and surface waters: Application as chemical markers for domestic wastewater. *Environ Sci Technol* 42:6354–6360.

Burgos-Aceves, M.A., Cohen, A., Smith, Y. and Faggio, C. 2018. MicroRNAs and their role on fish oxidative stress during xenobiotic environmental exposures. *Ecotoxicol Environ Saf* 148:995–1000.

Bush, K., Courvalin, P., Dantas, G., Davies, J., Eisenstein, B., Huovinen, P., et al. 2011. Tackling antibiotic resistance. *Nat Rev Microbiol* 9:894–896.

Cabello, F.C. 2006. Heavy use of prophylactic antibiotics in aquaculture: A growing problem for human and animal health and for the environment. *Environ Microbiol* 8:1137–1144.

Cahill, J.D., Furlong, E.T., Burkhardt, M.R., Kolpin, D. and Anderson, L.G. 2004. Determination of pharmaceutical compounds in surface- and ground-water samples by solid-phase extraction and high-performance liquid chromatography—electrospray ionization mass spectrometry. *J Chromatogr A* 1041:171–180.

Carmona, E., Andreu, V. and Picó, Y. 2014. Occurrence of acidic pharmaceuticals and personal care products in Turia River Basin: From waste to drinking water. *Sci Total Environ* 484:53–63.

Cars, O., Chandy, S.J., Mpundu, M., Peralta, A.Q., Zorzet, A. and So, A.D. 2021. Resetting the agenda for antibiotic resistance through a health systems perspective. *Lancet Glob Health* 9:e1022–e1027.

Carvalho, I.T. and Santos, L. 2016. Antibiotics in the aquatic environments: A review of the European scenario. *Environ Int* 94:736–757.

Chen, H.-C., Wang, P.-L. and Ding, W.-H. 2008. Using liquid chromatography—ion trap mass spectrometry to determine pharmaceutical residues in Taiwanese rivers and wastewaters. *Chemosphere* 72:863–869.

Chen, R.-A., Wu, W.-K., Panyod, S., Liu, P.-Y., Chuang, H.-L., Chen, Y.-H., et al. 2022. Dietary exposure to antibiotic residues facilitates metabolic disorder by altering the gut microbiota and bile acid composition. *mSystems* 7:e00172–e00122.

Choi, K., Kim, Y., Park, J., Park, C.K., Kim, M., Kim, H.S., et al. 2008. Seasonal variations of several pharmaceutical residues in surface water and sewage treatment plants of Han River, Korea. *Sci Total Environ* 405:120–128.

Chu, S. and Metcalfe, C.D. 2007. Analysis of paroxetine, fluoxetine and norfluoxetine in fish tissues using pressurized liquid extraction, mixed mode solid phase extraction cleanup and liquid chromatography—tandem mass spectrometry. *J Chromatogr A* 1163:112–118.

Daughton, C.G. 2001. Pharmaceuticals and personal care products in the environment: Overarching issues and overview. In *Pharmaceuticals and Care Products in the Environment*, eds. Daughton, C.G., Jones-Lepp, T.L., 2–38. ACS Symposium Series Vol. 791. Washington, DC: American Chemical Society.

Davies, J. and Davies, D. 2010. Origins and evolution of antibiotic resistance. *Microbiol Mol Biol Rev* 74:417–433.

Del Fiol, F.S., Balcão, V.M., Barberato-Fillho, S., Lopes, L.C. and Bergamaschi, C.C. 2018. Obesity: A new adverse effect of antibiotics? *Front Pharmacol* 9:1408.

Doğan, Y.N., Pamuk, Ş. and Gürler, Z. 2020. Chloramphenicol and sulfonamide residues in sea bream (*Sparus aurata*) and sea bass (*Dicentrarchus labrax*) fish from aquaculture farm. *Environ Sci Pollut Res* 27:41248–41252.

Doretto, K.M., Peruchi, L.M. and Rath, S. 2014. Sorption and desorption of sulfadimethoxine, sulfaquinoxaline and sulfamethazine antimicrobials in Brazilian soils. *Sci Total Environ* 476–477:406–414.

EFSA CONTAM Panel (EFSA Panel on Contaminants in the Food Chain), 2014. 2014. Scientific Opinion on Chloramphenicol in food and feed. *EFSA Journal* 2014; 12:3907, 145 pp.

Elmund, G.K., Morrison, S.M., Grant, D.W. and Nevins, M.P. 1971. Role of excreted chlortetracycline in modifying the decomposition process in feedlot waste. *Bull Environ Contam Toxicol* 6:129–132.

Evgenidou, E.N., Konstantinou, I.K. and Lambropoulou, D.A. 2015. Occurrence and removal of transformation products of PPCPs and illicit drugs in wastewaters: A review. *Sci Total Environ* 505:905–926.

Faggio, C., Pagano, M., Alampi, R., Vazzana, I. and Felice, M.R. 2016. Cytotoxicity, haemolymphatic parameters, and oxidative stress following exposure to sub-lethal concentrations of quaternium-15 in *Mytilus galloprovincialis*. *Aquat Toxicol* 180:258–265.

Faggio, C., Tsarpali, V. and Dailianis, S. 2018. Mussel digestive gland as a model tissue for assessing xenobiotics: An overview. *Sci Total Environ* 636:220–229.

Fauci, A.S. and Marston, H.D. 2014. The perpetual challenge of antimicrobial resistance. *JAMA* 311:1853–1854.

Fekadu, S., Alemayehu, E., Dewil, R. and Van der Bruggen, B. 2019. Pharmaceuticals in freshwater aquatic environments: A comparison of the African and European challenge. *Sci Total Environ* 654:324–337.

Fent, K. 2008. Effects of pharmaceuticals on aquatic organisms. In *Pharmaceuticals in the Environment: Sources, Fate, Effects and Risks*, ed. Kümmerer, K., 175–203. Berlin, Heidelberg: Springer.

Fernandes, D., Schnell, S. and Porte, C. 2011. Can pharmaceuticals interfere with the synthesis of active androgens in male fish? An in vitro study. *Mar Pollut Bull* 62:2250–2253.

Ferreira, A.P., de Lourdes, C. and da Cunha, N. 2005. Anthropic pollution in aquatic environment: Development of a caffeine indicator. *Int J Environ Health Res* 15:303–311.

Fonseca, V.F., Duarte, I.A., Duarte, B., Freitas, A., Pouca, A.S.V., Barbosa, J., et al. 2021. Environmental risk assessment and bioaccumulation of pharmaceuticals in a large urbanized estuary. *Sci Total Environ* 783:147021.

García, J., García-Galán, M.J., Day, J.W., Boopathy, R., White, J.R., Wallace, S., et al. 2020. A review of emerging organic contaminants (EOCs), antibiotic resistant bacteria (ARB), and antibiotic resistance genes (ARGs) in the environment: Increasing removal with wetlands and reducing environmental impacts. *Bioresour Technol* 307:123228.

Gavrilescu, M., Demnerová, K., Aamand, J., Agathos, S. and Fava, F. 2015. Emerging pollutants in the environment: Present and future challenges in biomonitoring, ecological risks and bioremediation. *N Biotechnol* 32:147–156.

Ge, H., Zhang, J., Gong, Y., Gupte, J., Ye, J., Weiszmann, J., et al. 2014. Fibroblast growth factor receptor 4 (FGFR4) deficiency improves insulin resistance and glucose metabolism under diet-induced obesity conditions. *J Biol Chem* 289:30470–30480.

Golovko, O., Kumar, V., Fedorova, G., Randak, T. and Grabic, R. 2014. Seasonal changes in antibiotics, antidepressants/psychiatric drugs, antihistamines and lipid regulators in a wastewater treatment plant. *Chemosphere* 111:418–426.

González-Alcaraz, M.N., Malheiro, C., Cardoso, D.N., Prodana, M., Morgado, R.G., van Gestel, C.A.M., et al. 2020. Bioaccumulation and toxicity of organic chemicals in terrestrial invertebrates. In *Bioavailability of Organic Chemicals in Soil and Sediment*, eds. Ortega-Calvo, J.J. and Parsons, J.R., 149–189. Cham: Springer International Publishing.

Gottschall, N., Topp, E., Edwards, M., Payne, M., Kleywegt, S., Russell, P., et al. 2013. Hormones, sterols, and fecal indicator bacteria in groundwater, soil, and subsurface drainage following a high single application of municipal biosolids to a field. *Chemosphere* 91:275–286.

Gros, M., Petrović, M., Ginebreda, A. and Barceló, D. 2010. Removal of pharmaceuticals during wastewater treatment and environmental risk assessment using hazard indexes. *Environ Int* 36:15–26.

Han, S., Choi, K., Kim, J., Ji, K., Kim, S., Ahn, B., et al. 2010. Endocrine disruption and consequences of chronic exposure to ibuprofen in Japanese medaka (*Oryzias latipes*) and freshwater cladocerans *Daphnia magna* and *Moina macrocopa*. *Aquat Toxicol* 98:256–264.

Hansen, B.G., Munn, S.J., De Bruijn, J., Pakalin, S., Luotamo, M., Berthault, F., et al., 2002. European Union risk assessment report—4-nonylphenol (branched) and nonylphenol. In *Environment and quality of life series*. https://echa.europa.eu/documents/10162/43080e23-3646-4ddf-836b-a248bd4225c6 (accessed July 15, 2022).

Hawker, J.I., Smith, S., Smith, G.E., Morbey, R., Johnson, A.P., Fleming, D.M., et al. 2014. Trends in antibiotic prescribing in primary care for clinical syndromes subject to national recommendations to reduce antibiotic resistance, UK 1995–2011: Analysis of a large database of primary care consultations. *J Antimicrob Chemother* 69:3423–3430.

He, S., Li, D., Wang, F., Zhang, C., Yue, C., Huang, Y., et al. 2022. Parental exposure to sulfamethazine and nanoplastics alters the gut microbial communities in the offspring of marine madaka (*Oryzias melastigma*). *J Hazard Mater* 423:127003.

Heberer, T., Reddersen, K. and Mechlinski, A. 2002. From municipal sewage to drinking water: Fate and removal of pharmaceutical residues in the aquatic environment in urban areas. *Water Sci Technol* 46:81–88.

Heberer, T., Schmidt-Bäumler, K. and Stan, H.-J. 1998. Occurrence and distribution of organic contaminants in the aquatic system in Berlin. Part I: Drug residues and other polar contaminants in Berlin surface and groundwater. *Acta Hydrochim Hydrobiol* 26:272–278.

Heberer, T. and Stan, H.J. 1997. Determination of clofibric acid and N-(phenylsulfonyl)-sarcosine in sewage, river and drinking water. *Int J Environ Anal Chem* 67:113–124.

Hernando-Amado, S., Coque, T.M., Baquero, F. and Martínez, J.L. 2019. Defining and combating antibiotic resistance from One Health and Global Health perspectives. *Nat Microbiol* 4:1432–1442.

Higashi, T. and Fukuhara, S. 2009. Antibiotic prescriptions for upper respiratory tract infection in Japan. *Intern Med* 48:1369–1375.

Holzbauer, S. and Chiller, T. 2006. Antimicrobial resistance in bacteria of animal origin. *Emerg Infect Dis* 12:1180–1181.

Huerta, B., Jakimska, A., Gros, M., Rodríguez-Mozaz, S. and Barceló, D. 2013a. Analysis of multi-class pharmaceuticals in fish tissues by ultra-high-performance liquid chromatography tandem mass spectrometry. *J Chromatogr A* 1288:63–72.

Huerta, B., Rodríguez-Mozaz, S. and Barcelo, D. 2013b. Chapter 6—analysis of pharmaceutical compounds in biota. In *Comprehensive Analytical Chemistry*, eds. Petrovic, M., Barcelo, D. and Pérez, S., 169–193. Poland: Elsevier.

Jiang, J.-J., Lee, C.-L. and Fang, M.-D. 2014. Emerging organic contaminants in coastal waters: Anthropogenic impact, environmental release and ecological risk. *Mar Pollut Bull* 85:391–399.

Kibbey, T.C.G., Paruchuri, R., Sabatini, D.A. and Chen, L. 2007. Adsorption of beta blockers to environmental surfaces. *Environ Sci Technol* 41:5349–5356.

Kim, J., Ohtani, H., Tsujimoto, M. and Sawada, Y. 2009. Quantitative comparison of the convulsive activity of combinations of twelve fluoroquinolones with five nonsteroidal antiinflammatory agents. *Drug Metab Pharmacokinet* 24:167–174.

K'Oreje, K.O., Vergeynst, L., Ombaka, D., De Wispelaere, P., Okoth, M., Van Langenhove, H., et al. 2016. Occurrence patterns of pharmaceutical residues in wastewater, surface water and groundwater of Nairobi and Kisumu city, Kenya. *Chemosphere* 149:238–244.

Koschorreck, J., Koch, C. and Rönnefahrt, I. 2002. Environmental risk assessment of veterinary medicinal products in the EU—a regulatory perspective. *Toxicol Lett* 131:117–124.

Kosonen, J. and Kronberg, L. 2009. The occurrence of antihistamines in sewage waters and in recipient rivers. *Environ Sci Pollut Res* 16:555–564.

Koutsouba, V., Heberer, T., Fuhrmann, B., Schmidt-Baumler, K., Tsipi, D. and Hiskia, A. 2003. Determination of polar pharmaceuticals in sewage water of Greece by gas chromatography—mass spectrometry. *Chemosphere* 51:69–75.

Kristofco, L.A. and Brooks, B.W. 2017. Global scanning of antihistamines in the environment: Analysis of occurrence and hazards in aquatic systems. *Sci Total Environ* 592:477–487.

Kruse, H. and Sørum, H. 1994. Transfer of multiple drug resistance plasmids between bacteria of diverse origins in natural microenvironments. *Appl Environ Microbiol* 60:4015–4021.

Krzeminski, P., Tomei, M.C., Karaolia, P., Langenhoff, A., Almeida, C.M.R., Felis, E., et al. 2019. Performance of secondary wastewater treatment methods for the removal of contaminants of emerging concern implicated in crop uptake and antibiotic resistance spread: A review. *Sci Total Environ* 648:1052–1081.

Kuch, H.M. and Ballschmiter, K. 2001. Determination of endocrine-disrupting phenolic compounds and estrogens in surface and drinking water by HRGC–(NCI)–MS in the picogram per liter range. *Environ Sci Technol* 35:3201–3206.

Kumar, S., Kaushik, G. and Villarreal-Chiu, J.F. 2016. Scenario of organophosphate pollution and toxicity in India: A review. *Environ Sci Pollut Res* 23:9480–9491.

Lambropoulou, D.A. and Nollet, L. 2014. *Transformation Products of Emerging Contaminants in the Environment: Analysis, Processes, Occurrence, Effects and Risks*. India: John Wiley & Sons.

Le, T.X., Munekage, Y. and Kato, S.-I. 2005. Antibiotic resistance in bacteria from shrimp farming in mangrove areas. *Sci Total Environ* 349:95–105.

Letsinger, S., Kay, P., Rodríguez-Mozaz, S., Villagrassa, M., Barceló, D. and Rotchell, J.M. 2019. Spatial and temporal occurrence of pharmaceuticals in UK estuaries. *Sci Total Environ* 678:74–84.

Li, W.C. 2014. Occurrence, sources, and fate of pharmaceuticals in aquatic environment and soil. *Environ Pollut* 187:193–201.

Lillehaug, A., Lunestad, B.T. and Grave, K. 2003. Epidemiology of bacterial diseases in Norwegian aquaculture—a description based on antibiotic prescription data for the ten-year period 1991 to 2000. *Dis Aquat Organ* 53:115–125.

Liu, S., Dong, G., Zhao, H., Chen, M., Quan, W. and Qu, B. 2018. Occurrence and risk assessment of fluoroquinolones and tetracyclines in cultured fish from a coastal region of northern China. *Environ Sci Pollut Res* 25:8035–8043.

Liu, X., Steele, J.C. and Meng, X.-Z. 2017. Usage, residue, and human health risk of antibiotics in Chinese aquaculture: A review. *Environ Pollut* 223:161–169.

Luan, Y., Chen, K., Zhao, J. and Cheng, L. 2022. Comparative study on synergistic toxicity of enrofloxacin combined with three antibiotics on proliferation of THLE-2 Cell. *Antibiotics* 11:394.

Marti, E., Variatza, E. and Balcazar, J.L. 2014. The role of aquatic ecosystems as reservoirs of antibiotic resistance. *Trends Microbiol* 22:36–41.

Maul, J.D., Schuler, L.J., Belden, J.B., Whiles, M.R. and Lydy, M.J. 2006. Effects of the antibiotic ciprofloxacin on stream microbial communities and detritivorous macroinvertebrates. *Environ Toxicol Chem* 25:1598–1606.

McDermott, P.F., Zhao, S., Wagner, D.D., Simjee, S., Walker, R.D. and White, D.G. 2002. The food safety perspective of antibiotic resistance. *Anim Biotechnol* 13:71–84.

McEachran, A.D., Shea, D., Bodnar, W. and Nichols, E.G. 2016. Pharmaceutical occurrence in groundwater and surface waters in forests land-applied with municipal wastewater. *Environ Toxicol Chem* 35:898–905.

Mennigen, J.A., Stroud, P., Zamora, J.M., Moon, T.W. and Trudeau, V.L. 2011. Pharmaceuticals as Neuroendocrine Disruptors: Lessons Learned from Fish on Prozac. *J Toxicol Env Health-Pt B-Crit Rev* 14:387–412.

Mezzelani, M., Gorbi, S., Da Ros, Z., Fattorini, D., d'Errico, G., Milan, M., et al. 2016. Ecotoxicological potential of non-steroidal anti-inflammatory drugs (NSAIDs) in marine organisms: Bioavailability, biomarkers and natural occurrence in Mytilus galloprovincialis. *Mar Environ Res* 121:31–39.

Mezzelani, M., Gorbi, S. and Regoli, F. 2018. Pharmaceuticals in the aquatic environments: Evidence of emerged threat and future challenges for marine organisms. *Mar Environ Res* 140:41–60.

Miller, E.L., Nason, S.L., Karthikeyan, K.G. and Pedersen, J.A. 2016. Root uptake of pharmaceuticals and personal care product ingredients. *Environ Sci Technol* 50:525–541.

Mohd Nasir, F.A., Praveena, S.M. and Aris, A.Z. 2019. Public awareness level and occurrence of pharmaceutical residues in drinking water with potential health risk: A study from Kajang (Malaysia). *Ecotoxicol Environ Saf* 185:109681.

Mrdjanović, J., Šolajić, S., Srđenović-Čonić, B., Bogdanović, V., Dea, K.-J., Kladar, N., et al. 2021. The oxidative stress parameters as useful tools in evaluating the DNA damage and changes in the complete blood count in hospital workers exposed to low doses of antineoplastic drugs and ionizing radiation. *Int J Environ Res Public Health* 18:8445.

Myers, S.S., Gaffikin, L., Golden, C.D., Ostfeld, R.S., H. Redford, K., H. Ricketts, T., et al. 2013. Human health impacts of ecosystem alteration. *Proc Natl Acad Sci U S A* 110:18753–18760.

Nakano, Y., Watari, T., Adachi, K., Watanabe, K., Otsuki, K., Amano, Y., et al. 2022. Survey of potentially inappropriate prescriptions for common cold symptoms in Japan: A cross-sectional study. *PLoS ONE* 17:e0265874.

Ngubane, N.P., Naicker, D., Ncube, S., Chimuka, L. and Madikizela, L.M. 2019. Determination of naproxen, diclofenac and ibuprofen in Umgeni estuary and seawater: A case of northern Durban in KwaZulu—Natal Province of South Africa. *Reg Stud Mar Sci* 29:100675.

Ojemaye, C.Y. and Petrik, L. 2019. Pharmaceuticals in the marine environment: A review. *Environ Rev* 27:151–165.

Okubo, Y., Nishi, A., Michels, K.B., Nariai, H., Kim-Farley, R.J., Arah, O.A., et al. 2022. The consequence of financial incentives for not prescribing antibiotics: A Japan's nationwide quasi-experiment. *Int J Epidemiol* 51:1645–1655.

Öllers, S., Singer, H.P., Fässler, P. and Müller, S.R. 2001. Simultaneous quantification of neutral and acidic pharmaceuticals and pesticides at the low-ng/l level in surface and waste water. *J Chromatogr A* 911:225–234.

Paíga, P., Lolić, A., Hellebuyck, F., Santos, L.H.M.L.M., Correia, M. and Delerue-Matos, C. 2015. Development of a SPE—UHPLC—MS/MS methodology for the determination of non-steroidal anti-inflammatory and analgesic pharmaceuticals in seawater. *J Pharm Biomed Anal* 106:61–70.

Pan, M. and Chu, L.M. 2016. Adsorption and degradation of five selected antibiotics in agricultural soil. *Sci Total Environ* 545–546:48–56.

Parolini, M. 2020. Toxicity of the Non-Steroidal Anti-Inflammatory Drugs (NSAIDs) acetylsalicylic acid, paracetamol, diclofenac, ibuprofen and naproxen towards freshwater invertebrates: A review. *Sci Total Environ* 740:140043.

Paxéus, N. 2004. Removal of selected non-steroidal anti-inflammatory drugs (NSAIDs), gemfibrozil, carbamazepine, b-blockers, trimethoprim and triclosan in conventional wastewater treatment plants in five EU countries and their discharge to the aquatic environment. *Water Sci Technol* 50:253–260.

Praveena, S.M., Mohd Rashid, M.Z., Mohd Nasir, F.A., Sze Yee, W. and Aris, A.Z. 2019. Occurrence and potential human health risk of pharmaceutical residues in drinking water from Putrajaya (Malaysia). *Ecotoxicol Environ Saf* 180:549–556.

Praveena, S.M., Mohd Rashid, M.Z., Mohd Nasir, F.A., Wee, S.Y. and Aris, A.Z. 2021. Occurrence, human health risks, and public awareness level of pharmaceuticals in tap water from Putrajaya (Malaysia). *Expo Health* 13:93–104.

Principi, N. and Esposito, S. 2016. Antibiotic administration and the development of obesity in children. *Int J Antimicrob Agents* 47:171–177.

Qian, M., Wang, J., Ji, X., Yang, H., Tang, B., Zhang, H., et al. 2021. Sub-chronic exposure to antibiotics doxycycline, oxytetracycline or florfenicol impacts gut barrier and induces gut microbiota dysbiosis in adult zebrafish (*Daino rerio*). *Ecotoxicol Environ Saf* 221:112464.

Rana, M.S., Lee, S.Y., Kang, H.J. and Hur, S.J. 2019. Reducing veterinary drug residues in animal products: A review. *Food Sci Anim Resour* 39:687–703.

Robicsek, A., Jacoby, G.A. and Hooper, D.C. 2006. The worldwide emergence of plasmid-mediated quinolone resistance. *Lancet Infect Dis* 6:629–640.

Rodil, R., Quintana, J.B., Concha-Graña, E., López-Mahía, P., Muniategui-Lorenzo, S. and Prada-Rodríguez, D. 2012. Emerging pollutants in sewage, surface and drinking water in Galicia (NW Spain). *Chemosphere* 86:1040–1049.

Rossknecht, H., Hetzenauer, H. and Ternes, T.A. 2001. Pharmaceuticals in Lake Constance. *Nachr Chem* 49:145–149.

Sanseverino, I., Navarro Cuenca, A., Loos, R., Marinov, D. and Lettieri, T. 2018. *State of the Art on the Contribution of Water to Antimicrobial Resistance, EUR 29592 EN, JRC114775*. Luxembourg: Publications Office of the European Union.

Scaria, J., Anupama, K.V. and Nidheesh, P.V. 2021. Tetracyclines in the environment: An overview on the occurrence, fate, toxicity, detection, removal methods, and sludge management. *Sci Total Environ* 771:145291.

Schmerold, I. and Ungemach, F. 2004. Antibiotics | use in animal husbandry. In *Encyclopedia of Meat Sciences*, ed. Jensen, W.K., 32–38. Oxford: Elsevier.

Schultz, M.M., Furlong, E.T., Kolpin, D.W., Werner, S.L., Schoenfuss, H.L., Barber, L.B., et al. 2010. Antidepressant pharmaceuticals in two U.S. Effluent-impacted streams: Occurrence and fate in water and sediment, and selective uptake in fish neural tissue. *Environ Sci Technol* 44:1918–1925.

Sharma, K. and Kaushik, G. 2017. NSAIDS in the environment: From emerging problem to green solution. *Ann Pharmacol Pharm* 2:1077.

Shi, J.L. and McCurry, D.L. 2020. Transformation of N-methylamine drugs during wastewater ozonation: Formation of nitromethane, an efficient precursor to halonitromethanes. *Environ Sci Technol* 54:2182–2191.

Simazaki, D., Kubota, R., Suzuki, T., Akiba, M., Nishimura, T. and Kunikane, S. 2015. Occurrence of selected pharmaceuticals at drinking water purification plants in Japan and implications for human health. *Water Res* 76:187–200.

Solliec, M., Roy-Lachapelle, A., Gasser, M.-O., Coté, C., Généreux, M. and Sauvé, S. 2016. Fractionation and analysis of veterinary antibiotics and their related degradation products in agricultural soils and drainage waters following swine manure amendment. *Sci Total Environ* 543:524–535.

Sorinolu, A.J., Tyagi, N., Kumar, A. and Munir, M. 2021. Antibiotic resistance development and human health risks during wastewater reuse and biosolids application in agriculture. *Chemosphere* 265:129032.

Spinks, A., Glasziou, P.P. and Del Mar, C.B. 2021. Antibiotics for treatment of sore throat in children and adults. *Cochrane Database Syst Rev*: CD000023.

Stanton, I.C., Murray, A.K., Zhang, L., Snape, J. and Gaze, W.H. 2020. Evolution of antibiotic resistance at low antibiotic concentrations including selection below the minimal selective concentration. *Commun Biol* 3:467.

Styszko, K., Proctor, K., Castrignanò, E. and Kasprzyk-Hordern, B. 2021. Occurrence of pharmaceutical residues, personal care products, lifestyle chemicals, illicit drugs and metabolites in wastewater and receiving surface waters of Krakow agglomeration in South Poland. *Sci Total Environ* 768:144360.

Sui, Q., Cao, X., Lu, S., Zhao, W., Qiu, Z. and Yu, G. 2015. Occurrence, sources and fate of pharmaceuticals and personal care products in the groundwater: A review. *Emerg Contam* 1:14–24.

Suzuki, S. and Hoa, P.T.P. 2012. Distribution of quinolones, sulfonamides, tetracyclines in aquatic environment and antibiotic resistance in Indochina. *Front Microbiol* 3:67.

Szalowska, E., van der Burg, B., Man, H.-Y., Hendriksen, P.J.M. and Peijnenburg, A.A.C.M. 2014. Model steatogenic compounds (amiodarone, valproic acid, and tetracycline) alter lipid metabolism by different mechanisms in mouse liver slices. *PLoS ONE* 9:e86795.

Tajik, H., Malekinejad, H., Razavi-Rouhani, S.M., Pajouhi, M.R., Mahmoudi, R. and Haghnazari, A. 2010. Chloramphenicol residues in chicken liver, kidney and muscle: A comparison among the antibacterial residues monitoring methods of Four Plate Test, ELISA and HPLC. *Food Chem Toxicol* 48:2464–2468.

Ternes, T.A. and Hirsch, R. 2000. Occurrence and behavior of X-ray contrast media in sewage facilities and the aquatic environment. *Environ Sci Technol* 34:2741–2748.

Ternes, T.A., Joss, A. and Siegrist, H. 2004. Peer reviewed: Scrutinizing pharmaceuticals and personal care products in wastewater treatment. *Environ Sci Technol* 38:392A–399A.

Thiele-Bruhn, S. 2003. Pharmaceutical antibiotic compounds in soils—a review. *J Plant Nutr Soil Sci-Z Pflanzenernahr Bodenkd* 166:145–167.

Thuy, H.T.T., Nga, L.P. and Loan, T.T.C. 2011. Antibiotic contaminants in coastal wetlands from Vietnamese shrimp farming. *Environ Sci Pollut Res* 18:835–841.

Tiseo, K., Huber, L., Gilbert, M., Robinson, T.P. and Van Boeckel, T.P. 2020. Global trends in antimicrobial use in food animals from 2017 to 2030. *Antibiotics* 9:918.

Tixier, C., Singer, H.P., Oellers, S. and Müller, S.R. 2003. Occurrence and fate of carbamazepine, clofibric acid, diclofenac, ibuprofen, ketoprofen, and naproxen in surface waters. *Environ Sci Technol* 37:1061–1068.

Treiber, F.M. and Beranek-Knauer, H. 2021. Antimicrobial residues in food from animal origin—A review of the literature focusing on products collected in stores and markets worldwide. *Antibiotics* 10:534.

Turani, B., Aliko, V. and Faggio, C. 2019. Amphibian embryos as an alternative model to study the pharmaceutical toxicity of cyclophosphamide and ibuprofen. *J Biol Res-Boll Soc Biol Sper* 92:72–76.

Valcárcel, Y., González Alonso, S., Rodríguez-Gil, J.L., Gil, A. and Catalá, M. 2011. Detection of pharmaceutically active compounds in the rivers and tap water of the Madrid Region (Spain) and potential ecotoxicological risk. *Chemosphere* 84:1336–1348.

Verlicchi, P., Al Aukidy, M. and Zambello, E. 2012. Occurrence of pharmaceutical compounds in urban wastewater: Removal, mass load and environmental risk after a secondary treatment—a review. *Sci Total Environ* 429:123–155.

Verlicchi, P. and Zambello, E. 2015. Pharmaceuticals and personal care products in untreated and treated sewage sludge: Occurrence and environmental risk in the case of application on soil—a critical review. *Sci Total Environ* 538:750–767.

VICH. 2005. *VICH GL38 Environmental Impact Assessments for Veterinary Medicinal Products—Phase II.* London UK: European Medicines Agency.

Vishnuraj, M.R., Kandeepan, G., Rao, K.H., Chand, S. and Kumbhar, V. 2016. Occurrence, public health hazards and detection methods of antibiotic residues in foods of animal origin: A comprehensive review. *Cogent Food Agric* 2:1235458.

Vrieze, A., Out, C., Fuentes, S., Jonker, L., Reuling, I., Kootte, R.S., et al. 2014. Impact of oral vancomycin on gut microbiota, bile acid metabolism, and insulin sensitivity. *J Hepatol* 60:824–831.

Wagil, M., Kumirska, J., Stolte, S., Puckowski, A., Maszkowska, J., Stepnowski, P., et al. 2014. Development of sensitive and reliable LC-MS/MS methods for the determination of three fluoroquinolones in water and fish tissue samples and preliminary environmental risk assessment of their presence in two rivers in northern Poland. *Sci Total Environ* 493:1006–1013.

Wang, J. and Gardinali, P.R. 2012. Analysis of selected pharmaceuticals in fish and the fresh water bodies directly affected by reclaimed water using liquid chromatography-tandem mass spectrometry. *Anal Bioanal Chem* 404:2711–2720.

Wang, Y., Zhang, W., Mhungu, F., Zhang, Y., Liu, Y., Li, Y., et al. 2021. Probabilistic risk assessment of dietary exposure to chloramphenicol in Guangzhou, China. *Int J Environ Res Public Health* 18:8805.

Webb, S., Ternes, T., Gibert, M. and Olejniczak, K. 2003. Indirect human exposure to pharmaceuticals via drinking water. *Toxicol Lett* 142:157–167.

Wee, S.Y. and Aris, A.Z. 2017. Endocrine disrupting compounds in drinking water supply system and human health risk implication. *Environ Int* 106:207–233.

Wee, S.Y., Ismail, N.A.H., Haron, D.E.M., Yusoff, F.M., Praveena, S.M. and Aris, A.Z. 2022. Pharmaceuticals, hormones, plasticizers, and pesticides in drinking water. *J Hazard Mater* 424:127327.

Weigel, S., Berger, U., Jensen, E., Kallenborn, R., Thoresen, H. and Hühnerfuss, H. 2004. Determination of selected pharmaceuticals and caffeine in sewage and seawater from Tromsø/Norway with emphasis on ibuprofen and its metabolites. *Chemosphere* 56:583–592.

Weinberger, J. and Klaper, R. 2014. Environmental concentrations of the selective serotonin reuptake inhibitor fluoxetine impact specific behaviors involved in reproduction, feeding and predator avoidance in the fish *Pimephales promelas* (fathead minnow). *Aquat Toxicol* 151:77–83.

Wernersson, A.-S., Carere, M., Maggi, C., Tusil, P., Soldan, P., James, A., et al. 2015. The European technical report on aquatic effect-based monitoring tools under the water framework directive. *Environ Sci Eur* 27:7.

Wilkins, L.J., Monga, M. and Miller, A.W. 2019. Defining dysbiosis for a cluster of chronic diseases. *Sci Rep* 9:12918.

Wu, Q., Li, Z. and Hong, H. 2013. Adsorption of the quinolone antibiotic nalidixic acid onto montmorillonite and kaolinite. *Appl Clay Sci* 74:66–73.

Xu, M., Li, H., Li, S., Li, C., Li, J. and Ma, Y. 2020. The presence of tetracyclines and sulfonamides in swine feeds and feces: Dependence on the antibiotic type and swine growth stages. *Environ Sci Pollut Res* 27:43093–43102.

Yang, Q., Gao, Y., Ke, J., Show, P.L., Ge, Y., Liu, Y., et al. 2021. Antibiotics: An overview on the environmental occurrence, toxicity, degradation, and removal methods. *Bioengineered* 12:7376–7416.

Zhang, C., Barron, L. and Sturzenbaum, S. 2021a. The transportation, transformation and (bio)accumulation of pharmaceuticals in the terrestrial ecosystem. *Sci Total Environ* 781:146684.

Zhang, H., Luo, Y., Wu, L., Huang, Y. and Christie, P. 2015a. Residues and potential ecological risks of veterinary antibiotics in manures and composts associated with protected vegetable farming. *Environ Sci Pollut Res* 22:5908–5918.

Zhang, Q., Cheng, J. and Xin, Q. 2015b. Effects of tetracycline on developmental toxicity and molecular responses in zebrafish (*Danio rerio*) embryos. *Ecotoxicology* 24:707–719.

Zhang, R., Yang, S., An, Y., Wang, Y., Lei, Y. and Song, L. 2022. Antibiotics and antibiotic resistance genes in landfills: A review. *Sci Total Environ* 806:150647.

Zhang, Y.-L., Lin, S.-S., Dai, C.-M., Shi, L. and Zhou, X.-F. 2014. Sorption—desorption and transport of trimethoprim and sulfonamide antibiotics in agricultural soil: Effect of soil type, dissolved organic matter, and pH. *Environ Sci Pollut Res* 21:5827–5835.

Zhang, Y.T., Chen, H., He, S., Wang, F., Liu, Y., Chen, M., et al. 2021b. Subchronic toxicity of dietary sulfamethazine and nanoplastics in marine medaka (*Oryzias melastigma*): Insights from the gut microbiota and intestinal oxidative status. *Ecotoxicol Environ Saf* 226:112820.

Zhi, D., Yang, D., Zheng, Y., Yang, Y., He, Y., Luo, L., et al. 2019. Current progress in the adsorption, transport and biodegradation of antibiotics in soil. *J Environ Manage* 251:109598.

Zhou, J., Yun, X., Wang, J., Li, Q. and Wang, Y. 2022. A review on the ecotoxicological effect of sulphonamides on aquatic organisms. *Toxicol Rep* 9:534–540.

Zhou, S., Di Paolo, C., Wu, X., Shao, Y., Seiler, T.-B. and Hollert, H. 2019. Optimization of screening-level risk assessment and priority selection of emerging pollutants—the case of pharmaceuticals in European surface waters. *Environ Int* 128:1–10.

Zicarelli, G., Multisanti, C.R., Falco, F. and Faggio, C. 2022. Evaluation of toxicity of Personal Care Products (PCPs) in freshwaters: Zebrafish as a model. *Environ Toxicol Pharmacol* 94:103923.

9 Toxicity and Adverse Effects of Veterinary Pharmaceuticals in Animals

Muhammad Adil, Mavara Iqbal, Shamsa Kanwal, and Ghazanfar Abbas

CONTENTS

9.1 INTRODUCTION

The significance of animals for humankind has been primarily reflected in terms of their consumption as food by ancient human beings. Accordingly, the traditional interaction between the cave-dwelling humans and wild animals represented a characteristic predator-prey relationship. Nevertheless, the domestication of wild animals and birds associated with the advent of human civilization markedly evolved their pivotal role in human society. Thus, in addition to food production, several species of animals and birds are currently reared for companionship, exhibitions, competitions, biomedical research and draught purposes (Pond et al. 2011). This growing and multifaceted utilization of animals and birds also called for parallel development in the provision of health and husbandry practices. The progressive advancement in human medicine paved the way for

DOI: 10.1201/9781003361091-9

designing better diagnostic, therapeutic and preventive measures against the diseases of animals and birds. The currently available, wide range of pharmaceutical agents with diverse physico-chemical and pharmacological attributes are further expanded through the gradual introduction of new and more effective drugs, to fulfill the medical needs of veterinary patients. Veterinary pharmaceuticals are generally used for the treatment (therapy), control (metaphylaxis) or prevention (prophylaxis) of disease conditions, or stimulation of growth rate (growth promotion) in target species. Typical veterinary drugs encompass the antimicrobials, anthelmintics, ectoparasiticides and antineoplastics as well as the agents used for relieving inflammation, and modifying the functions of digestive, respiratory, cardiovascular, nervous, renal and endocrine systems. Commonly formulated as solutions, boluses, feed additives, implants, suspensions, powders, sprays and injectable dosage forms, these drugs are usually administered via oral, topical, parenteral and inhalatory routes. Before they are approved, authorized or licensed, veterinary medicinal products must normally pass three major criteria including quality, efficacy and safety (Woodward 2000). For prognostic purposes, pharmacological and toxicological evaluations have formerly been included in the safety assessment of human and veterinary pharmaceuticals.

9.2 BASIC ASPECTS OF DRUG-INDUCED TOXICITY IN ANIMALS

9.2.1 Types of Drug-Induced Toxicity

Toxicity is defined as the amount (dose) of a drug that leads to detrimental effects in exposed animals under specific conditions. Whereas, the pathological state attributed to a poisonous substance is termed as toxicosis, poisoning or intoxication (Osweiler 1996). Based upon the frequency of exposure to causative agent and the duration taken for the onset of deleterious effects, toxicity is classified into acute, subacute, subchronic and chronic types (Table 9.1). Acute toxicity usually occurs due to the accidental ingestion of drugs by the animals in substantially greater amounts. Similarly, the inadvertent intravenous injection of a highly toxic drug in an excessive dose can also lead to acute toxicity. Drug-mediated anaphylaxis is an example of acute toxicity. The other forms of toxicity are commonly caused by drugs having cumulative effect. Carcinogenesis, mutagenesis and teratogenesis are the common examples of drug-induced chronic toxicity. Depending upon the nature of exposure, some drugs such as organophosphates can cause either acute or chronic forms of toxicity in treated animals.

9.2.2 Factors Affecting the Drug-Induced Toxicity

Generally, both overdosage and shorter dosing interval result in adverse drug reactions owing to increased plasma concentration and saturation of biodisposition. The toxicity of some drugs has been linked with the improper route of administration. For example, magnesium sulphate is usually given by mouth to stimulate intestinal motility for the treatment of constipation; however, it may lead to hypotension, cardiac arrest and even sudden death following intravenous injection.

TABLE 9.1
Types of Toxicity

Type of toxicity	Frequency of exposure	Duration taken for the onset of detrimental effects
Acute toxicity	Single/multiple	24 hours
Subacute toxicity	Multiple	Up to 1 month
Subchronic toxicity	Multiple	1–3 months
Chronic toxicity	Multiple	More than 3 months

The effects of certain drugs also vary with their duration of use; for instance, single exposure to organophosphate insecticides leads to acute toxicity characterized by hypersalivation, convulsions, dyspnea and muscle fasciculation whereas repeated exposure gives rise to delayed toxicity in the form of neuropathy (Sandhu and Rampal 2006). Medication errors including incorrect combination and administration of drugs, can induce iatrogenic (treatment-induced) diseases such as drenching pneumonia, inflammation, sloughing and abscessation of the injection site, and even nerve damage leading to paralysis of the dependent body part. Lipophilic and non-polar vehicles enhance the toxicity of drugs by facilitating their absorption and penetration across the biological membranes. Hydrophobic drugs including aminoglycosides, tetracyclines and quinolones can cross the placental barrier and cause teratogenic effects in the developing fetus. Besides, certain drugs exhibit high affinity for specific tissue proteins and hence the long-term retention of protein-bound drug molecules can disrupt the morphological and/or physiological attributes of target tissues. Drug interactions resulting from the concurrent administration of drugs having similar toxicity profiles augment the magnitude of overall toxicity. Thus, the co-administration of potentially nephrotoxic drugs such as gentamicin and frusemide may enhance the degree of kidney damage. Similarly, the anticoagulant effect of warfarin is potentiated in the form of hemorrhages by the concomitant use of aspirin.

Interspecies differences in the susceptibility to adverse drug reactions exist on account of altered digestive, metabolic, renal and reproductive functions. For instance, cats and fish, being deficient in glucuronide conjugatory mechanism, are unable to properly metabolize aspirin and paracetamol and may consequently suffer from severe intoxication. Unlike most species of domestic animals, rabbits are comparatively resistant to the toxic effects of *Atropa belladonna* due to the presence of atropinase enzyme in their plasma and liver (Harrison et al. 2006). Equines and turkeys are more vulnerable to the adverse effects of ionophore anticoccidial agents. Unlike the adult animals, their neonatal and geriatric counterparts are more prone to drug-induced toxic effects on account of underdeveloped hepatic drug-metabolizing systems and improper renal function. Drugs requiring extensive biotransformation (such as macrolides and amphenicols) can lead to poisoning in young animals due to improper metabolic conversion and resultant accumulation of unmetabolized drug molecules. Dehydrated animals and those subjected to starvation may suffer from harmful effects caused by hydrophilic and lipophilic drugs, respectively. The high fat content and low volume of extracellular fluid enhance the susceptibility of obese animals to detrimental effects caused by hydrophilic drugs such as aminoglycosides. Generally, testosterone protects the male animals from adverse drug reactions through the induction of drug-metabolizing enzymes (Osweiler 1996). However, the nephrotoxic effects of cisplatin were significantly higher in male Wistar rats as compared to their female counterparts (Nematbakhsh et al. 2013).

Pathological conditions, particularly those involving the vital body organs and systems (such as liver, kidneys and cardiovascular system), enhance the susceptibility of target animals to drug-induced toxicity. Hepatic dysfunction leads to adverse drug reactions through the impairment of normal biotransformation and resultant retention of unmetabolized drug(s) in the animal body. Additionally, liver diseases also diminish the production of antioxidants (like glutathione), which protect the body tissues from drug-induced oxidative injury. Animals with pre-existing renal insufficiency are predisposed to more profound kidney damage caused by potentially nephrotoxic drugs. Moreover, inflammatory diseases such as meningitis and encephalitis increase the neurotoxic effects of drugs via disrupting the permeability of the blood-brain barrier (Sandhu and Rampal 2006).

Environmental factors, including altitude, temperature, atmospheric pressure and light, also affect the biological response of living organisms to pharmacological agents. For example, the toxicity potential of digitalis, phenobarbitone and strychnine increases with low oxygen pressure and high altitude. Unlike the pesticides that uncouple oxidative phosphorylation, organochlorines and pyrethroids are more toxic at low ambient temperature during winter. Post-treatment exposure of animals to sunlight may cause photosensitivity reaction manifested by erythema, edema, serum exudation, blistering and scab formation. Drugs, such as phenothiazine derivatives and demeclocycline or their metabolites, exhibiting photodynamic activity have been implicated in causing photosensitization of lightly pigmented skin in animals (Moore 2002).

Managemental factors, particularly feeding and handling practices, affect the outcome of drug administration in animals. Water deprivation and even inadequate water supply can predispose the carnivorous animals treated with sulphonamides to crystalluria (Sandhu and Rampal 2006). Organophosphate poisoning typically results from the licking of topically applied drugs by the animals that are left unattended or due to lack of essential preventive measures like the application of a muzzle. Overcrowding and external noise aggravate the toxicity of amphetamine and other CNS stimulants.

9.2.3 Mechanistic Insight into Drug-Induced Toxicity

Generally, drug-mediated toxicity occurs through any of the following mechanisms (Figure 9.1). Drug-induced toxicity may occur as a result of the metabolic conversion of relatively non-toxic parent drugs into toxic metabolites, known as bioactivation, such as the biotransformation of paracetamol (acetaminophen) into hepatotoxic metabolite N-acetyl-p-benzoquinone imine (Siroka and Svobodova 2013). Mechanism-based toxicity arises due to extended/exaggerated pharmacological action of the drug or its metabolites on the target site. For instance, the overdosing or prolonged administration of a non-steroidal anti-inflammatory drug leads to gastric hyperacidity and oliguria through the inhibition of prostaglandin synthesis in the stomach and kidneys, respectively.

Off-target toxicity results from the action of the drug or its metabolites on an alternative target site. The antihistaminic effect of terfenadine occurs due to blockage of H_1 receptors, whereas its binding with human Ether-à-go-go-Related Gene (hERG) receptors leads to cardiac arrhythmia (Guengerich 2013).

Drug-induced hypersensitivity (allergic reaction) is caused by the action of the drug or its metabolites as hapten and consequent induction of antibodies production. Several drugs including penicillins, cephalosporins and sulphonamides have been implicated in causing hypersensitivity reactions in animals. Immunologically mediated allergic reactions may occur in hypersensitive animals or birds previously exposed to penicillins, cephalosporins or sulphonamides. Furthermore, drugs with similar chemical structures (such as penicillins and cephalosporins) can also lead to cross-hypersensitivity. Idiosyncrasy refers to a genetically determined adverse drug reaction that cannot be predicted from the pharmacological profile of the drug. Genetic deficiency of drug-metabolizing enzymes can predispose the individual animal to idiosyncratic drug reactions (Siroka and Svobodova

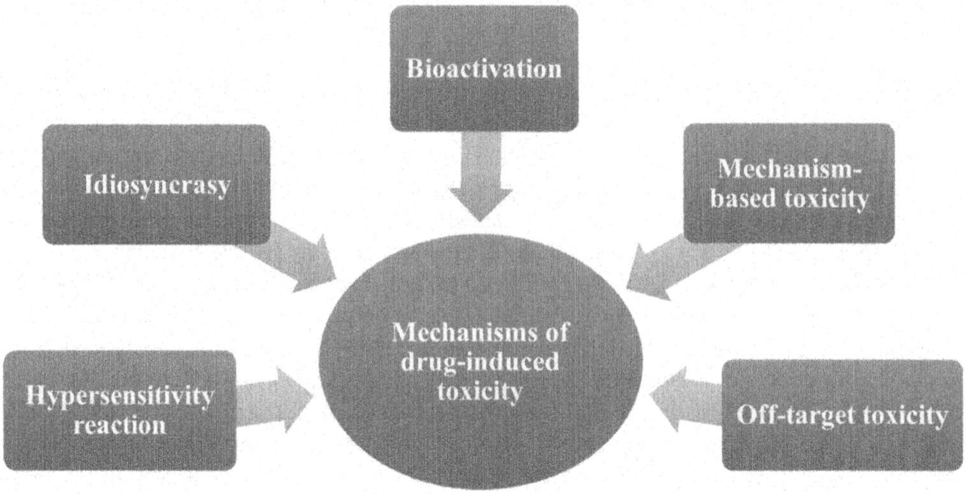

FIGURE 9.1 Mechanisms of drug-induced toxicity.

2013). Idiosyncratic blood dyscrasia and apnea have been linked with the genetic deficiency of glucose-6-phosphate dehydrogenase and pseudocholinesterase enzymes in animals receiving sulphonamides and suxamethonium, respectively (Trepanier 2004). Moreover, pyridine-containing sulphonamides have been associated with keratoconjunctivitis sicca in some dogs (Collins et al. 1986).

9.3 TOXICITY AND MAJOR DETRIMENTAL EFFECTS OF VETERINARY DRUGS IN ANIMALS

Drug-induced intoxications in animals and birds primarily occur on account of inappropriate dosage regimen, off-label use of medicines, inadvertent ingestion, negligence or intentional poisonings (Siroka and Svobodova 2013). Almost 10–30% of the overall drug-induced poisonings are documented to occur in animals, with companion animals (such as dogs and cats) being more commonly affected than farm animals (Kupper et al. 2010). Anti-inflammatory drugs, antiparasitics and antimicrobials represent the most frequently reported causes of adverse drug reactions (Xavier and Kogika 2002).

9.3.1 Drug-Induced Toxicity in Ruminants

Animals with a fermenting gastrointestinal tract such as ruminants and horses depend upon the beneficial microflora for the digestion of plant-derived cellulose, and are therefore susceptible to dysbiosis, indigestion and diarrhea following the systemic administration of broad-spectrum antimicrobial drugs. Chloramphenicol-treated calves exhibited more severe diarrhea than those receiving ampicillin, neomycin or tetracycline (Rollin et al. 1986). Hypersensitivity reactions have been recorded in large animals after the administration of different antibiotics including chloramphenicol, penicillin and tetracycline (Sakar 1993, Krishna Bhat et al. 1995, Thirunavukkarasu et al. 1995). Nephrotoxicity and ototoxicity were observed in neomycin-treated calves (Crowell et al. 1981), whereas, accidental overdosage of doxycycline in calves led to pulmonary edema and cardiotoxicity (Yeruham et al. 2002). Localized allergic reactions are caused by chloramphenicol and oxytetracycline, whereas, urticaria and anaphylaxis have been associated with penicillin administration in cattle (Brisbane 1963). Intravenous injection of oxytetracycline sometimes leads to bradycardia, reduced cardiac output and collapse in cattle and calves (Gross et al. 1981, Gyrd-Hansen et al. 1981). Besides, diarrhea, ataxia, reduced milk production and ketosis were manifested by dairy cattle following the consumption of feeds containing low concentrations of tylosin and lincomycin (Crossman and Poyser 1981, Rice et al. 1983). Nitrofurazone and tiamulin administration caused neurological signs such as mydriasis, ataxia, nystagmus and muscle tremors in calves (Lister and Fisher 1970, Ziv et al. 1983).

Organophosphate intoxication frequently occurs in ruminants and is typically marked by hypersalivation, lacrimation, frequent urination, diarrhea, bronchoconstriction, miosis, bradycardia, delirium, muscle fasciculation and even death due to respiratory muscle paralysis. Sheep and goats treated with closantel (imidathiazole anthelmintic) suffered from blindness, retinopathy and optic neuropathy (Button et al. 1987, Gill et al. 1999, Barlow et al. 2002). Additionally, cases of salinomycin, monensin and lasalocid intoxications have been described in horses, sheep and calves, respectively (Aleman et al. 2007, França et al. 2009, Oruç et al. 2011).

9.3.2 Drug-Induced Toxicity in Equines and Camels

Gastrointestinal disturbance, particularly antibiotic-induced diarrhea and *clostridium difficile* colitis have been reported in horses exposed to rifampicin and erythromycin (Stratton-Phelps et al. 2000). Likewise, the parenteral use of lincomycin has been linked with the incidence of acute diarrhea in treated equines (Mansmann 1975). The concomitant administration of sulphonamides and diaminopyrimidines with vitamin E and folic acid to pregnant mares produced congenital anomalies in foals (Toribio et al. 1998). Besides, hemolytic anemia in horses has been ascribed to co-trimoxazole

(Thomas and Livesey 1998). Severe and acute anaphylactic reactions occurred in horses following the intravenous injections of oxytetracycline and trimethoprim (Alexander and Collett 1975, Bowen and McMullan 1975). Ataxia, depression, mydriasis, reduced pupillary reflexes, transient blindness and muscle fasciculation were noticed in ivermectin-intoxicated zebras and horses (Hautekeete et al. 1998, Swor et al. 2009). Moxidectin also leads to similar adverse effects following its overdosage in horses (Khan et al. 2002). Amitraz toxicity is characterized by dyspnea, difficulty in swallowing and chewing, reduced cutaneous sensibility and stridor in horses (Duarte et al. 2003). Phenylbutazone-treated horses exhibited neutropenia and reduced mineral apposition in bones (McConnico et al. 2008). The opioid analgesic etorphine caused muscle tremors, tachycardia and hypertension in donkeys and horses (Dobbs and Ling 1972, Van Laun 1977).

Camels exhibit relatively mild and transient type of ivermectin intoxication in terms of hypersalivation and dropping of the lower lip (Ali 1988). The antitrypanosomal drug isometamidium caused serious adverse effects including salivation, frequent urination, diarrhea, paralysis of hind legs and recumbency in camels via inhibiting the plasma cholinesterase enzyme (Ali and Hassan 1986). Another antiprotozoal drug, furazolidone, led to irritability, head tilting, recumbency and trembling when administered to camels in high doses (Ali 1988). Likewise, convulsions, salivation, hyperesthesia, diaphoresis, frequent urination and diarrhea were evident in diminazene-treated camels (Homeida et al. 1981). Mild hypotension, sinus arrhythmia, relaxation of neck, hyperglycemia, drooping of the lower lip and hematological changes have been linked with xylazine administration in camels (Custer et al. 1977, Peshin et al. 1980). The intravenous general anesthetic chloral hydrate caused respiratory acidosis and marked tachycardia in camels (Sharma et al. 1983).

9.3.3 Drug-Induced Toxicity in Companion Animals

Benzimidazole-induced bone marrow toxicity has been documented in various species of animals including cats, dogs and porcupines (Stokol et al. 1997, Gary et al. 2004, Weber et al. 2006). Alopecia, lethargy and occasionally toxic epidermal necrolysis have been described in dogs receiving thiabendazole (Siroka and Svobodova 2013). Dachshunds are relatively more vulnerable to thiabendazole toxicity than other breeds of dogs (Plumb 1999). Levamisole, the commonly used nematocidal drug may lead to allergic cutaneous reactions, pulmonary edema, convulsions, respiratory muscle paralysis, asphyxia and even death in dogs (Plumb 1999, Hsu 2008). Dogs of Collie breed are highly susceptible to ivermectin toxicity on account of their poorly functional blood brain barrier (Mealey 2004, Geyer and Janko 2012). Ivermectin and doramectin toxicity in dogs is usually characterized by ataxia, bradycardia, disorientation, hypersalivation, mydriasis, coma, retinal edema and transient blindness (Yas-Natan et al. 2003, Kenny et al. 2008). Whereas, agitation, limb paresis, decrease or lack of ocular reflexes, tremors and blindness have been described in ivermectin-affected cats (Siroka and Svobodova 2013). Dogs having MDR1 gene mutation are predisposed to loperamide toxicity, characterized by sedation, pancreatitis, paralytic ileus and megacolon (Sartor et al. 2004). Cats are extremely vulnerable to adverse effects of pyrethrins and pyrethroids, owing to the deficiency of glucuronoyl transferase enzyme. Adverse effects of pyrethroids include ataxia, seizures, mydriasis, pyrexia, temporary blindness, hyperesthesia and tremors (Boland and Angles 2010). Besides, arrhythmia, respiratory distress and hypothermia have been documented in cats treated with the acaricidal drug amitraz (Andrade et al. 2007).

Fluoroquinones have been associated with arthropathy in young puppies (Sandhu and Rampal 2006, Stahlmann and Lode 2010), whereas penicillins (such as ampicillin and amoxicillin) and oxytetracycline induce the onset of hypersensitivity reactions in dogs (Srinivasan and Kumar 1991). Moreover, blood dyscrasia, epilepsy and thyroid dysfunction in dogs have been linked with the toxicity of chloramphenicol, penicillin and sulphonamides, respectively (Currie et al. 1970, Baig and Sharma 2002, Daminet and Ferguson 2003). Many antibiotics including cephalexin, potentiated sulphonamides, gentamicin and chloramphenicol may give rise to toxic epidermal necrolysis in dogs (Roosje 1991). The antifungal drug itraconazole caused hepatotoxicity, vasculitis and skin lesions in dogs and cats (Siroka and Svobodova 2013). Peripheral polyneuropathy, paresis and death (in severe cases) occur due to salinomycin poisoning in cats (Van der Linde-Sipman et al. 1999). Likewise, accidental lasalocid poisoning has been reported in dogs (Segev et al. 2004). Metronidazole has been implicated in causing several adverse effects including neurotoxicity (Caylor and Cassimatis 2001, Olson et al. 2005) and genotoxicity in cats (Sekis et al. 2009) and neurological disruption in dogs (Wright and Tyler 2003).

Chemotherapy-induced vomiting has been reported in dogs receiving methotrexate, cisplatin and 5-fluorouracil, and cats treated with cyclophosphamide and doxorubicin (Lee et al. 1973, Gylys et al. 1979). Being highly toxic in cats, cisplatin caused acute pulmonary edema followed by death (Knapp et al. 1988). Cats are predominantly susceptible to the myelosuppressive effect of thioguanine, necessitating the dose adjustment (Henness et al. 1977). Doxorubicin resulted in alopecia, myelosuppression, nephrotoxicity and cardiomyopathy in cats (Bristow et al. 1980, Cotter et al. 1985). Pulmonary fibrosis may occur following the chronic administration of busulfan in dogs (Witschi et al. 1987). Dogs treated with procarbazine and L-asparaginase exhibited neutropenia and hemorrhagic pancreatitis, respectively (Henry and Marlow 1973). The common adverse effects of vincristine in felines include weight loss, generalized weakness, leukopenia and convulsions (Johnson et al. 1963).

Cholestasis and hepatotoxicity have been described in dogs receiving the combination of phenytoin with primidone and phenobarbitone (Bunch 1993). The high toxicity of barbiturates to Greyhounds is attributable to relatively low content of adipose tissue (Gupta 2019). Whereas, dogs of brachycephalic breeds are relatively more vulnerable to acepromazine-induced hypotension (Lemke 2004). Dogs subjected to the overdosage of antihistaminic drug diphenhydramine displayed disorientation, tachycardia, dryness of mucous membranes and nystagmus (Goldfrank et al. 2002). Similarly, ventricular dysrhythmia and myocardial infarction were observed in dogs due to overdose of phenylpropanolamine and ephedrine (Peterson and Talcott 2013). Hepatotoxicity, enhanced bleeding tendency and gastric lesions in dogs have been ascribed to repeated administration of carprofen (Luna et al. 2007). Anemia, pancytopenia and thrombocytopenia were recorded in phenylbutazone-treated dogs (Weiss and Klausner 1990). Hepatic damage and renal failure occurred in a cat subjected to the overdosage of nimesulide (Borku et al. 2008). Characteristic signs of acetaminophen poisoning in cats include inappetence, emesis, facial edema, hypoxia, dyspnea, methemoglobinemia, hypothermia and cyanosis (Siroka and Svobodova 2013). Similar to cats, ferrets also lack the glucuronidation capacity that is critical for the detoxification of acetaminophen and thus may suffer from acute hepatic necrosis, methemoglobinemia and, sporadically, acute renal failure (Wickstrom and Eason 1999, Court 2001). Moreover, ibuprofen-intoxicated ferrets manifested gastrointestinal disturbances including anorexia, abdominal pain, diarrhea and melena, and renal signs such as polydipsia and polyuria (Richardson and Balabuszko 2001, Dunayer 2008).

9.3.4 Drug-Induced Toxicity in Laboratory Animals

Rabbits and Guinea pigs are quite sensitive to the adverse effects of fipronil including anorexia, ptyalism, seizures, hypothermia, lethargy, emaciation and depression (Webster 1999, Petritz and Chen 2018). The toxicity of ionophore anticoccidial drug maduramycin led to extensive mortality in rabbits (Martino et al. 2009). Rabbits are also vulnerable to the adverse effects of benzimidazoles including inappetence, lethargy, epistaxis, diarrhea, abdominal hemorrhages and mucosal discoloration (Graham et al. 2014). Besides, altered levels of progesterone and estradiol with subsequent reproductive disturbance were recorded in fipronil-treated rats (Ohi et al. 2004). Mucopurulent oculo-nasal discharge and pododermatitis with progressive abscessation of the hock joint have been ascribed to steroid-induced immunosuppression in rabbits (Petritz and Chen 2018). Unlike Guinea pigs, dogs and ferrets, other species including rabbits, mice and rats may suffer from marked lymphopenia on exposure to exogenous glucocorticoids (Cohn 1991, O'Malley 2005). Ketoprofen-poisoned rats demonstrated gastric ulceration, abdominal pain, mucosal discoloration, melena and lethargy (Petritz and Chen 2018).

9.3.5 Drug-Induced Toxicity in Birds

The toxicity of benzimidazoles occurs in many avian species including doves, pigeons, vultures, pelicans and storks, and is characterized by anemia, agranulocytosis, heteropenia and myelodysplasia (Howard et al. 2002, Weber et al. 2002, Bonar et al. 2003, Lindemann et al. 2016). Acute respiratory distress and mortality occurred in captive kiwis following the administration of levamisole (Gartrell et al. 2005). Sporadic reports of ivermectin intoxication have been documented in pigeons (Chen et al. 2013, Li et al. 2013). The antifungal agent itraconazole is highly toxic to African gray parrots and may cause inappetence, depression and even death in case of acute poisoning (Orosz et al. 1996).

Involution of thymus, spleen and bursa was accompanied by the suppression of both humoral and cell-mediated immunities in corticosteroid-treated birds (Lumeij et al. 1994, Leili and Scanes 1998). Monensin supplementation in the feed gave rise to diarrhea, ataxia, reduced water consumption, paralysis, feed rejection and mortality in a broiler flock (Zavala et al. 2011). While reduced egg production, abnormal gait, early embryonic mortality, decreased fertility and hatchability of eggs have been recorded in monensin-affected layers (Perelman et al. 1993). Likewise, salinomycin-intoxicated birds reflected muscular weakness, inability to stand, dyspnea, stiffness, drowsiness, recumbency and mortality (Andreasen and Schleifer 1995).

Wild birds commonly suffer from acute respiratory failure and subsequent death due to carbamate and organophosphate poisoning through contaminated water or carcasses of previously treated dead animals (Fleischli et al. 2004). Mallard ducklings and pheasants are highly prone to organophosphate-induced delayed neurotoxicity (Brown and Julian 2003). Predatory birds (raptors) suffer from huge mortality when exposed to organophosphate and carbamate insecticides following the direct ingestion of poison via scavenging on previously treated carcasses or through intoxicated prey (Hill and Mendenhall 1980, Reece and Handson 1982). Disorientation, ataxia, convulsions, muscle fasciculation, weight loss and death are the characteristic signs of organochlorine toxicosis in waterfowl (Walker 2003). Similarly, acute pyrethroid toxicosis leads to restlessness, sudden movements and feather pecking in exposed birds (Zwart 1988). Extensive decline in the vulture population of India, Pakistan and Nepal occurred due to acute renal failure following the consumption of flesh from improperly disposed, diclofenac-treated, dead cattle (Oaks et al. 2004). Consequently, diclofenac was banned for use in veterinary practice in 2006 and meloxicam was recommended as an eco-safe alternative (Swarup et al. 2007).

9.3.6 Drug-Induced Toxicity in Amphibians, Fish and Reptiles

Amphibians can be exposed to the pharmaceutically active substances contained in municipal or domestic waste water, as well as emissions from the drug manufacturing companies (Creusot et al. 2014). Exposure to high levels of progesterone and estrogens at critical larval stages interferes with the gonadal differentiation in amphibians. Likewise, exposure to selective serotonin re-uptake inhibitors, fluoxetine or sertraline for 10 weeks reduced the weight at metamorphosis (Conners et al. 2009). Besides, the ovarian differentiation was interrupted in amphibian larvae treated with aromatase inhibitors such as fadrozole and flavone (Mackenzie et al. 2003).

When released into water, the benzimidazole anthelmintics represent a risk of developmental toxicity to fish and aquatic invertebrates (Carlsson et al. 2011, Sasagawa et al. 2016). Sulphonamides such as sulphadimethoxine and sulphamethoxazole have been documented to persist in surface water across the world (Yang et al. 2004). Exposure to sulphadimidine, sulphadiazine and sulfamethoxazole has been linked with tachycardia and poor spontaneous swimming capacity in zebrafish (Lin et al. 2014). Sulfamethazine retarded the growth of juvenile fish, and disrupted the redox potential of zebrafish embryos (Ji et al. 2012, Yan et al. 2018). Besides, sulphamethoxazole enhanced the expression level of inflammatory cytokines and caused mortality in zebrafish (Zhou et al. 2016) and deteriorated the nutrient metabolism and innate immunity in tilapia (Limbu et al. 2018). Both acute as well as chronic exposure to oxytetracycline and erythromycin induced chromosomal aberrations and DNA damage leading to genotoxic effects in rainbow trout (Rodrigues et al. 2016, 2017). Likewise, delayed hatching, intestinal damage, hemato-biochemical disturbances and reduced swimming tendency in fish have also been attributed to oxytetracycline (Ambili et al. 2013, Oliveira et al. 2013). Macrolides, including tylosin, clarithromycin, tilmicosin and azithromycin, led to oxidative stress, cardiotoxicity and developmental anomalies in zebrafish embryos (Yan et al. 2019). Concurrent administration of different fluoroquinolones, including enrofloxacin, ciprofloxacin, norfloxacin and ofloxacin, attenuated the hatching and larval development in zebrafish (Wang et al. 2014). Mortality has been recorded in fish receiving ciprofloxacin in combination with fluoxetine and ibuprofen (Richards et al. 2004). Histopathological lesions were evident in the liver and gills of adult zebrafish treated with maduramicin (Ni et al. 2019). Amoxicillin treatment perturbed the level of antioxidant enzymes and triggered oxidative stress in carp and adult zebrafish (Oliveira et al. 2013, Elizalde-Velázquez et al. 2017).

Nephrotoxicity and ototoxicity have been linked with aminoglycosides including amikacin and gentamicin (Montali et al. 1979). Whereas, metronidazole resulted in ataxia, seizures, nystagmus, opisthotonus and tremors in reptiles (Bennett 1996). Reptiles are also susceptible to organophosphate intoxication manifested as hypersalivation, ataxia, muscle twitching, loss of righting reflex, coma respiratory distress. Secondary toxicity in lizards has been associated with the feeding of fipronil-contaminated insects (Peveling and Demba 2003). Anorexia, ocular irritation and altered defecation intervals were observed in amitraz-exposed tortoises (Burridge et al. 2002).

9.3.7 Toxicity of Veterinary Drugs Affecting Invertebrates

Pharmacologically active metabolites and/or residues of antiparasitic drugs, released in the urine and feces of treated animals, not only contaminate the soil and water but can also affect the aquatic and terrestrial invertebrates. Almost 98% of macrocyclic lactones excreted in the feces persist as active metabolic products or unmetabolized drugs (Horvat et al. 2012). Similarly, around 96–98% of the topically applied deltamethrin is released in the form of residues through the fecal route (Mahefarisoa et al. 2021). The toxicity of avermectins including ivermectin and abamectin has been extensively documented in dung beetles (*Euoniticellus fulvus*), honey bees (*Apis mellifera*) and bumble bees (Lumaret et al. 1993, Marletto et al. 2003, Iwasa et al. 2005, Sánchez-Bayo 2012). Additionally, the toxic effects of deltamethrin have also been reported in dung-feeding beetles, alfalfa leaf-cutting bees (*Megachile rotundata*) and honey bees (Jordan et al. 2006, Belzunces et al. 2012, Piccolomini et al. 2018). Aquatic invertebrates and fish are highly prone to fipronil-induced teratogenic effects (Stehr et al. 2006).

Several commonly used NSAIDs, including aspirin, paracetamol, ibuprofen, diclofenac and naproxen, contaminate the aquatic ecosystems and may bring about a serious risk to freshwater invertebrates. The growth curves of two invertebrate species (*Plationus patulus* and *Moina macrocopa*) were negatively affected by paracetamol (Sarma et al. 2014). Regeneration defects of eyes and tails in freshwater planarians (*Dugesia japonica*) and DNA fragmentation and lipid peroxidation in crustaceans (*Daphnia magna*) have been attributed to aspirin treatment (Gómez-Oliván et al. 2014, Zhang et al. 2019). Besides, increasing concentrations of paracetamol elicited apparent changes in the cellular redox status of freshwater clam, *Corbicula fluminea* (Brandão et al. 2014). Reduced egg production and mortality were recorded in *Daphnia magna* exposed to diclofenac and ibuprofen (Du et al. 2016). The bivalve *Dreissena polymorpha* exhibited cytotoxic and genotoxic effects when treated with diclofenac (Parolini et al. 2009). Exposure to ibuprofen and naproxen has been linked with impaired regeneration and contraction of tentacles in cnidarians *Hydra vulgaris* and *Hydra magnipapillata* (Quinn et al. 2008, Yamindago et al. 2019).

9.4 THERAPEUTIC MANAGEMENT OF DRUG-INDUCED TOXICITY IN ANIMALS

The clinical cases of drug-induced toxicity can be therapeutically managed using the following six-step approach.

9.4.1 Restoration and Maintenance of Vital Body Functions

Stabilization of the patient should be carried out through the maintenance of optimal cardiovascular, respiratory, metabolic and neurological functions. Epinephrine and corticosteroids are capable to normalize the blood pressure in hypotensive patients. Atropine or lidocaine are effective for the treatment of bradycardia and tachycardia, respectively (Osweiler 1996). CNS depressants like diazepam, phenobarbitone or phenothiazine tranquilizers can be used to control seizures. Respiratory stimulants (analeptics) including doxapram and nikethamide help to restore spontaneous respiration. Sodium lactate or sodium bicarbonate are effective for counteracting metabolic acidosis. In some cases, the patient may require necessary intervention to relieve hyperthermia or hypothermia.

9.4.2 Prevention of Further Exposure

Elimination of further exposure through the shifting of the animal(s) from the affected site, cleansing of contaminated body part(s) with soap and water and changing the feed or water source is recommended. Proper washing of contaminated hair several times with a non-irritant shampoo and even the trimming of affected hair can be necessary in some cases.

9.4.3 Evasion of Systemic Absorption

Preclusion of further systemic absorption by means of emesis, gastric lavage or adsorption therapy may be required in some cases of poisonings. Apomorphine and xylazine are generally used for the induction of emesis in dogs and cats, respectively (Osweiler 1996). Whereas, oral administration of mustard oil leads to vomiting in birds. Conversely, ruminants, equines, rabbits and rodents are unable to vomit and therefore emetics are contraindicated in these animals (Beasley 1999). Likewise, emetics are not recommended in poisonings caused by corrosive drugs. Gastric lavage is an alternative way of gastrointestinal decontamination using normal saline or water. Adsorption therapy is carried out by means of a non-specific chelating agent such as activated charcoal.

9.4.4 Enhancement of Drug Elimination

Facilitation of drug excretion through diuresis, ion trapping or catharsis can be effective in certain intoxications. Diuretics such as frusemide and mannitol are typically used to enhance the elimination of poisonous drugs in intoxicated animals and birds (Bischoff 2018). The ion trapping phenomenon involves the use of urinary alkalinizer (sodium bicarbonate) or urinary acidifier (ammonium chloride) to facilitate the ionization and excretion of acidic and basic drugs, respectively. Whereas, catharsis can be induced using sorbitol or sodium sulphate.

9.4.5 Use of Specific Reversal Agents (antidotes)

Reversal of the unfavorable clinical manifestations should be attempted using specific antidotes. However, specific antidotes are only available against a few drugs, and the effectiveness of antidotal therapy requires timely application, prior to the systemic absorption of the toxic drug. Table 9.2 represents a list of antidotes that can be useful in neutralizing the detrimental effects caused by some veterinary drugs. The lack of specific antidote for a toxic drug necessitates the anticipation of symptomatic treatment.

TABLE 9.2
List of Common Antidotes Used against Drug-Induced Toxicities in Animals

Poisonous drug(s)	Antidote(s)
Anticholinergic drugs	Physostigmine
Benzodiazepines	Flumazenil
Cardiac glycosides	Potassium chloride
Coumarin anticoagulants	Vitamin K_1
Depolarizing muscle relaxants	Anti-cholinesterase agent
Imidathiazoles	Atropine + hexamethonium
Imidocarb	Atropine
Macrocyclic lactones	Physostigmine
Non-depolarizing muscle relaxants	Aminopyridine, sugammadex
Opioid analgesics	Naloxone
Organophosphate insecticides	Atropine + pralidoxime
Paracetamol (acetaminophen)	Acetylcysteine
Xylazine	Yohimbine

9.4.6 Palliative Therapy

Supportive care can be provided (if required) in terms of fluid administration and nutritional supplementation. Plasma extenders or lactated Ringer's solution are recommended for compensating fluid loss (Osweiler 1996). Moreover, the requirement for whole blood transfusion should be evaluated in severe cases.

9.5 CONCLUSIONS AND RECOMMENDATIONS

Owing to the remarkable diversity among animals, the occurrence of adverse drug reactions is not uncommon. Although the majority of these detrimental effects are predictable from the pharmacological profile of the concerned drug(s), others are unlikely and primarily linked with variability in terms of species, age, breed, gender or physiological status of the animal. Moreover, overdosing, improper dosing intervals, drug interactions, incompatibilities, polypharmacy and other types of medication errors also contribute to the incidence of drug-induced toxicity. Environmental contamination resulting from the passage of drug residues in soil, air and water can induce ecotoxic effects in exposed animals.

Accordingly, rational prescribing and prudent use of veterinary drugs along with the avoidance of medication errors can help to minimize the probability of drug-induced toxicity. Prescribing guidelines described in the relevant formularies can provide a valuable source of the most up-to-date information for practicing veterinarians. Drugs with relatively high toxicity should be used only when the therapeutic benefits outweigh the potential harmful consequences. Proper discarding of expired and unused pharmaceutical products is also requisite to ensure the safety of animals and their environment. Current advancements in the fields of pharmacogenetics and precision medicine are projected to circumvent the occurrence of hypersensitivities and idiosyncratic drug reactions through designing personalized therapy. The discovery and development of alternative, non-toxic veterinary drugs and highly effective antidotes may diminish the likelihood of adverse drug reactions in animals.

REFERENCES

Aleman, Monica, K Gary Magdesian, Tracy S Peterson, and Francis D Galey. 2007. "Salinomycin toxicosis in horses." *Journal of the American Veterinary Medical Association* no. 230 (12):1822–1826.

Alexander, F, and RA Collett. 1975. "Trimethoprim in the horse." *Equine Veterinary Journal* no. 7 (4):203–206.

Ali, BH. 1988. "A survey of some drugs commonly used in the camel." *Veterinary Research Communications* no. 12 (1):67–75.

Ali, BH, and T Hassan. 1986. "Some observations on the toxicosis of isometamidium chloride (samorin) in camels." *Veterinary and Human Toxicology* no. 28 (5):424–426.

Ambili, TR, M Saravanan, M Ramesh, DB Abhijith, and RK Poopal. 2013. "Toxicological effects of the antibiotic oxytetracycline to an Indian major carp Labeo rohita." *Archives of Environmental Contamination and Toxicology* no. 64 (3):494–503.

Andrade, SF, M Sakate, CB Laposy, SF Valente, VM Bettanim, LT Rodrigues, and J Marcicano. 2007. "Effects of experimental amitraz intoxication in cats." *Arquivo Brasileiro de Medicina Veterinária e Zootecnia* no. 59:1236–1244.

Andreasen, James R, and John H Schleifer. 1995. "Salinomycin toxicosis in male breeder turkeys." *Avian Diseases*, 638–642.

Baig, J, and MC Sharma. 2002. "Kalicharan (2002) Chloramphenicol administration in dogs: Bone marrow and pathological alterations." *Indian Journal of Veterinary Research* no. 11:24–30.

Barlow, AM, JA Sharpe, and EA Kincaid. 2002. "Blindness in lambs due to inadvertent closantel overdose." *The Veterinary Record* no. 151 (1):25–26.

Beasley, V. 1999. "Absorption, distribution, metabolism, and elimination: Differences among species (9-Aug-1999)." In *Veterinary toxicology*, edited by Beasley V, 1–19. Ithaca, NY: IVIS.

Belzunces, Luc P, Sylvie Tchamitchian, and Jean-Luc Brunet. 2012. "Neural effects of insecticides in the honey bee." *Apidologie* no. 43 (3):348–370.

Bennett, RA. 1996. "Neurology." In *Reptile medicine and surgery*, edited by Mader DR. Philadelphia, PA: WB Saunders.

Bischoff, K. 2018. "Toxicity of drugs of abuse." In *Veterinary toxicology: Basic and clinical principles*, edited by Gupta RC, 385–408. Oxford: Academic Press.

Boland, Lara A, and John M Angles. 2010. "Feline permethrin toxicity: Retrospective study of 42 cases." *Journal of Feline Medicine and Surgery* no. 12 (2):61–71.

Bonar, Christopher J, Albert H Lewandowski, and Jordan Schaul. 2003. "Suspected fenbendazole toxicosis in 2 vulture species (Gyps africanus, Torgos tracheliotus) and marabou storks (Leptoptilos crumeniferus)." *Journal of Avian Medicine and Surgery* no. 17 (1):16–19.

Borku, M, M Guzel, Mehmet Karakurum, K Ural, and S Aktas. 2008. "Nimesulide-induced acute biliary tract injury and renal failure in a kitten: A case report." *Veterinarni Medicina* no. 53 (3).

Bowen, JM, and WC McMullan. 1975. "Influence of induced hypermagnesemia and hypocalcemia on neuromuscular blocking property of oxytetracycline in the horse." *American Journal of Veterinary Research* no. 36 (7):1025–1028.

Brandão, Fátima Pinto, Joana Luísa Pereira, Fernando Gonçalves, and Bruno Nunes. 2014. "The impact of paracetamol on selected biomarkers of the mollusc species Corbicula fluminea." *Environmental Toxicology* no. 29 (1):74–83.

Brisbane, WP. 1963. "Antibiotic reactions in cattle." *The Canadian Veterinary Journal* no. 4 (9):234.

Bristow, Michael R, W Scott Sageman, Robert H Scott, Margaret E Billingham, Robert E Bowden, Robert S Kernoff, George H Snidow, and John R Daniels. 1980. "Acute and chronic cardiovascular effects of doxorubicin in the dog: The cardiovascular pharmacology of drug-induced histamine release." *Journal of Cardiovascular Pharmacology* no. 2 (5):487–515.

Brown, Thomas P, and Richard J Julian. 2003. Other toxins and poisons. In *Diseases of poultry*, edited by Saif YM, 1133–1159. Ames, IA: Iowa State Press.

Bunch, Susan E. 1993. "Hepatotoxicity associated with pharmacologic agents in dogs and cats." *Veterinary Clinics of North America: Small Animal Practice* no. 23 (3):659–670.

Burridge, Michael J, Trevor F Peter, Sandra A Allan, and Suman M Mahan. 2002. "Evaluation of safety and efficacy of acaricides for control of the African tortoise tick (Amblyomma marmoreum) on leopard tortoises (Geochelone pardalis)." *Journal of Zoo and Wildlife Medicine: Official Publication of the American Association of Zoo Veterinarians* no. 33 (1):52–57.

Button, C, I Jerrett, P Alexander, and W Mizon. 1987. "Blindness in kids associated with overdosage of closantel." *Australian Veterinary Journal* no. 64 (7):226–226.

Carlsson, Gunnar, Johan Patring, Erik Ullerås, and Agneta Oskarsson. 2011. "Developmental toxicity of albendazole and its three main metabolites in zebrafish embryos." *Reproductive Toxicology* no. 32 (1):129–137.
Caylor, Kevin B, and Marina K Cassimatis. 2001. "Metronidazole neurotoxicosis in two cats." *Journal of the American Animal Hospital Association* no. 37 (3):258–262.
Chen, Li-Jie, Bao-Hong Sun, Jian ping Qu, Shiwen Xu, and Shu Li. 2013. "Avermectin induced inflammation damage in king pigeon brain." *Chemosphere* no. 93 (10):2528–2534.
Cohn, Leah A. 1991. "The influence of corticosteroids on host defense mechanisms." *Journal of Veterinary Internal Medicine* no. 5 (2):95–104.
Collins, BK, CP Moore, and JH Hagee. 1986. "Sulfonamide-associated keratoconjunctivitis sicca and corneal ulceration in a dysuric dog." *Journal of the American Veterinary Medical Association* no. 189 (8):924–926.
Conners, Deanna E, Emily D Rogers, Kevin L Armbrust, Jeong-Wook Kwon, and Marsha C Black. 2009. "Growth and development of tadpoles (Xenopus laevis) exposed to selective serotonin reuptake inhibitors, fluoxetine and sertraline, throughout metamorphosis." *Environmental Toxicology and Chemistry* no. 28 (12):2671–2676.
Cotter, SM, PJ Kanki, and M Simon. 1985. "Renal disease in five tumor-bearing cats treated with adriamycin." *The Journal of the American Animal Hospital Association (USA)* no. 21:405–409.
Court, MH. 2001. "Acetaminophen UDP-glucuronosyltransferase in ferrets: Species and gender differences, and sequence analysis of ferret UGT1A6." *Journal of Veterinary Pharmacology and Therapeutics* no. 24 (6):415–422.
Creusot, Nicolas, Selim Aït-Aïssa, Nathalie Tapie, Patrick Pardon, François Brion, Wilfried Sanchez, Eric Thybaud, Jean-Marc Porcher, and Hélène Budzinski. 2014. "Identification of synthetic steroids in river water downstream from pharmaceutical manufacture discharges based on a bioanalytical approach and passive sampling." *Environmental Science & Technology* no. 48 (7):3649–3657.
Crossman, PJ, and MR Poyser. 1981. "Effect of inadvertently feeding tylosin and tylosin with dimetridazole to dairy cows." *The Veterinary Record* no. 108 (13):285.
Crowell, WA, TJ Divers, TD Byars, AE Marshall, KE Nusbaum, and L Larsen. 1981. "Neomycin toxicosis in calves." *American Journal of Veterinary Research* no. 42 (1):29–34.
Currie, TT, Nancy J Hayward, Pamela C Campbell, G Westlake, and J Williams. 1970. "Epilepsy in dogs caused by large doses of penicillin and concurrent brain damage." *British Journal of Experimental Pathology* no. 51 (5):492.
Custer, Randy, Lynn Kramer, Suzanne Kennedy, and R Mitchell Bush. 1977. "Hematologic effects of xylazine when used for restraint of Bactrian camels." *Journal of the American Veterinary Medical Association* no. 171 (9):899–901.
Daminet, Sylvie, and Duncan C Ferguson. 2003. "Influence of drugs on thyroid function in dogs." *Journal of Veterinary Internal Medicine* no. 17 (4):463–472.
Dobbs, HE, and CM Ling. 1972. "Use of etorphine/acepromazine in the horse and donkey." *Veterinary Record* no. 91 (2):40–41.
Du, Juan, Cheng-Fang Mei, Guang-Guo Ying, and Mei-Ying Xu. 2016. "Toxicity thresholds for diclofenac, acetaminophen and ibuprofen in the water flea Daphnia magna." *Bulletin of Environmental Contamination and Toxicology* no. 97 (1):84–90.
Duarte, Marcos Dutra, Paulo Vargas Peixoto, Pedro Soares Bezerra Júnior, Krishna Duro de Oliveira, Alexandre Paulino Loretti, and Carlos Hubinger Tokarnia. 2003. "Natural and experimental poisoning by amitraz in horses and donkey: Clinical aspects." *Pesquisa Veterinária Brasileira* no. 23:105–118.
Dunayer, Eric. 2008. "Toxicology of ferrets." *Veterinary Clinics of North America: Exotic Animal Practice* no. 11 (2):301–314.
Elizalde-Velázquez, Armando, Héctor Martínez-Rodríguez, Marcela Galar-Martínez, Octavio Dublán-García, Hariz Islas-Flores, Juana Rodríguez-Flores, Gregorio Castañeda-Peñalvo, Isabel Lizcano-Sanz, and Leobardo Manuel Gómez-Oliván. 2017. "Effect of amoxicillin exposure on brain, gill, liver, and kidney of common carp (Cyprinus carpio): The role of amoxicilloic acid." *Environmental Toxicology* no. 32 (4):1102–1120.
Fleischli, Margaret A, JC Franson, NJ Thomas, DL Finley, and W Riley. 2004. "Avian mortality events in the United States caused by anticholinesterase pesticides: A retrospective summary of National Wildlife Health Center records from 1980 to 2000." *Archives of Environmental Contamination and Toxicology* no. 46 (4):542–550.
França, Ticiana N, Vivian A Nogueira, Elise M Yamasaki, Saulo A Caldas, Carlos H Tokarnia, and Paulo V Peixoto. 2009. "Accidental monensin poisoning in sheep in Rio de Janeiro state." *Pesquisa Veterinária Brasileira* no. 29:743–746.

Gartrell, BD, MR Alley, and AH Mitchell. 2005. "Fatal levamisole toxicosis of captive kiwi (Apteryx mantelli)." *New Zealand Veterinary Journal* no. 53 (1):84–86.

Gary, Anthony T, Marie E Kerl, Charles E Wiedmeyer, Susan E Turnquist, and Leah A Cohn. 2004. "Bone marrow hypoplasia associated with fenbendazole administration in a dog." *Journal of the American Animal Hospital Association* no. 40 (3):224–229.

Geyer, Joachim, and Christina Janko. 2012. "Treatment of MDR1 mutant dogs with macrocyclic lactones." *Current Pharmaceutical Biotechnology* no. 13 (6):969–986.

Gill, PA, RW Cook, JG Boulton, WR Kelly, B Vanselow, and LA Reddacliff. 1999. "Optic neuropathy and retinopathy in closantel toxicosis of sheep and goats." *Australian Veterinary Journal* no. 77 (4):259–261.

Goldfrank, Lewis R, Neal E Flomenbaum, Neal A Lewin, Mary Ann Howland, Robert S Hoffman, and Lewis S Nelson. 2002. *Goldfrank's toxicologic emergencies*. New York: McGraw-Hill.

Gómez-Oliván, Leobardo Manuel, Marcela Galar-Martínez, Sandra García-Medina, Analleli Valdés-Alanís, Hariz Islas-Flores, and Nadia Neri-Cruz. 2014. "Genotoxic response and oxidative stress induced by diclofenac, ibuprofen and naproxen in Daphnia magna." *Drug and Chemical Toxicology* no. 37 (4):391–399.

Graham, Jennifer E, Michael M Garner, and Drury R Reavill. 2014. "Benzimidazole toxicosis in rabbits: 13 cases (2003 to 2011)." *Journal of Exotic Pet Medicine* no. 23 (2):188–195.

Gross, DR, KT Dodd, JD Williams, and HR Adams. 1981. "Adverse cardiovascular effects of oxytetracycline preparations and vehicles in intact awake calves." *American Journal of Veterinary Research* no. 42 (8):1371–1377.

Guengerich, F Peter. 2013. "Cytochrome P450 activation of toxins and hepatotoxicity." In *Drug-induced liver disease*, 15–33. New York, NY: Elsevier.

Gupta, PK. 2019. "General principles of toxicology." In *Concepts and applications in veterinary toxicology*, 1–26. New York, NY: Springer.

Gylys, JA, KM Doran, and JP Buyniski. 1979. "Antagonism of cisplatin induced emesis in the dog." *Research Communications in Chemical Pathology and Pharmacology* no. 23 (1):61–68.

Gyrd-Hansen, N, F Rasmussen, and M Smith. 1981. "Cardiovascular effects of intravenous administration of tetracycline in cattle." *Journal of Veterinary Pharmacology and Therapeutics* no. 4 (1):15–25.

Harrison, Patrick K, John EH Tattersall, and Ed Gosden. 2006. "The presence of atropinesterase activity in animal plasma." *Naunyn-Schmiedeberg's Archives of Pharmacology* no. 373 (3):230–236.

Hautekeete, Larissa A, Safdar Ali Khan, and WS Hales. 1998. "Ivermectin toxicosis in a zebra." *Veterinary and Human Toxicology* no. 40 (1):29–31.

Henness, AM, GH Theilen, and JP Lewis. 1977. "Clinical investigation of doxorubicin, daunomycin, and 6-thioguanine in normal cats." *American Journal of Veterinary Research* no. 38 (4):521–524.

Henry, MC, and M Marlow. 1973. "Preclinical toxicologic study of procarbazine (NSC-77213)." *Cancer Chemotherapy Reports. Part 3* no. 4 (1):97–102.

Hill, Elwood F, and Vivian M Mendenhall. 1980. "Secondary poisoning of barn owls with famphur, an organophosphate insecticide." *The Journal of Wildlife Management* no. 44 (3):676–681.

Homeida, AM, EA El Amin, SEI Adam, and MM Mahmoud. 1981. "Toxicity of diminazene aceturate (Berenil) to camels." *Journal of Comparative Pathology* no. 91 (3):355–360.

Horvat, AJM, S Babić, DM Pavlović, D Ašperger, S Pelko, M Kaštelan-Macan, M Petrović, and AD Mance. 2012. "Analysis, occurrence and fate of anthelmintics and their transformation products in the environment." *TrAC Trends in Analytical Chemistry* no. 31:61–84.

Howard, Lauren L, Rebecca Papendick, Ilse H Stalis, Jack L Allen, Meg Sutherland-Smith, Jeffery R Zuba, Daniel L Ward, and Bruce A Rideout. 2002. "Fenbendazole and albendazole toxicity in pigeons and doves." *Journal of Avian Medicine and Surgery* no. 16 (3):203–210.

Hsu, Walter H. 2008. *Handbook of veterinary pharmacology*. John Wiley & Sons.

Iwasa, Mitsuhiro, Tomokazu Nakamura, Kyoko Fukaki, and Nobuo Yamashita. 2005. "Nontarget effects of ivermectin on coprophagous insects in Japan." *Environmental Entomology* no. 34 (6):1485–1492.

Ji, Kyunghee, Sunmi Kim, Sunyoung Han, Jihyun Seo, Sangwoo Lee, Yoonsuk Park, Kyunghee Choi, Young-Lim Kho, Pan-Gyi Kim, and Jeongim Park. 2012. "Risk assessment of chlortetracycline, oxytetracycline, sulfamethazine, sulfathiazole, and erythromycin in aquatic environment: Are the current environmental concentrations safe?" *Ecotoxicology* no. 21 (7):2031–2050.

Johnson, Irving S, James G Armstrong, Marvin Gorman, and J Paul Burnett Jr. 1963. "The vinca alkaloids: A new class of oncolytic agents." *Cancer Research* no. 23 (8_Part_1):1390–1427.

Jordan, Chihiya, Tafara Gadzirayi Christopher, and Mut Edward. 2006. "Effect of three different treatment levels of deltamethrin on the numbers of dung beetles in dung pats." *African Journal of Agricultural Research* no. 1 (3):74–77.

Kenny, Patrick J, Karen M Vernau, Birgit Puschner, and David J Maggs. 2008. "Retinopathy associated with ivermectin toxicosis in two dogs." *Journal of the American Veterinary Medical Association* no. 233 (2):279–284.

Khan, Safdar A, Dee Ann Kuster, and Steve R Hansen. 2002. "A review of moxidectin overdose cases in equines from 1998 through 2000." *Veterinary and Human Toxicology* no. 44 (4):232–235.

Knapp, Deborah W, Ralph C Richardson, Patty L Bonney, and Kevin Hahn. 1988. "Cisplatin therapy in 41 dogs with malignant tumors." *Journal of Veterinary Internal Medicine* no. 2 (1):41–46.

Krishna Bhat, K, KG Udupa, and N Prakash. 1995. "Adverse drug reaction in a chloramphenicol sodium succinate treated cow." *Indian Journal of Veterinary Medicine* no. 15:51–51.

Kupper, Jaqueline, Hanspeter Naegeli, and Meret Wehrli Eser. 2010. "Common poisonings in the horse." *Praktische Tierarzt* no. 91 (6):492–498.

Lee, Cheng-Chun, Th R Castles, and Loren D Kintner. 1973. "Single-dose toxicity of cyclophosphamide (NSC-26271) in dogs and monkeys." *Cancer Chemotherapy Reports. Part 3* no. 4 (1):51–76.

Leili, Steven, and Colin G Scanes. 1998. "The effects of glucocorticoids (dexamethasone) on insulin-like growth factor-I, IGF-binding proteins, and growth in chickens." *Proceedings of the Society for Experimental Biology and Medicine* no. 218 (4):329–333.

Lemke, Kip A. 2004. "Perioperative use of selective alpha-2 agonists and antagonists in small animals." *The Canadian Veterinary Journal* no. 45 (6):475.

Li, Ming, Tian-Zi You, Wen-Jun Zhu, Jian-Ping Qu, Ci Liu, Bing Zhao, Shi-Wen Xu, and Shu Li. 2013. "Antioxidant response and histopathological changes in brain tissue of pigeon exposed to avermectin." *Ecotoxicology* no. 22 (8):1241–1254.

Limbu, Samwel M, Li Zhou, Sheng-Xiang Sun, Mei-Ling Zhang, and Zhen-Yu Du. 2018. "Chronic exposure to low environmental concentrations and legal aquaculture doses of antibiotics cause systemic adverse effects in Nile tilapia and provoke differential human health risk." *Environment International* no. 115:205–219.

Lin, Tao, Shilin Yu, Yanqiu Chen, and Wei Chen. 2014. "Integrated biomarker responses in zebrafish exposed to sulfonamides." *Environmental Toxicology and Pharmacology* no. 38 (2):444–452.

Lindemann, Dana M, David Eshar, Jerome C Nietfeld, and In Joong Kim. 2016. "Suspected fenbendazole toxicity in an American white pelican (Pelecanus erythrorhynchos)." *Journal of Zoo and Wildlife Medicine* no. 47 (2):681–685.

Lister, EE, and LJ Fisher. 1970. "Establishment of the toxic level of nitrofurazone for young liquid-fed calves." *Journal of Dairy Science* no. 53 (10):1490–1495.

Lumaret, Jean-Pierre, Eduardo Galante, C Lumbreras, J Mena, Michel Bertrand, JL Bernal, JF Cooper, Nassera Kadiri, and Deirdre Crowe. 1993. "Field effects of ivermectin residues on dung beetles." *Journal of Applied Ecology*:428–436.

Lumeij, JT, BW Ritchie, GJ Harrison, and LR Harrison. 1994. "Avian medicine: Principles and application." In *Endocrinology*, 599–601. Lake Worth, FL: Wingers Publishers.

Luna, Stelio PL, Ana C Basílio, Paulo VM Steagall, Luciana P Machado, Flávia Q Moutinho, Regina K Takahira, and Cláudia VS Brandão. 2007. "Evaluation of adverse effects of long-term oral administration of carprofen, etodolac, flunixin meglumine, ketoprofen, and meloxicam in dogs." *American Journal of Veterinary Research* no. 68 (3):258–264.

Mackenzie, Constanze A, Michael Berrill, Chris Metcalfe, and Bruce D Pauli. 2003. "Gonadal differentiation in frogs exposed to estrogenic and antiestrogenic compounds." *Environmental Toxicology and Chemistry: An International Journal* no. 22 (10):2466–2475.

Mahefarisoa, KL, N Simon Delso, V Zaninotto, ME Colin, and JM Bonmatin. 2021. "The threat of veterinary medicinal products and biocides on pollinators: A one health perspective." *One Health* no. 12:100237.

Mansmann, Richard A. 1975. "Antimicrobial therapy in horses." *The Veterinary Clinics of North America* no. 5 (1):81–99.

Marletto, Franco, Augusto Patetta, and Aulo Manino. 2003. "Laboratory assessment of pesticide toxicity to bumblebees." *Bulletin of Insectology* no. 56 (1):155–158.

Martino, PE, E Parrado, R Sanguinetti, C Espinoza, R Debenedetti, NM Di Benedetto, C Cisterna, and P Gomez. 2009. "Massive mortality in rabbits by maduramicin poisoning." *World Rabbit Science* no. 17 (1):45–48.

McConnico, Rebecca S, Timothy W Morgan, Cathleen C Williams, Jeremy D Hubert, and Rustin M Moore. 2008. "Pathophysiologic effects of phenylbutazone on the right dorsal colon in horses." *American Journal of Veterinary Research* no. 69 (11):1496–1505.

Mealey, KL. 2004. "Therapeutic implications of the MDR-1 gene." *Journal of Veterinary Pharmacology and Therapeutics* no. 27 (5):257–264.

Montali, RJ, M Bush, and JM Smeller. 1979. "The pathology of nephrotoxicity of gentamicin in snakes." *Veterinary Pathology* no. 16 (1):108–115.

Moore, Douglas E. 2002. "Drug-induced cutaneous photosensitivity." *Drug Safety* no. 25 (5):345–372.

Nematbakhsh, Mehdi, Shadi Ebrahimian, Mona Tooyserkani, Fatemeh Eshraghi-Jazi, Ardeshir Talebi, and Farzaneh Ashrafi. 2013. "Gender difference in Cisplatin-induced nephrotoxicity in a rat model: Greater intensity of damage in male than female." *Nephro-urology Monthly* no. 5 (3):818.

Ni, Han, Lin Peng, Xiuge Gao, Hui Ji, Junxiao Ma, Yanping Li, and Shanxiang Jiang. 2019. "Effects of maduramicin on adult zebrafish (Danio rerio): Acute toxicity, tissue damage and oxidative stress." *Ecotoxicology and Environmental Safety* no. 168:249–259.

Oaks, J Lindsay, Martin Gilbert, Munir Z Virani, Richard T Watson, Carol U Meteyer, Bruce A Rideout, HL Shivaprasad, Shakeel Ahmed, Muhammad Jamshed Iqbal Chaudhry, and Muhammad Arshad. 2004. "Diclofenac residues as the cause of vulture population decline in Pakistan." *Nature* no. 427 (6975):630–633.

Ohi, M, PR Dalsenter, AJM Andrade, and AJ Nascimento. 2004. "Reproductive adverse effects of fipronil in Wistar rats." *Toxicology Letters* no. 146 (2):121–127.

Oliveira, Rhaul, Sakchai McDonough, Jessica CL Ladewig, Amadeu MVM Soares, António JA Nogueira, and Inês Domingues. 2013. "Effects of oxytetracycline and amoxicillin on development and biomarkers activities of zebrafish (Danio rerio)." *Environmental Toxicology and Pharmacology* no. 36 (3):903–912.

Olson, EJ, SC Morales, AS McVey, and DW Hayden. 2005. "Putative metronidazole neurotoxicosis in a cat." *Veterinary Pathology* no. 42 (5):665–669.

O'Malley, Bert W. 2005. "A life-long search for the molecular pathways of steroid hormone action." *Molecular Endocrinology* no. 19 (6):1402–1411.

Orosz, Susan E, Donita L Frazier, Edward C Schroeder, Sherry K Cox, Dorcas O Schaeffer, Sonia Doss, and Patrick J Morris. 1996. "Pharmacokinetic properties of itraconazole in blue-fronted Amazon parrots (Amazona aestiva aestiva)." *Journal of Avian Medicine and Surgery*:168–173.

Oruç, Hasan Hüseyin, IT Cangul, Murat Cengiz, and R Yilmaz. 2011. "Acute lasalocid poisoning in calves associated with off-label use." *Journal of Veterinary Pharmacology and Therapeutics* no. 34 (2):187–189.

Osweiler, GD. 1996. *Toxicology. The national veterinary medical series.* Philadelphia, PA: Williams & Wilkins.

Parolini, Marco, Andrea Binelli, Daniele Cogni, Consuelo Riva, and Alfredo Provini. 2009. "An in vitro biomarker approach for the evaluation of the ecotoxicity of non-steroidal anti-inflammatory drugs (NSAIDs)." *Toxicology in Vitro* no. 23 (5):935–942.

Perelman, B, M Pirak, and B Smith. 1993. "Effects of the accidental feeding of lasalocid sodium to broiler breeder chickens." *The Veterinary Record* no. 132 (11):271–273.

Peshin, PK, JM Nigam, SC Singh, and BA Robinson. 1980. "Evaluation of xylazine in camels." *Journal of the American Veterinary Medical Association* no. 177 (9):875–878.

Peterson, Michael E, and Patricia A Talcott. 2013. *Small animal toxicology.* USA: Elsevier Saunders.

Petritz, Olivia A, and Sue Chen. 2018. "Therapeutic contraindications in exotic pets." *Veterinary Clinics: Exotic Animal Practice* no. 21 (2):327–340.

Peveling, Ralf, and Sy Amadou Demba. 2003. "Toxicity and pathogenicity of Metarhizium anisopliae var. acridum (Deuteromycotina, Hyphomycetes) and fipronil to the fringe-toed lizard Acanthodactylus dumerili (Squamata: Lacertidae)." *Environmental Toxicology and Chemistry: An International Journal* no. 22 (7):1437–1447.

Piccolomini, Alyssa M, Shavonn R Whiten, Michelle L Flenniken, Kevin M O'Neill, and Robert KD Peterson. 2018. "Acute toxicity of permethrin, deltamethrin, and etofenprox to the alfalfa leafcutting bee." *Journal of Economic Entomology* no. 111 (3):1001–1005.

Plumb, DC. 1999. *Veterinary drug handbook*, 3rd ed. Ames, IA: Iowa State University, 750.

Pond, W, F Bazer, and B Rollin. 2011. "Contributions of farm and laboratory animals to society." In *Animal welfare in animal agriculture*, 30–53. Boca Raton, FL: CRC Press.

Quinn, Brian, François Gagné, and Christian Blaise. 2008. "The effects of pharmaceuticals on the regeneration of the cnidarian, Hydra attenuata." *Science of the Total Environment* no. 402 (1):62–69.

Reece, RL, and P Handson. 1982. "Observations on the accidental poisoning of birds by organophosphate insecticides and other toxic substances." *The Veterinary Record* no. 111 (20):453–455.

Rice, DA, CH McMurray, and JF Davidson. 1983. "Ketosis in dairy cows caused by low levels of lincomycin in concentrate feed." *Veterinary Record* no. 113 (21):495–496.

Richards, Sean M, Christian J Wilson, David J Johnson, Dawn M Castle, Monica Lam, Scott A Mabury, Paul K Sibley, and Keith R Solomon. 2004. "Effects of pharmaceutical mixtures in aquatic microcosms." *Environmental Toxicology and Chemistry: An International Journal* no. 23 (4):1035–1042.

Richardson, Jill A, and Rachel A Balabuszko. 2001. "Ibuprofen ingestion in ferrets: 43 cases January 1995–March 2000." *Journal of Veterinary Emergency and Critical Care* no. 11 (1):53–58.
Rodrigues, Sara, Sara C Antunes, Alberto T Correia, and Bruno Nunes. 2016. "Acute and chronic effects of erythromycin exposure on oxidative stress and genotoxicity parameters of Oncorhynchus mykiss." *Science of the Total Environment* no. 545:591–600.
Rodrigues, Sara, Sara C Antunes, Alberto T Correia, and Bruno Nunes. 2017. "Rainbow trout (Oncorhynchus mykiss) pro-oxidant and genotoxic responses following acute and chronic exposure to the antibiotic oxytetracycline." *Ecotoxicology* no. 26 (1):104–117.
Rollin, RE, KN Mero, PB Kozisek, and RW Phillips. 1986. "Diarrhea and malabsorption in calves associated with therapeutic doses of antibiotics: Absorptive and clinical changes." *American Journal of Veterinary Research* no. 47 (5):987–991.
Roosje, PJ. 1991. "Your dog has TEN . . . Toxic epidermal necrolysis in the dog." *Tijdschrift voor diergeneeskunde* no. 116 (21):1072–1075.
Sakar, D. 1993. "Some unfavoured activities of tetracycline antibiotics with regard to the occurrence of early allergic reaction in cattle." *Praxis Veterinaria-Zagreb-* no. 41:87–108.
Sánchez-Bayo, F. 2012. "Insecticides mode of action in relation to their toxicity to non-target organisms." *Journal of Environmental & Analytical Toxicology* no. 4:S4–002.
Sandhu, HS, and S Rampal. 2006. *Essentials of veterinary pharmacology and therapeutics*, 1st ed. New Delhi: Kalyani Publishers.
Sarma, SSS, Brenda Karen González-Pérez, Rosa Martha Moreno-Gutiérrez, and S Nandini. 2014. "Effect of paracetamol and diclofenac on population growth of Plationus patulus and Moina macrocopa." *Journal of Environmental Biology* no. 35 (1):119.
Sartor, Laura L, Steven A Bentjen, Lauren Trepanier, and Katrina L Mealey. 2004. "Loperamide toxicity in a collie with the MDR1 mutation associated with ivermectin sensitivity." *Journal of Veterinary Internal Medicine* no. 18 (1):117–118.
Sasagawa, Shota, Yuhei Nishimura, Tetsuo Kon, Yukiko Yamanaka, Soichiro Murakami, Yoshifumi Ashikawa, Mizuki Yuge, Shiko Okabe, Koki Kawaguchi, and Reiko Kawase. 2016. "DNA damage response is involved in the developmental toxicity of mebendazole in zebrafish retina." *Frontiers in Pharmacology* no. 7:57.
Segev, G, G Baneth, B Levitin, I Aroch, and I Shlosberg. 2004. "Accidental poisoning of 17 dogs with lasalocid." *Veterinary Record* no. 155 (6):174–176.
Sekis, Ivana, Kerry Ramstead, Mark Rishniw, Wayne S Schwark, Sean P McDonough, Richard E Goldstein, Mark Papich, and Kenneth W Simpson. 2009. "Single-dose pharmacokinetics and genotoxicity of metronidazole in cats." *Journal of Feline Medicine and Surgery* no. 11:60e68
Sharma, SK, Jit Singh, PK Peshin, and AP Singh. 1983. "Evaluation of chloral hydrate anaesthesia in camels." *Zentralblatt für Veterinärmedizin Reihe A* no. 30 (8):674–681.
Siroka, Z, and Z Svobodova. 2013. "The toxicity and adverse effects of selected drugs in animals—overview." *Polish Journal of Veterinary Sciences* no. 16 (1):181–191.
Srinivasan, SR, and Ajith Kumar. 1991. "Adverse reaction to oxytetracycline in a dog: A case report." *Indian Veterinary Journal* no. 68 (2):159–160.
Stahlmann, Ralf, and Hartmut Lode. 2010. "Safety considerations of fluoroquinolones in the elderly." *Drugs & Aging* no. 27 (3):193–209.
Stehr, Carla M, Tiffany L Linbo, John P Incardona, and Nathaniel L Scholz. 2006. "The developmental neurotoxicity of fipronil: Notochord degeneration and locomotor defects in zebrafish embryos and larvae." *Toxicological Sciences* no. 92 (1):270–278.
Stokol, T, JF Randolph, S Nachbar, C Rodi, and SC Barr. 1997. "Development of bone marrow toxicosis after albendazole administration in a dog and cat." *Journal of the American Veterinary Medical Association* no. 210 (12):1753–1756.
Stratton-Phelps, Meri, W David Wilson, and Ian A Gardner. 2000. "Risk of adverse effects in pneumonic foals treated with erythromycin versus other antibiotics: 143 cases (1986–1996)." *Journal of the American Veterinary Medical Association* no. 217 (1):68–73.
Swarup, DPRC, RC Patra, V Prakash, R Cuthbert, D Das, P Avari, DJ Pain, RE Green, AK Sharma, and M Saini. 2007. "Safety of meloxicam to critically endangered Gyps vultures and other scavenging birds in India." *Animal Conservation* no. 10 (2):192–198.
Swor, Tamara M, Jamie L Whittenburg, and M Keith Chaffin. 2009. "Ivermectin toxicosis in three adult horses." *Journal of the American Veterinary Medical Association* no. 235 (5):558–562.
Thirunavukkarasu, PS, M Subramanian, and K Vasu. 1995. "Adverse reaction to oxytetracycline in cow." *Indian Journal of Veterinary Medicine* no. 15:105.

Thomas, Helen L, and Michael A Livesey. 1998. "Immune-mediated hemolytic anemia associated with trimethoprim-sulphamethoxazole administration in a horse." *The Canadian Veterinary Journal* no. 39 (3):171.

Toribio, Ramiro E, Fairfield T Bain, Dawn R Mrad, NT Messer, Rani S Sellers, and Kenneth W Hinchcliff. 1998. "Congenital defects in newborn foals of mares treated for equine protozoal myeloencephalitis during pregnancy." *Journal of the American Veterinary Medical Association* no. 212 (5):697–701.

Trepanier, LA. 2004. "Idiosyncratic toxicity associated with potentiated sulfonamides in the dog." *Journal of Veterinary Pharmacology and Therapeutics* no. 27 (3):129–138.

Van der Linde-Sipman, JS, TSGAM Van Den Ingh, JJ Van Nes, H Verhagen, JGTM Kersten, AC Beynen, and R Plekkringa. 1999. "Salinomycin-induced polyneuropathy in cats: Morphologic and epidemiologic data." *Veterinary Pathology* no. 36 (2):152–156.

Van Laun, T. 1977. "Anaesthetising donkeys." *The Veterinary Record* no. 100 (18):391.

Walker, CH. 2003. "Neurotoxic pesticides and behavioural effects upon birds." *Ecotoxicology* no. 12 (1):307–316.

Wang, Huili, Baoguang Che, Ailian Duan, Jingwen Mao, Randy A Dahlgren, Minghua Zhang, Hongqin Zhang, Aibing Zeng, and Xuedong Wang. 2014. "Toxicity evaluation of β-diketone antibiotics on the development of embryo-larval zebrafish (Danio rerio)." *Environmental Toxicology* no. 29 (10):1134–1146.

Weber, Martha A, Michele A Miller, Donald L Neiffer, and Scott P Terrell. 2006. "Presumptive fenbendazole toxicosis in North American porcupines." *Journal of the American Veterinary Medical Association* no. 228 (8):1240–1242.

Weber, Martha A, Scott P Terrell, Donald L Neiffer, Michele A Miller, and Barbara J Mangold. 2002. "Bone marrow hypoplasia and intestinal crypt cell necrosis associated with fenbendazole administration in five painted storks." *Journal of the American Veterinary Medical Association* no. 221 (3):417–369.

Webster, M. 1999. "Product warning: Frontline." *Australian Veterinary Journal* no. 77 (3):202.

Weiss, DJ, and JS Klausner. 1990. "Drug-associated aplastic anemia in dogs: Eight cases (1984–1988)." *Journal of the American Veterinary Medical Association* no. 196 (3):472–475.

Wickstrom, ML, and CT Eason. 1999. "Literature search for mustelid-specific toxicants." *Science for Conservation E* no. 127:57–65.

Witschi, Hanspeter, Gayle Godfrey, Ed Frome, and Robert C Lindenschmidt. 1987. "Pulmonary toxicity of cytostatic drugs: Cell kinetics." *Toxicological Sciences* no. 8 (2):253–262.

Woodward, KN. 2000. "Regulation of veterinary drugs." *General and Applied Toxicology*, 1633–1652.

Wright, Kathryn H, and John W Tyler. 2003. "Recognizing metronidazole toxicosis in dogs." *Veterinary Medicine* no. 98 (5):410–418.

Xavier, Fabiana Galtarossa, and Márcia Mery Kogika. 2002. "Common causes of poisoning in dogs and cats in a Brazilian veterinary teaching hospital from 1998 to 2000." *Veterinary and Human Toxicology* no. 44 (2):115–116.

Yamindago, Ade, Nayun Lee, Seonock Woo, and Seungshic Yum. 2019. "Transcriptomic profiling of Hydra magnipapillata after exposure to naproxen." *Environmental Toxicology and Pharmacology* no. 71:103215.

Yan, Zhaoyang, Xiaoyong Huang, Yangzhouyun Xie, Meirong Song, Kui Zhu, and Shuangyang Ding. 2019. "Macrolides induce severe cardiotoxicity and developmental toxicity in zebrafish embryos." *Science of the Total Environment* no. 649:1414–1421.

Yan, Zhengyu, Qiulian Yang, Weili Jiang, Jilai Lu, Zhongrun Xiang, Ruixin Guo, and Jianqiu Chen. 2018. "Integrated toxic evaluation of sulfamethazine on zebrafish: Including two lifespan stages (embryo-larval and adult) and three exposure periods (exposure, post-exposure and re-exposure)." *Chemosphere* no. 195:784–792.

Yang, Shinwoo, Jongmun Cha, and Kenneth Carlson. 2004. "Quantitative determination of trace concentrations of tetracycline and sulfonamide antibiotics in surface water using solid-phase extraction and liquid chromatography/ion trap tandem mass spectrometry." *Rapid Communications in Mass Spectrometry* no. 18 (18):2131–2145.

Yas-Natan, E, M Shamir, S Kleinbart, and I Aroch. 2003. "Doramectin toxicity in a collie." *The Veterinary Record* no. 153 (23):718.

Yeruham, I, S Perl, D Sharony, and Y Vishinisky. 2002. "Doxycycline toxicity in calves in two feedlots." *Journal of Veterinary Medicine, Series B* no. 49 (8):406–408.

Zavala, Guillermo, Douglas A Anderson, James F Davis, and Louise Dufour-Zavala. 2011. "Acute monensin toxicosis in broiler breeder chickens." *Avian Diseases* no. 55 (3):516–521.

Zhang, Siqi, Danielle Hagstrom, Patrick Hayes, Aaron Graham, and Eva-Maria S Collins. 2019. "Multi-behavioral endpoint testing of an 87-chemical compound library in freshwater planarians." *Toxicological Sciences* no. 167 (1):26–44.

Zhou, Li-Jun, Qinglong L Wu, Bei-Bei Zhang, Yong-Gang Zhao, and Bi-Ying Zhao. 2016. "Occurrence, spatiotemporal distribution, mass balance and ecological risks of antibiotics in subtropical shallow Lake Taihu, China." *Environmental Science: Processes & Impacts* no. 18 (4):500–513.

Ziv, G, SL Levisohn, B Bar-Moshe, A Bor, and S Soback. 1983. "Clinical pharmacology of tiamulin in ruminants." *Journal of Veterinary Pharmacology and Therapeutics* no. 6 (1):23–32.

Zwart, P. 1988. "Poisoning caused by the pyrethroid compound deltamethrin in Gould's amadines (Chloebia gouldiae). Description of a case." *Tijdschrift Voor Diergeneeskunde* no. 113 (18):1009–1010.

10 Accumulation, Uptake Pathways, and Toxicity of Pharmaceuticals into Plants and Soil

Neetu Singh and Surender Singh Yadav

CONTENTS

10.1 INTRODUCTION

Modern technologies are driving the development of the world with rapid speed. At the same time, this technological advancement is also associated with producing several serious environmental contaminants like heavy metals, nanomaterials, harmful microbes, etc. Besides these environmental contaminants, many pharmaceuticals (e.g., antibiotics, herbicides, pesticides, veterinary drugs) and personal care products like sunscreens, soaps, lotions, creams, etc., which we use for glorifying our daily lives, are also environmental contaminants of emerging concern (Wu et al., 2015). Since the industrial revolution, thousands of new pharmaceuticals have come into existence to ease our lives and are accumulating in the environment. The accumulation of pharmaceutical and cosmeceutical products is increasing daily (Dodgen et al., 2015). These pharmaceutical contaminants are widely distributed in our surroundings and are very toxic in nature. These compounds possess potential ecotoxicological impacts on aquatic as well as terrestrial lives (Bhoyar et al., 2023; Christou et al., 2017b; Zhang et al., 2021).

The antibiotic residue and inorganic salts from these pharmaceutical compounds are imposing serious ill effects on biological survival and disturbing smooth environmental functioning

DOI: 10.1201/9781003361091-10

(Guo et al., 2016). Exposure to these pharmaceutical contaminants is responsible for causing various diseases, resulting in a great loss of life and an imbalance in the whole ecosystem (Tripathi et al., 2017). These chemical compounds negatively affect proper cell functioning and reproductive ability of aquatic as well as terrestrial organisms (Dodgen et al., 2015; Svobodníková et al., 2019). These contaminants are also responsible for the degradation of soil quality and associated soil microbiota, ultimately leading to plant deterioration (Christou et al., 2017b).

The risk assessment from the accumulation of these pharmaceuticals and their residues within a reasonable time frame is somewhat difficult to determine because every year, new pharmaceuticals are coming into existence at a very high rate. According to a study, the total consumption of antibiotics was estimated at 1–2 lakh tons on a global scale. Approximately 15,000 tons of antibiotics are released into the European environment alone per annum (Bartrons & Peñuelas, 2017). Though all these pharmaceuticals are not newly designed, some are already present in the environment due to their continuous exposure over a long time frame. Most of the time, anthropogenic activities remain the major source of the emergence and dispersion of these pharmaceutical contaminants (Bartrons & Peñuelas, 2017). Now the question arises, how can these pharmaceutical contaminants be reduced from the environment?

Plants are one of the natural pollution indicators or biomonitors. They can trace even a very small number of environmental pollutants. Plants mostly accumulate these chemical compounds from reclaimed water (used for irrigation) and contaminated soil (essential for growth and support), which results in contaminated food crops. The accumulation of these pharmaceuticals imposes toxicity not only on the plants but also on the associated microbiota. The intakes of such plants also pose a risk to humans and ecosystems. They may disturb living beings' normal metabolism and behavior (Dodgen et al., 2013). Only a few studies are documented on the pharmaceutical uptake by plants. These studies are also restricted to a few pharmaceutical compounds and plant types. Besides, detecting these chemical compounds also requires precise and advanced analytical tools. Therefore, there is an urgent need to overcome this paucity of tracing these chemical compounds. In this regard, an attempt was made to cover different aspects of pharmaceutical accumulation like in soil, plants, their uptake pathways, phytotoxicity, and decontamination techniques.

10.2 PHARMACEUTICALS OF EMERGING CONCERN

In our day-to-day life, we are facing a lot of problems related to physical and mental illness. To resolve all these problems, we primarily depend upon modern allopathic medications which have emerged as one of the major sources of pharmaceutical accumulation. These pharmaceuticals are considered persistent or pseudo-persistent environmental pollutants (Christou et al., 2017a; Puckowski et al., 2016): antibiotics (roxithromycin, erythromycin, and ketoconazole), antiepileptics (carbamazepine), antidepressants (sertraline and fluoxetine), antipyretics, antidiabetics, analgesics, anti-inflammatories (diclofenac, naproxen, and ibuprofen), antiseptics, diuretics, lipid regulators, β-blockers (propranolol), antiprotozoals, steroid hormones (17α-ethinylestradiol); drugs used for the treatment of pulmonary disorders, erectile dysfunction, hypertension, psychiatric disorders; and veterinary drugs. Though these pharmaceuticals play a major role in maintaining human and livestock well-being for a short period of time, in combination they also impose long-term negative effects (Dodgen et al., 2013). The active constituents of these medicines, whose accumulation is increasing day by day, are of emerging concern (Bartrons & Peñuelas, 2017; Liu et al., 2015). Some of these emerging pharmaceutical contaminants are listed in Table 10.1.

TABLE 10.1
Emerging Pharmaceuticals Contaminants

S. No.	Class of pharmaceuticals	Examples	S. No.	Class of pharmaceuticals	Examples
1.	Antidiabetics	Metformin	8.	Antidepressants	Diazepam and fluoxetine
2.	Antiestrogenics	Tamoxifen	9.	Psychostimulants	Caffeine and paraxanthine
3.	Antiepileptics	Carbamazepine, codeine, diclofenac, 4-aminoantipyrine, antipyrine	10.	Phytosanitary products	Clofibric acid
4.	Antihistamines	Diphenhydramine	11.	Antiseptics	Triclosan, chlorophene
5.	Antiprotozoals	Quinacrine dihydrochloride	12.	Erectile dysfunction and pulmonary arterial hypertension	Sildenafil
6.	Diuretics	Furosemide, hydrochlorothiazide, amidotrizoic acid, diatrizoate, iothalamic acid	13.	Lipid regulators	Acebutolol, atenolol, atorvastatin, bezafibrate, fenofibric acid, gemfibrozil
7.	Veterinary and human medications	Azithromycin, chlortetracycline, clarithromycin, ciprofloxacin, doxycycline, enrofloxacin erythromycin, tetracycline, trimethoprim, tylosin, sulfadimethoxine, salinomycin, etc.	14.	Analgesics and anti-inflammatories	Codeine, diclofenac, fenoprofen, ibuprofen, indomethacin, ketoprofen, ketorolac, paracetamol, phenylbutazone, naproxen, clofibric acid, etc.

10.3 THE MAJOR SOURCE OF PHARMACEUTICAL CONTAMINANTS IN SOIL AND PLANTS

The major source of pharmaceutical contaminants is anthropogenic activities associated with municipal, agricultural, industrial garbage, and wastewater (Dodgen et al., 2013; Wu et al., 2012; Wu et al., 2015). Globally, agricultural practices require 67% of total water withdrawals and 86% of total water consumption (Miller et al., 2016). Due to water scarcity, the irrigation of vegetables and fruit crops with treated wastewater has become the most common phenomenon (Malchi et al., 2014; Pan et al., 2014). Such practices allow the accumulation of pharmaceuticals in the soil, dietary crops, and their translocation into the food web and ultimately to the human body. Veterinary pharmaceuticals are also one of the major sources of pharmaceutical contamination. About 90% of antibiotics are released by the animals in their feces (as shown in Table 10.1). The pharmaceuticals accumulate in the soil up to mg/kg levels. For instance, the biosolids applied to the soil at a rate of 5 kg/m^2 lead to triclosan accumulation in the farm soil at a concentration of 0.77–0.95 mg/kg (Li et al., 2019; Wu et al., 2012). The persistency of residuals in water bodies develops antibiotic-resistant pathogens known as "superbugs" (Bhoyar et al., 2023; Yuan et al., 2020).

The human body also acts as a source of pharmaceutically active ingredients because most pharmaceutical compounds remain poorly absorbed or not completely metabolized; therefore, 30–90% of pharmaceuticals are released into the environment (Christou et al., 2017b). The major pathways of these compounds that come from the human body to the environment are excretion (urine and feces), vomiting, sweating, and so on (Daughton & Ruhoy, 2013). These antibiotics are released back in a concise time frame after their application. They are leached in their parent form or metabolites (Zhang et al., 2014). The exposure of the human body to such pharmaceuticals will subsequently lead to human complications (Christou et al., 2017b; Pan et al., 2014; Prosser & Sibley, 2015; Rand-Weaver et al., 2013; Wang et al., 2014). Recently, carbamazepine and its metabolites were also reported in human urine after consumption of freshly produced crops irrigated with reclaimed water (Li et al., 2019).

Besides, veterinary pharmaceuticals, including antiparasitics, antifungals, antibiotics, antiseptics, anti-inflammatory drugs, anesthetics, sedatives, hormones, etc., also play a significant role in environmental pollution (Bartrons & Peñuelas, 2017). These veterinary contaminants reach the environment through the improper disposal of containers, manure, unused drugs or livestock feed, and inappropriate manufacturing processes (Bártíková et al., 2016). The rise of population, prosperity, and inappropriate use of drugs has also energized the tremendous overproduction of numerous drugs. Sometimes, most of the drugs remain unused and expired, and are directly released into the environment (Christou et al., 2017b). The different sources of environmental contaminants are discussed in Table 10.2.

10.3.1 Pharmaceutical Occurrence in Soil

Soil is a living, dynamic entity that is greatly influenced by the application of pharmaceutically active compounds. Once they are introduced into the soil, soil goes through a lot of chemical, physical, and biological changes. These pharmaceutically active compounds are retained in the topmost layer of soil or may reach deeper soil layers. Ultimately these compounds reach the groundwater table where they interact with soil microbiota and are absorbed by the plants (Christou et al., 2017b). Sewage sludge is one of the major factors for soil contamination. There are mainly three types of sewage sludges: primary, secondary andtertiary. The sludge containing 1–4 % solids, formed by the combination of these three types of sludge is known as raw sludge. They are profoundly originating from wastewater-treated plants. Globally, 180 pharmaceutically active compounds and 45 metabolites

TABLE 10.2
Overview of Different Sources of Environmental Contaminants

Sources of environmental contaminants		
Natural	**Anthropogenic**	**Metals and other dusts**
Dust storms	Diesel and petrol	Coal and coal ash
Forest fires	Cigarette and smoke	Organic dust
Volcanoes	Buildings demolition	Lead and copper
Ocean and water evaporation	Cosmeceutical products	Polymer fumes
Organisms	Indoor pollution	Asbestos

Source: Tripathi et al. (2017).

have been reported in sewage sludge but their concentration varies substantially according to the usage, sludge matrix, sludge application frequency, precipitation and run-off, geographical location, soil characteristics, and conditions (Mejias et al., 2021). Their range varies from μg/kg to mg/kg. The antimicrobials (triclosan and triclocarban), and antibiotics (fluoroquinolones) were detected in major concentrations of 1200–3000 ng/g dm and 50–550 ng/g dm, respectively. Besides, other antibiotics including tetracycline, ibuprofen, diclofenac, azithromycin, diphenhydramine, fluoxetine, miconazole, gemfibrozil, sulfamethoxazole, oxytetracycline have also been reported frequently but at lower concentrations (Mejias et al., 2021).

10.3.2 Pharmaceutical Accumulation in Plants

Nowadays, we are facing numerous challenges in our daily life due to the increased accumulation of pharmaceutical contaminants in the environment. Plants play a major role in combating these pollutants. These pharmaceuticals are accumulated in significant amounts in many plant species at varying concentrations. Plants mainly acquire these pharmaceuticals from animals' manure, contaminated soil, and wastewater through their roots (Miller et al., 2016). The uptake pathways of these pharmaceuticals in plants are not fully known to date but it is assumed that most of the pharmaceuticals are mainly consumed through roots and aerial parts (Boxall et al., 2012; Goldstein et al., 2014; Miller et al., 2016; Tanoue et al., 2012). In roots, the uptake is facilitated either by mass flow or diffusion of dissolved contaminants. Neutral compounds are diffused across the root cell membrane by partition-coefficient while ionized compounds make their way by electrostatic interactions of the ionic fraction. In aerial parts, these contaminants get deposited from volatile compounds or aerosols. The aerial parts also get direct exposure to these contaminants during irrigation, specifically with wastewater (Bartrons & Peñuelas, 2017). It was also noticed that the accumulation of these compounds is more pronounced in roots as compared to other plant parts (Dodgen et al., 2013). The overall pharmaceutical uptake pathway of plants is depicted in Figure 10.1.

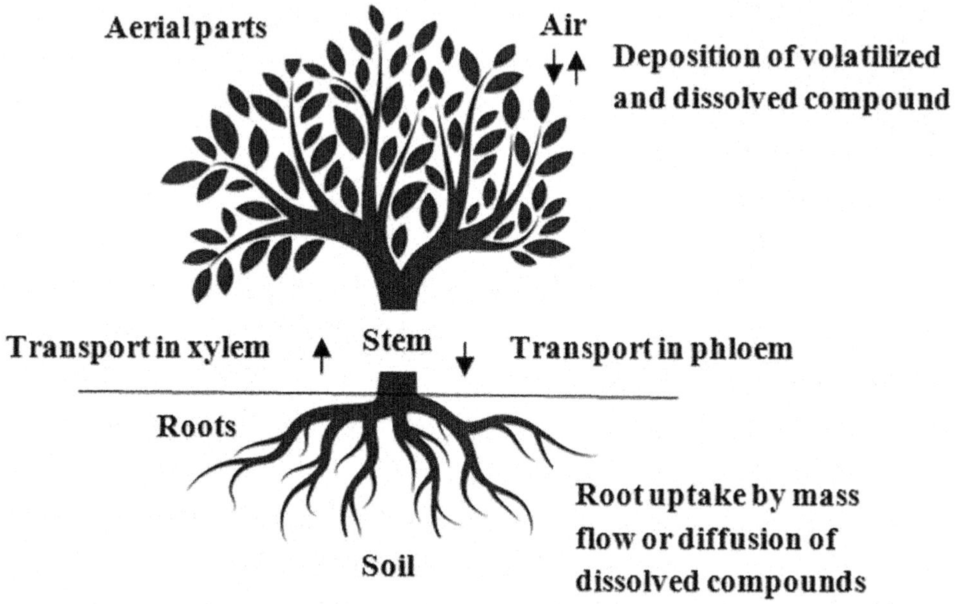

FIGURE 10.1 Overview of different uptake pathways of pharmaceutical accumulation by plants

Source: Bartrons & Peñuelas (2017).

10.3.2.1 Pharmaceutical Uptake Pathway by Roots

Roots play a major role in pharmaceutical uptake, translocation, and accumulation in different plant parts. Each molecule has to pass through the different layers of roots. The epidermis is the outermost layer of root cells. Small pharmaceutical molecules or those dissolved in water first of all have to pass through the epidermis of growing root tip cells and root hairs. In mature root cells, the permeability of another layer, the exodermis, resists the compounds passing through it. But the young root cells lack exodermis that resultantly enhances the permeabilityof root cell epidermis. After passing through the epidermis, the solute or water molecules reach the cortex. From there, these molecules make their way to vascular tissue through the endodermis. Once entered into the vascular tissues, the molecules can be translocated to other aboveground and belowground parts through the xylem and phloem. The compounds which fail to reach vascular tissues remain un-translocated. The presence of Casparian strips in the endodermis acts as the hydrophobic barrier in the transport of a cocktail of antibiotics, so it was suggested that one of these compounds should follow the symplastic pathway (Miller et al., 2016). The pharmaceutical uptake pathway of roots is shown in Figure 10.2. There are three different pathways through which water and solute molecules can move from the soil pore to the vasculature systems (Miller et al., 2016). These pathways are as follows:

- **Transmembrane pathway:** The compounds have to pass between cells through cell walls and membranes.
- **Symplastic pathway:** The compounds move between the cells through interconnecting plasmodesmata.
- **Apoplastic pathway:** In this pathway, compounds move along the cell walls through the intercellular space.

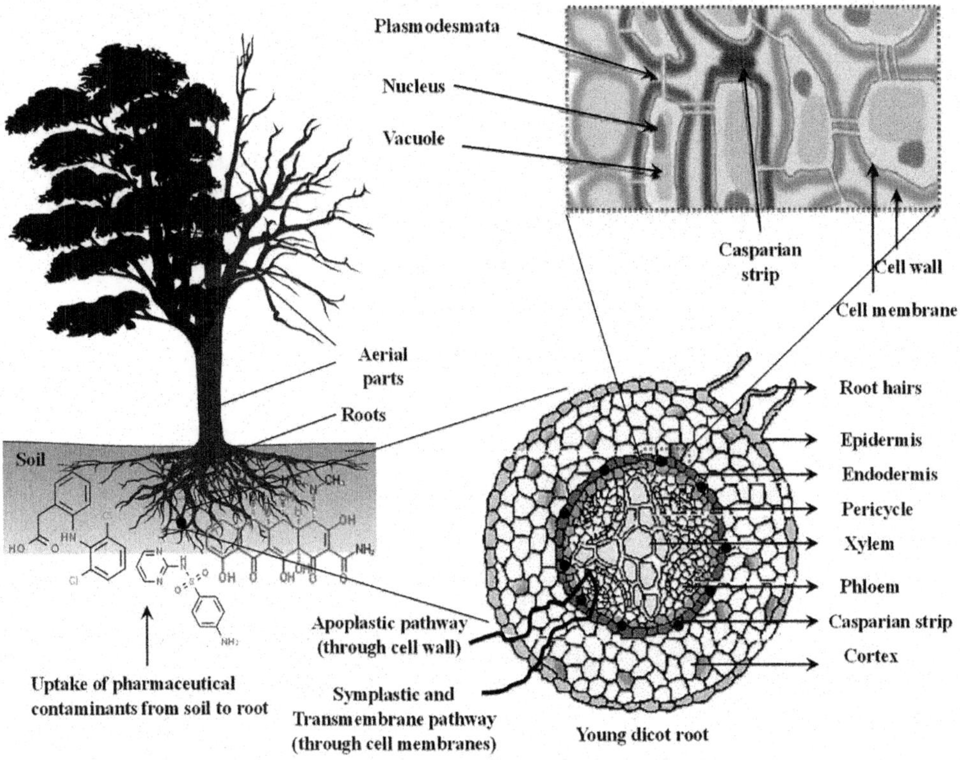

FIGURE 10.2 Overview of pharmaceutical uptake pathway by roots.

10.4 FACTORS AFFECTING THE PHARMACEUTICAL UPTAKE BY PLANTS

The uptake, absorbance, accumulation, translocation, and transformations of these pharmaceutical compounds not only depend upon their physiochemical properties (hydrophobicity, ionization behavior, charge, and so on) but are also influenced by the physiological behavior of the plant, soil properties (like pH, mineral concentration, dissolved organic content, cation exchange capacity, and soil texture), type of plant species, water quality, exposure to the contaminant, their duration, types of growth media, and environmental factors (Dodgen et al., 2013; Liu et al., 2013; Miller et al., 2016; Park & Huwe, 2016; Wu et al., 2013). Cationic and neutral contaminants have almost similar accumulation in roots and leaves whereas, in anionic form, they get more accumulated in roots than leaves. It was also noticed that organic compounds present in soil pore water are easily bioavailable for plant uptake (Li et al., 2019). However, most of the phytotoxic studies related to antibiotics accumulation were conducted in hydroponic laboratory conditions using unrealistic concentrations of antibiotics. These studies, however, were not able to fully justify the antibiotics' phytotoxicity but they were able to provide preliminary data regarding such studies (Michelini et al., 2013; Pan & Chu, 2016).

For instance, the antibiotics (oxytetracycline and norfloxacin) uptake from the contaminated soil required six to eight days to reach the maximum plant tissue concentration in soybean plants whereas it required only two days for antibiotic uptake when plants were grown in spiked deionized water. The difference occurred due to the higher availability of antibiotics in water as compared to the soil (Christou et al., 2017b). In another study by Carter et al. (2014), the uptake of five different pharmaceuticals (carbamazepine, diclofenac, fluoxetine, propranolol, sulfamethazine) and one personal care product (i.e., triclosan) was evaluated using *Raphanus sativus* (radish) and *Lolium perenne* (ryegrass). It was found that carbamazepine was the most accumulated compound in both the plants (i.e., 52 μg/g and 33 μg/g in radish and ryegrass, respectively) while sulfamethazine (<0.01 μg/g) was the least taken up compound. The difference in uptake concentration of different chemicals was assigned to their different extent of hydrophobicity and extent of ionization of each chemical in soil. Results have also indicated that the uptake of chemicals was specific to plant species. Malchi et al. (2014) quantified the pharmaceutical uptake by carrot and sweet potatoes grown in the treated wastewater. They found that in both plants the nonionic pharmaceutical compounds including caffeine, carbamazepine, and lamotrigine were in higher concentrations as compared to the ionic pharmaceutical compounds, namely bezafibrate, metoprolol, diclofenac, clofibric acid, gemfibrozil, sildenafil, sulfamethoxazole, naproxen, ketoprofen, ibuprofen, and so on. They also noticed that pharmaceutical accumulation was greater in leaves than in roots.

10.5 TOXICITY OF PHARMACEUTICAL ACCUMULATION

As we discussed earlier, there are so many sources of pharmaceutical accumulation. Antibiotics are one of the major sources of environmental contamination. It has also been reported that lots of veterinary drugs and hormones are also phytotoxic in nature. These medications impose toxic effects upon non-targeted objects, including soil and plants (Bártíková et al., 2016). These compounds disturb the proper functioning of soil microbiota. Some of the plants, like rice, lettuce, soybeans, wheat, sweet oat, lettuce, etc., are affected badly by antibiotics. They lead to increased production of reactive oxygen species which ultimately cause oxidative stress (Liu et al., 2013). Wang et al. (2014) reported that the total concentration of tetracyclines was in the range of 12.7–145.2 μg/kg. The parent compound (tetracycline) was found in higher concentrations than their by-products. They also detected the total highest concentration of fluoroquinolones (i.e., 79.2 μg/kg) in random soil samples. In another study by Grossberger et al. (2014), the concentration of sulfamethoxazole was detected in the soil irrigated with wastewater for a single growing period of carrot crop. The concentration was 0.12–0.28 μg/kg, depending upon the soil type. These accumulated pharmaceutical contaminants will ultimately reach the plants through the soil and result in different toxic symptoms in different types of crops. The phytotoxicity symptoms of different antibiotics on different plant species are discussed in Table 10.3.

TABLE 10.3
Phytotoxicity Symptoms of Antibiotics

S. No.	Class of comp.	Name of comp.	Botanical name of plant sp.	Common name of plant sp.	Phytotoxicity symptoms
1.	Tetracyclines	Chlortetracycline	*Phaseolus vulgaris*	Common bean	Retarded root growth and development
			Cichorium endivia	Sweet oat	Stop seed germination
			Cucumis sativus	Cucumber	
			Oryza sativa	Rice	
			Lactuca sativa	Lettuce	Root length decreased
			Daucus carota	Carrot	
		Doxycycline	*Triticum aestivum*	Wheat	Chlorophyll content and photosynthesis rate is reduced
		Oxytetracycline	*Phaseolus vulgaris*	Common bean	Retarded root growth and development
			Cichaorium endivia	Sweet oat	Seed germination is halted
			Cucumis sativus	Cucumber	
			Oryza sativa	Rice	
			Phragmites australis	Common reed	Decreased chlorophyll content and root activity
			Medicago sativa	Alfalfa	Retarded plant growth
			Lemna minor	Common duckweed	
			Daucus carota	Carrot	Root length and plant growth reduced
			Lactuca sativa	Lettuce	
		Tetracycline	*Lolium perenne*	Perennial Ryegrass	Reduction in root biomass and phosphorous assimilation
			Euphorbia pulcherrima	Poinsettia	Suppression of free branching
			Triticum aestivum	Wheat	Reduced chlorophyll content and photosynthesis
			Daucus carota	Carrot	Decreased root length
2.	Sulfonamides	Sulfadiazine	*Triticum aestivum*	Wheat	Retarded root and shoot elongation
			Brassica campestris	Mustard	
			Cyphomandrabetacea	Tree tomato	
			Salix fragilis	Crack willow	Suppress plant growth by root alterations

S. No.	Class of comp.	Name of comp.	Botanical name of plant sp.	Common name of plant sp.	Phytotoxicity symptoms
		Sulfadimidine	*Zea mays*	Maize	
		Sulfamethoxazole	*Cichaorium endivia*	Sweet oat	Retarded seed germination
			Cucumis sativus	Cucumber	
			Oryza sativa	Rice	Reduced seed germination and plant growth
			Lemnagibba	Swollen duckweed	Plant growth and development is decreased
			Daucus carota	Carrot	Plant growth stops
3.	Macrolides	Tylosin	*Cichaorium endivia*	Sweet oat	Seed germination is reduced
			Daucus carota	Carrot	Root length decreased
		Erythromycin	*Triticum aestivum*	Wheat	Photosynthesis and chlorophyll content is reduced
4.	Fluoroquinolones	Ciprofloxacin	*Triticum aestivum*	Wheat	Low photosynthesis and chlorophyll content
			Lemnagibba	Swollen duckweed	Reproduction rate decreases and chlorosis
			Daucus carota	Carrot	Suppression of plant growth and development
			Phragmites australis	Common reed	Low chlorophyll content and root activity
		Enrofloxacin	*Lactuca sativa*	Lettuce	Retarded plant growth and development
			Triticum aestivum	Wheat	Shortens root and shoot elongation
5.	β-lactams	Amoxicillin			Photosynthesis decreases
			Daucus carota	Carrot	Plant growth halts
		Cephotaxim	*Antirrhinum majus*	Snapdragon	
6.	Lincosamides	Lincomycin	*Daucus carota*	Carrot	Root length decreases
7.	Polyether ionophores	Salinomycin	*Brassica rapa*	Wild mustard	Lowers the plant growth and development
		Monensin	*Gossypium hirsutum*	American tetraploid	
		Maduramycin	*Amaranthus hypochondriacus*	Prince-of-Wales feather	
8.	Phenicols	Florfenicol	*Lemna minor*	Common duckweed	Plant growth decreases
9.	Other antibiotics	Trimethoprim	*Cichaorium endivia*	Sweet oat	Stops seed germination
		Novobiocin	*Oryza sativa*	Rice	

Source: Bártíková et al. (2016)

10.5.1 Phytotoxicity of Pharmaceutical Accumulation

Sometimes, the recommended doses of pharmaceuticals can result in unintended and unwelcome consequences for the environment (Daughton & Ruhoy, 2013). Plants are one of these unintended objects that suffer a lot due to pharmaceutical accumulation. The accumulated pharmaceuticals from plants enter the different trophic levels of the food chain and ultimately reach human beings. The accumulation of pharmaceutical contaminants, however, reduces the contaminants from the environment but they impose negative effects upon the plants (Christou et al., 2017a). The negative impacts may affect plants directly, like the reduction of photosynthetic pigments, reduced foliage density and size, reduced root and stem length, and overall decreased growth and development of the whole plant. The negative effects may be indirect, like damage to beneficial plant microbiota, lower rate of germination, tissue deformation, disturbances in redox homeostasis, imbalance of nutrient cycle, and other stress-related phenomena (Bartrons & Peñuelas, 2017).

These negative impacts of antibiotic accumulation were also supported by Liu et al. (2013). They conducted a study to evaluate the effect of ciprofloxacin, oxytetracycline, and sulfamethazine on *Phragmites australis*. They used different antibiotic concentrations like 0, 0.1, 1, 10, 100, and 1000 μg/L. They noticed that at high antibiotic concentrations (i.e., >10 μg/L), it was toxic to the plant. This concentration exerted a bad effect on root activity as well as chlorophyll content, whereas at lower concentrations antibiotic-induced hormesis on roots and chlorophyll was observed. Christou et al. (2017b) reported the accumulation of diclofenac in fruits due to continuous irrigation with wastewater. Bruzzoniti et al. (2014) stated the involvement of oxidative stress generated reactive oxygen species in the degradation of oxytetracycline in hairy root cultures of *Helianthus annus*. The phytotoxicity symptoms of pharmaceutical accumulation within plants are depicted in Figure 10.3.

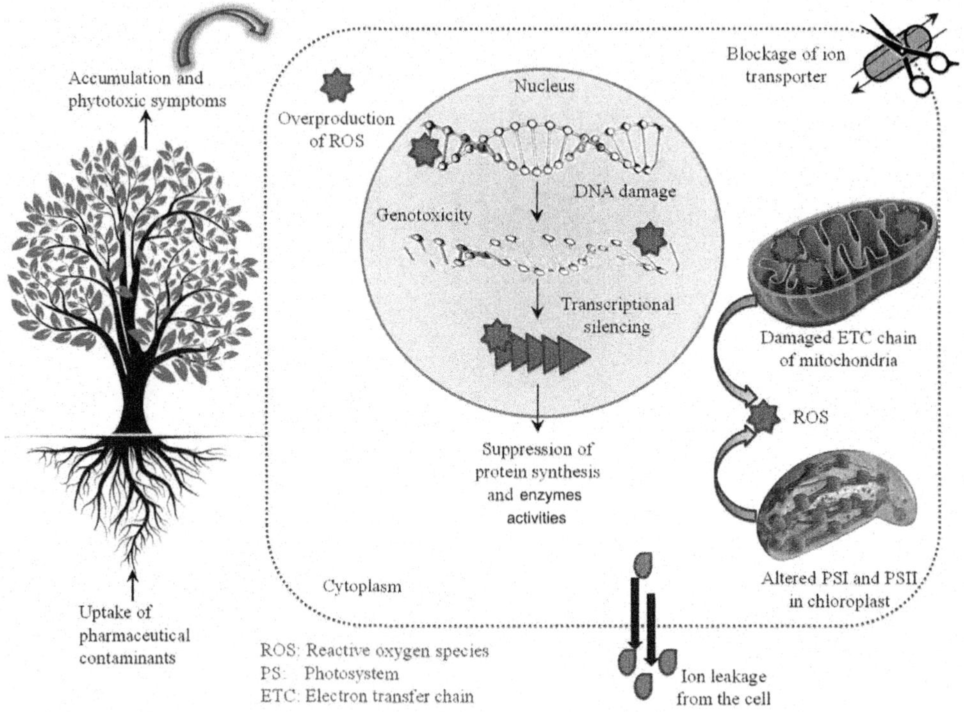

FIGURE 10.3 Phytotoxic symptoms of pharmaceutical accumulation.

10.6 DECONTAMINATION OF PHARMACEUTICAL ACCUMULATION

Plants are sessile in nature so they cannot be translocated to other non-contaminated places. However, there is no specific method for the removal of pharmaceutical accumulation from the plants but their source can be reduced to some extent. For instance, contaminated water should be decontaminated before irrigation so that the pharmaceutical accumulation can be reduced to a great extent (Bartrons & Peñuelas, 2017; Dodgen et al., 2015). Besides, the presence of antibiotic accumulation in plants can be traced using the combination of liquid chromatography with tandem mass spectroscopy. Bioremediation, sorption, membrane bioreactor, sequencing batch reactor, anaerobic-anoxic-oxic and moving-bed biofilm reactor, coagulation/disk filter, coagulation/sedimentation, rapid-coagulation sedimentation, ultraviolet light and powdered activated carbon etc. are some of the promising methods used for pharmaceutical removal.

10.7 ADVANTAGES AND DISADVANTAGES OF PHARMACEUTICAL ACCUMULATION

The agricultural use of organic pharmaceutical wastes including pharmaceutical sewage sludge is not always supposed to be harmful. Sometimes, it also exerts beneficial effects on the soil and plant microbiota in lower doses due to the rich content of macronutrients and low content of heavy metals and pharmaceutical contaminants. But the longer exposure to these contaminants will surely disturb the soil's properties and functionality. The beneficial effect of pharmaceutical sewage sludge exposure to soil was noticed only for a short period of 60 days. The most significant effects were the quality enhancement of soil organic matter; increased availability of plant nutrients like nitrogen, phosphorous, potassium, magnesium, calcium; and high germination index. The lower dose of pharmaceutical sewage sludge also prevents heavy metal accumulation in soil and also avoids soil salinization increasing exchangeable sodium (Cucina et al., 2019).

10.8 CONCLUSION AND FUTURE PERSPECTIVES

Pharmaceuticals ease our daily lives to a great extent by treating numerous ailments and maladies, so we can't imagine our lives without these pharmaceuticals. Therefore, there are continuous efforts to develop new and novel pharmaceuticals that possess a high therapeutic quotient at a lower dose. Besides, pharmaceutical compounds and their accumulation also pose a major threat not only to human beings but to all other forms of biota including plants and microbes. Though the accumulation of these pharmaceuticals in mobile objects can be avoided to some extent by translocating them, it becomes very difficult in the case of sessile objects like plants. Therefore, plants act as bioindicators of this accumulation, so they can be used for phytoremediation. They can accumulate these compounds from ng to µg/kg (Rand-Weaver et al., 2013). The fate of these contaminants' accumulation in plants is still unknown. But their high concentration might be toxic to plants, soil and plant microbiota, nutrient cycling, and a balanced ecosystem. The dietary intake of these pharmaceutical-contaminated plants also poses a great risk to living beings. Therefore, innovative decontaminated techniques are needed for the future use and reuse of water for drinking and agronomical cultures to avoid pharmaceutical contamination. Moreover, studies can also be focused on reducing drug concentration without compromising their therapeutic abilities, so ultimately the drug circulation in the environment can be reduced up to a great extent. Furthermore, emphasis should be given to field study using realistic conditions to evaluate the actual antibiotic accumulation and their phytotoxicity instead of hydroponic cultures/laboratory experimentation.

ACKNOWLEDGMENTS

The authors would like to thank the Council of Scientific and Industrial Research (CSIR), New Delhi and DST-SERB, and DST-FIST, New Delhi for providing the financial assistance.

REFERENCES

Bártíková, H., Podlipná, R., & Skálová, L. (2016). Veterinary drugs in the environment and their toxicity to plants. *Chemosphere, 144*, 2290–2301. https://doi.org/10.1016/j.chemosphere.2015.10.137.

Bartrons, M., & Peñuelas, J. (2017). Pharmaceuticals and personal-care products in plants. *Trends in Plant Science, 22*(3), 194–203. https://doi.org/10.1016/j.tplants.2016.12.010.

Bhoyar, T., Vidyasagar, D., & Umare, S. S. (2023). Mitigating phytotoxicity of tetracycline by metal-free 8-hydroxyquinoline functionalized carbon nitride photocatalyst. *Journal of Environmental Sciences, 125*, 37–46. https://doi.org/10.1016/j.jes.2021.10.032.

Boxall, A. B., Rudd, M. A., Brooks, B. W., Caldwell, D. J., Choi, K., Hickmann, S., & Van Der Kraak, G. (2012). Pharmaceuticals and personal care products in the environment: What are the big questions? *Environmental Health Perspectives, 120*(9), 1221–1229.

Bruzzoniti, M. C., Checchini, L., De Carlo, R. M., Orlandini, S., Rivoira, L., & Del Bubba, M. (2014). QuEChERS sample preparation for the determination of pesticides and other organic residues in environmental matrices: A critical review. *Analytical and Bioanalytical Chemistry, 406*(17), 4089–4116.

Carter, L. J., Harris, E., Williams, M., Ryan, J. J., Kookana, R. S., & Boxall, A. B. (2014). Fate and uptake of pharmaceuticals in soil—plant systems. *Journal of Agricultural and Food Chemistry, 62*(4), 816–825. https://doi.org/10.1021/jf404282y.

Christou, A., Agüera, A., Bayona, J. M., Cytryn, E., Fotopoulos, V., Lambropoulou, D., & Fatta-Kassinos, D. (2017a). The potential implications of reclaimed wastewater reuse for irrigation on the agricultural environment: The knowns and unknowns of the fate of antibiotics and antibiotic resistant bacteria and resistance genes—a review. *Water Research, 123*, 448–467. https://doi.org/10.1016/j.watres.2017.07.004.

Christou, A., Karaolia, P., Hapeshi, E., Michael, C., & Fatta-Kassinos, D. (2017b). Long-term wastewater irrigation of vegetables in real agricultural systems: Concentration of pharmaceuticals in soil, uptake and bioaccumulation in tomato fruits and human health risk assessment. *Water Research, 109*, 24–34. https://doi.org/10.1016/j.watres.2016.11.033.

Cucina, M., Ricci, A., Zadra, C., Pezzolla, D., Tacconi, C., Sordi, S., & Gigliotti, G. (2019). Benefits and risks of long-term recycling of pharmaceutical sewage sludge on agricultural soil. *Science of The Total Environment, 695*, 133762. https://doi.org/10.1016/j.scitotenv.2019.133762.

Daughton, C. G., & Ruhoy, I. S. (2013). Lower-dose prescribing: Minimizing "side effects" of pharmaceuticals on society and the environment. *Science of the Total Environment, 443*, 324–337. https://doi.org/10.1016/j.scitotenv.2012.10.092.

Dodgen, L. K., Li, J., Parker, D., & Gan, J. J. (2013). Uptake and accumulation of four PPCP/EDCs in two leafy vegetables. *Environmental Pollution, 182*, 150–156. https://doi.org/10.1016/j.envpol.2013.06.038.

Dodgen, L. K., Ueda, A., Wu, X., Parker, D. R., & Gan, J. (2015). Effect of transpiration on plant accumulation and translocation of PPCP/EDCs. *Environmental Pollution, 198*, 144–153. https://doi.org/10.1016/j.envpol.2015.01.002.

Goldstein, M., Shenker, M., & Chefetz, B. (2014). Insights into the uptake processes of wastewater-borne pharmaceuticals by vegetables. *Environmental Science & Technology, 48*(10), 5593–5600. https://doi.org/10.1021/es5008615.

Grossberger, A., Hadar, Y., Borch, T., & Chefetz, B. (2014). Biodegradability of pharmaceutical compounds in agricultural soils irrigated with treated wastewater. *Environmental Pollution, 185*, 168–177. https://doi.org/10.1016/j.envpol.2013.10.038.

Guo, W. Q., Zheng, H. S., Li, S., Du, J. S., Feng, X. C., Yin, R. L., & Chang, J. S. (2016). Removal of cephalosporin antibiotics 7-ACA from wastewater during the cultivation of lipid-accumulating microalgae. *Bioresource Technology, 221*, 284–290. https://doi.org/10.1016/j.biortech.2016.09.036.

Li, Y., Sallach, J. B., Zhang, W., Boyd, S. A., & Li, H. (2019). Insight into the distribution of pharmaceuticals in soil-water-plant systems. *Water Research, 152*, 38–46. https://doi.org/10.1016/j.watres.2018.12.039.

Liu, J., Lu, G., Xie, Z., Zhang, Z., Li, S., & Yan, Z. (2015). Occurrence, bioaccumulation and risk assessment of lipophilic pharmaceutically active compounds in the downstream rivers of sewage treatment plants. *Science of the Total Environment, 511*, 54–62. https://doi.org/10.1016/j.scitotenv.2014.12.033.

Liu, L., Liu, Y. H., Liu, C. X., Wang, Z., Dong, J., Zhu, G. F., & Huang, X. (2013). Potential effect and accumulation of veterinary antibiotics in *Phragmites australis* under hydroponic conditions. *Ecological Engineering, 53*, 138–143. https://doi.org/10.1016/j.ecoleng.2012.12.033.

Malchi, T., Maor, Y., Tadmor, G., Shenker, M., & Chefetz, B. (2014). Irrigation of root vegetables with treated wastewater: Evaluating uptake of pharmaceuticals and the associated human health risks. *Environmental Science & Technology, 48*(16), 9325–9333. https://doi.org/10.1021/es5017894.

Mejias, C., Martín, J., Santos, J. L., Aparicio, I., & Alonso, E. (2021). Occurrence of pharmaceuticals and their metabolites in sewage sludge and soil: A review on their distribution and environmental risk assessment. *Trends in Environmental Analytical Chemistry, 30*, e00125. https://doi.org/10.1016/j.teac.2021.e00125.

Michelini, L., La Rocca, N., Rascio, N., & Ghisi, R. (2013). Structural and functional alterations induced by two sulfonamide antibiotics on barley plants. *Plant Physiology and Biochemistry, 67*, 55–62. https://doi.org/10.1016/j.plaphy.2013.02.027.

Miller, E. L., Nason, S. L., Karthikeyan, K. G., & Pedersen, J. A. (2016). Root uptake of pharmaceuticals and personal care product ingredients. *Environmental Science & Technology, 50*(2), 525–541. https://doi.org/10.1021/acs.est.5b01546.

Pan, M., & Chu, L. M. (2016). Adsorption and degradation of five selected antibiotics in agricultural soil. *Science of the Total Environment, 545*, 48–56. https://doi.org/10.1016/j.scitotenv.2015.12.040.

Pan, M., Wong, C. K., & Chu, L. M. (2014). Distribution of antibiotics in wastewater-irrigated soils and their accumulation in vegetable crops in the Pearl River Delta, Southern China. *Journal of Agricultural and Food Chemistry, 62*(46), 11062–11069. https://doi.org/10.1021/jf503850v.

Park, J. Y., & Huwe, B. (2016). Effect of pH and soil structure on transport of sulfonamide antibiotics in agricultural soils. *Environmental Pollution, 213*, 561–570. https://doi.org/10.1016/j.envpol.2016.01.089.

Prosser, R. S., & Sibley, P. K. (2015). Human health risk assessment of pharmaceuticals and personal care products in plant tissue due to biosolids and manure amendments, and wastewater irrigation. *Environment International, 75*, 223–233. https://doi.org/10.1016/j.envint.2014.11.020.

Puckowski, A., Mioduszewska, K., Łukaszewicz, P., Borecka, M., Caban, M., Maszkowska, J., & Stepnowski, P. (2016). Bioaccumulation and analytics of pharmaceutical residues in the environment: A review. *Journal of Pharmaceutical and Biomedical Analysis, 127*, 232–255. https://doi.org/10.1016/j.jpba.2016.02.049.

Rand-Weaver, M., Margiotta-Casaluci, L., Patel, A., Panter, G. H., Owen, S. F., & Sumpter, J. P. (2013). The read-across hypothesis and environmental risk assessment of pharmaceuticals. *Environmental Science & Technology, 47*(20), 11384–11395. https://doi.org/10.1021/es402065a.

Svobodníková, L., Kummerová, M., Zezulka, Š., & Babula, P. (2019). Possible use of a Nicotiana tabacum 'Bright Yellow 2' cell suspension as a model to assess phytotoxicity of pharmaceuticals (diclofenac). *Ecotoxicology and Environmental Safety, 182*, 109369. https://doi.org/10.1016/j.ecoenv.2019.109369.

Tanoue, R., Sato, Y., Motoyama, M., Nakagawa, S., Shinohara, R., & Nomiyama, K. (2012). Plant uptake of pharmaceutical chemicals detected in recycled organic manure and reclaimed wastewater. *Journal of Agricultural and Food Chemistry, 60*(41), 10203–10211. https://doi.org/10.1021/jf303142t.

Tripathi, D. K., Tripathi, A., Singh, S., Singh, Y., Vishwakarma, K., Yadav, G., & Chauhan, D. K. (2017). Uptake, accumulation and toxicity of silver nanoparticle in autotrophic plants, and heterotrophic microbes: A concentric review. *Frontiers in Microbiology, 8*, 7. https://doi.org/10.3389/fmicb.2017.00007.

Wang, F. H., Qiao, M., Lv, Z. E., Guo, G. X., Jia, Y., Su, Y. H., & Zhu, Y. G. (2014). Impact of reclaimed water irrigation on antibiotic resistance in public parks, Beijing, China. *Environmental Pollution, 184*, 247–253. https://doi.org/10.1016/j.envpol.2013.08.038.

Wu, C., Spongberg, A. L., Witter, J. D., & Sridhar, B. M. (2012). Transfer of wastewater associated pharmaceuticals and personal care products to crop plants from biosolids treated soil. *Ecotoxicology and Environmental Safety, 85*, 104–109. https://doi.org/10.1016/j.ecoenv.2012.08.007.

Wu, Q., Li, Z., Hong, H., Li, R., & Jiang, W. T. (2013). Desorption of ciprofloxacin from clay mineral surfaces. *Water Research, 47*(1), 259–268. https://doi.org/10.1016/j.watres.2012.10.010.

Wu, X., Dodgen, L. K., Conkle, J. L., & Gan, J. (2015). Plant uptake of pharmaceutical and personal care products from recycled water and biosolids: A review. *Science of the Total Environment, 536*, 655–666. https://doi.org/10.1016/j.scitotenv.2015.07.129.

Yuan, Q., Zhang, D., Yu, P., Sun, R., Javed, H., Wu, G., & Alvarez, P. J. (2020). Selective adsorption and photocatalytic degradation of extracellular antibiotic resistance genes by molecularly-imprinted graphitic carbon nitride. *Environmental Science & Technology, 54*(7), 4621–4630. *https://doi.org/10.1021/acs.est.9b06926.*

Zhang, T., Li, N., Chen, G., Xu, J., Ouyang, G., & Zhu, F. (2021). Stress symptoms and plant hormone-modulated defense response induced by the uptake of carbamazepine and ibuprofen in Malabar spinach (*Basella alba* L.). *Science of the Total Environment, 793*, 148628. https://doi.org/10.1016/j.scitotenv.2021.148628.

Zhang, Y. L., Lin, S. S., Dai, C. M., Shi, L., & Zhou, X. F. (2014). Sorption—desorption and transport of trimethoprim and sulfonamide antibiotics in agricultural soil: Effect of soil type, dissolved organic matter, and pH. *Environmental Science and Pollution Research, 21*(9), 5827–5835.

11 Ecotoxicological and Risk Assessment of Pharmaceutical Chemicals in the Aquatic Environment

Wen-Jun Shi, Li Yao, and Jian-Liang Zhao

CONTENTS

11.1 INTRODUCTION

Pharmaceuticals are designed to treat diseases targeting certain organs. However, the uptake of pharmaceuticals into nontarget tissues from the environment may pose potential harm to organisms. Therefore, although the levels of pharmaceuticals in the aquatic environment are generally at trace levels, they still show ecotoxicological effects to nontarget aquatic organisms (Duarte et al. 2020). There are many classes of pharmaceuticals currently in use, so their toxicological effects to aquatic organisms also vary widely. Also, the same pharmaceutical often exhibits different toxicological effects to different aquatic organisms. For examples, steroid drugs may display endocrine-disrupting effects on aquatic vertebrates, mammals and humans at low dose, but usually have less endocrine-disrupting effects to aquatic invertebrates. Therefore, comprehensive knowledge of ecotoxicological effects of pharmaceuticals is the basis for the control of pharmaceuticals in the aquatic environment.

In this chapter, the ecotoxicological effects of common types of pharmaceuticals in fish, such as hormones, antibiotics, psychotropic chemicals, and NSAIDs, are summarized. The risk assessment method of pharmaceuticals and their ecological risks in aquatic environments are elaborated.

DOI: 10.1201/9781003361091-11

11.2 ECOTOXICOLOGICAL EFFECTS OF PHARMACEUTICALS

11.2.1 Steroid Hormones

Steroid hormones are a class of typical endocrine disruptors, and mainly interfere with the endocrine system of organisms. Estrogens are commonly used drugs in hormone replacement therapy and oral contraception. Due to the strong biological activity of estrogen, an increased number of studies has shown estrogenic effects on the growth, reproduction, immunity and metabolism in aquatic organisms at low concentrations in fish, such as zebrafish (*Danio rerio*), three-spined stickleback (*Gasterosteus aculeatus*), Japanese medaka (*Oryzias latipes*), fathead minnows (*Pimephales promelas*) and western mosquitofish (*Gambusia affinis affinis*) (Leet et al. 2011). Importantly, it has been reported that EE2 interferes with biological immune function by changing immune regulatory gene transfection and leukocyte activity in male rainbow trout (*Oncorhynchus mykiss*) (Massart et al. 2014).

Progesterone levonorgestrel (LNG), norethisterone (NET) and pregnenolone (NET) inhibit the spawning of the fathead minnow and Japanese medaka (Fent 2015; Runnalls et al. 2013; Zeilinger et al. 2009). In addition, 5 ng/L LNG, 306 ng/L progesterone (P4) and 500 ng/L dydrogesterone (DDG) all increase the frequency of atresia cells in the ovary of zebrafish and rainbow trout (*Salmo gairdneri* Richardson) for 21 days (Zhao et al. 2015; Zucchi et al. 2014; Shi et al., 2018). Oocyte atresia plays an important role in regulating the tissue balance in the ovary (Fent, 2015). The more atresia cells in the ovary indicate that the ovarian function is degenerating after progesterone exposure (Fent, 2015). In the testis, LNG at 100 ng/L promotes the sperm maturation and results in male secondary characteristics in female minnows for 14 days (Frankel et al. 2017; Runnalls et al. 2013). Importantly, 5 ng/L LNG and 500 ng/L DDG cause male bias in zebrafish (Shi et al. 2018, 2021).

TABLE 11.1
Main Endocrine-Disrupting Effects of Estrogens and Progestins

Pharmaceuticals	Organisms	Exposure time	Concentrations	Main effects	References
Estrogen	Zebrafish; three-spined stickleback; Japanese medaka; fathead minnow; frog	Acute; subchronic; chronic	ng/L to μg/L levels	Female bias; feminization; decrease egg production; elevated VTG levels; altered steroid levels; changed transcription of genes in HPG axis; influenced growth	Leet et al. 2011; Hahlbeck et al. 2004; Hagino 2001
Progestin	Zebrafish; three-spined stickleback; fathead minnow; western mosquitofish; roach	Acute; subchronic; chronic	ng/L to μg/L levels	Male bias; masculinization; decreased egg production; altered gonadal development; changed transcription of genes in HPG axis; influenced growth	Fent 2015 Hua et al. 2015; Zucchi et al. 2014; Zeilinger et al. 2009; Runnalls et al. 2013; Zhao et al. 2015

11.2.2 Antibiotics

Antibiotics display a certain impact on aquatic organisms. Antibiotics will inhibit the growth of algae, photosynthesis, and the growth and reproduction of water fleas. Oxolinic acid is commonly used in aquaculture, which can affect the reproductive ability of aquatic crustacean *Daphnia magna* at very low concentrations (Wollenberger et al. 2000). Zounkova et al. (2011) investigated the toxicity of oxytetracycline to six model organisms with EC50 ranging from 0.22 to 86 mg/L, of which *Pseudomonas putida* was the most sensitive organism (Zounkova et al. 2011). Kim et al. (2007) investigated the toxicity of six sulfonamides and trimethoprim against *Vibrio fischeri*, *Daphnia magna* and medaka, and found that *Daphnia magna* were the most sensitive (Kim et al. 2007).

In recent years, toxicological tests at the molecular level have also been carried out to study the mechanism of toxicity of antibiotics to aquatic organisms. Erythromycin, ciprofloxacin and sulfamethoxazole all had certain effects on the antioxidant system of carbon synthesis, ascorbic acid glutathione cycle, lutein cycle, antioxidant enzyme activity, thereby inhibiting the antioxidant effect of Pseudokirchneriella subcapitata (Nie et al. 2013). Antibiotics in the environment can also cause the problem of antibiotic resistance genes (ARGs). ARGs frequently occur in the Aquatic environments (Marti et al. 2014). Hence, compared with the various toxic effects of antibiotics on aquatic organisms, the spread of resistance genes is more concerning (He et al. 2016).

11.2.3 Psychotropic Chemicals

Selective serotonin reuptake inhibitors (SSRIs) and new psychoactive substances (NPSs) are typical emerging contaminants of interest. SSRIs can affect the development in embryos of fish (Cunha et al. 2016; Kalichak et al. 2016). Fluoxetine ranging from 0.52 to 276.63 μg/L not only increases the deformity rate, but also decreases the hatching time, survival rate, heart rate and body length in zebrafish embryos for 80 hours post fertilization (hpf) (Cunha et al. 2016; Kalichak et al. 2016). SSRIs affect the growth and development via influencing the 5-hydroxytryptamine system in fish (Sourbron et al. 2016). In addition, recent studies report that SSRIs strongly affect the movement, aggressive behavior, anxiety behavior and social behavior of mosquitofish and wild guppies (*Poecilia reticulata*) (Martin et al. 2019; Saaristo et al. 2017). However, the toxicological mechanism of SSRIs on behavior is still unclear. Furthermore, SSRIs at a range from 5.2 ng/L to 1104 ng/L also have endocrine-disrupting effects in male fathead minnows for 21 days (Schultz et al. 2011).

NPSs have a strong toxic effect on development and behavior (Félix et al. 2017; Gao et al. 2022). For instance, cannabis substances (THC and CBD) at mg/L level also significantly affect the hatchability of zebrafish embryos, even leading to death for 96 hpf or 120 hpf (Ahmed et al. 2018; Carty et al. 2018). The molecular polarity of NPSs is gradually reduced, and can easily cross the blood-brain barrier resulting in obvious excitation effect. A large number of evidence shows that long-term use or abuse of NPSs will cause serious damage in the nervous system (Robinson et al. 2019; Chen et al. 2016). Studies have shown that sensitivities of behavioral phenotypes are 10~1000 times more than acute lethal (Hellou 2011). Cannabis THC or CBD reduces the response to escape and sound stimulation in zebrafish larvae (Ahmed et al. 2018). It is worth noting that the behavioral changes in zebrafish are accompanied with the alteration of neurotransmitters (Ostroumov and Dani 2018). Until now, NPSs probably induce the neurotoxicity via dopamine pathways (Self and Nestler 1998). However, the specific mechanism of behavior changes needs further study.

11.2.4 Other Drugs

The mechanism of NSAIDs is mainly through the inhibition of prostaglandin (PG) cyclooxygenase, which prevents arachidonic acid transforming into PG, resulting in anti-inflammatory, analgesic and antipyretic effects (Vane and Botting 1998). At present, the effects of NSAIDs on fish mainly concentrate on the tissue, physiological and biochemical levels (Näslund et al. 2017). For example,

TABLE 11.2
Main Neurotoxicity Induced by Antidepressants and New Psychoactive Substance

Pharmaceutical	Organisms	Exposure time	Concentrations	Main effects	References
Antidepressants	Western mosquitofish; zebrafish; fathead minnows	Acute; subchronic; chronic	μg/L to mg/L	Influenced behavior; inhibited growth and development	Henry and Black 2008; Yang et al. 2021; Schultz et al. 2011
New psychoactive substance	Japanese medaka; zebrafish	Acute; subchronic; chronic	μg/L to mg/L	Altered swimming speeds; increased oxidative stress; changed behavior; embryo death; influenced heart rate; influenced hatching	Wang et al. 2020; Liao et al. 2018; Khor et al. 2011; Gao et al. 2022

320 μg/L diclofenac increased the renal hematopoietic tissue in three-spined stickleback for 28 days (Näslund et al. 2017). Ibuprofen also has certain effects on biological reproduction (Flippin et al. 2007; Han et al. 2010). For example, 10 μg/L ibuprofen reduces the spawning rate, but increases the number of eggs in Japanese medaka for 42 days (Flippin et al. 2007).

11.3 RISK ASSESSMENT

11.3.1 Risk Assessment Method of Pharmaceuticals in the Aquatic Environment

The ecological risk assessment is centered on the use of probabilistic distribution to describe exposure concentrations and toxicological responses of organisms to the chemicals of concern (Suter 2008). The processes include environmental exposure assessment, ecotoxicological effect assessment and ecological risk characterization. There are two ways to assess the environmental exposure, including the predicted environmental concentration (PEC) calculated by the modeling use data and the measured environmental concentration (MEC) obtained from the sampling and instrument analysis.

Before the ecotoxicological effect assessment, the ecological dataset of pharmaceuticals shall be established to summarize all collected ecotoxicity data, which include test species, test endpoint, duration, test effect, references, and Klimisch code (Caldwell et al. 2008). The common effects include LC50, EC50, NOEC and LOEC, etc. Among them, the tested organisms in the ecotoxicity data should be the biological species existing in the local aquatic environment as far as possible. Then, the derivation of predicted non-effect concentrations (PNECs) was necessary using the collected toxicity data. Table 11.3 shows the European Technical Guide for Risk Assessment adopting the assessment factor (AF) method to derive the PNEC (EC 2003).

Risk quotients (RQ) are the most commonly used to describe the risk level of chemicals to aquatic organisms (EC 2003). The RQ values are calculated as the ratio of exposure concentrations to PNECs using worst-case assumptions (EMEA 2006). The risk rating criteria are: $0.01 < RQ < 0.1$, low risk; $0.1 \leq RQ < 1$, medium risk; $RQ \geq 1$, high risk.

TABLE 11.3
Values of Assessment Factor (AF) for Derivation of PNEC

No.	Available ecotoxicity data	Assessment factor
1	The acute $L(E)C_{50}$ data of at least one species in three trophic levels (*Daphnia*, fish and algae)	1000
2	The chronic NOEC data of one species (*Daphnia* or fish)	100
3	The chronic NOEC data of two species representing two trophic levels (generally any two species of fish, *Daphnia* and algae)	50
4	The chronic NOEC data for three species representing at least three trophic levels (generally fish, *Daphnia* and algae)	10
5	The species sensitivity distribution curve (SSD) was used for chronic NOEC data of three branches and eight families	1~5
6	Field toxicity data or ecosystem simulation	depends on circumstances

Note: $L(E)C_{50}$: the semi-lethal concentration; NOEC: the maximum observed no effect concentration.

11.3.2 Risks of Pharmaceuticals in the Aquatic Environment

Based on the risk assessment strategy, many pharmaceuticals displayed low to high ecological risks in the aquatic environment. For hormones, Goeury et al. (2019) assessed the risks of priority endocrine disruptors in surface waters of Quebec (Canada), the results indicating that most samples were assigned the minimal risk class (RQ < 0.01), but β-estradiol and ethinylestradiol showed exceedances to the PNEC with the maximum RQ values of 5.6 and 5.5, respectively. The findings are in line with field observations in the Mille Iles River, where feminization of wild and caged invertebrates was previously reported (Bouchard et al. 2009; Gagné et al. 2011). Except for progesterone, most steroids (e.g., estrone, 17β-estradiol, estriol, androsterone, epiandrosterone, progesterone, cortisol and cortisone) have median RQs below 1 in surface water and sediment in China, indicating low to moderate risks for aquatic and sediment-dwelling livings. Meanwhile, the RQs of steroids span over a wide range and some of them are very high. Often these high risks (or concentrations) are observed in small tributaries with low dilution factors and proximity to emission sources (Zhong et al. 2021).

Antibiotics displayed high ecological effects to aquatic organisms in the environment. Near the disposal point of the antibiotics into wastewater, the RQ values (>1) show high risk to the environment (Zhang et al. 2017). Not only the aquaculture and wastewater are contaminated but the surface water is also contaminated by antibiotics. Recent studies on risk assessment in China show high risk to algae from norfloxacin, tetracycline, sulfonamide, ciprofloxacin and ofloxacin (Li et al. 2017). Studies on tap water revealed that it is also not safe from the antibiotics; about 17 types of antibiotics were present in tap water according to research in China. Out of 17, only four showed high levels (thiamphenicol, dimetridazole, sulfamethazine and clarithromycin (Leung et al. 2013). For eight selected antibiotics (e.g., trimethoprim, ciprofloxacin, ofloxacin, erythromycin, amoxicillin, oxytetracycline, tetracycline, sulfamethoxazole) in various types of water and with organisms, except for trimethoprim, all studied antibiotics showed a high risk to the aquatic environment in at least one of the presented Environmental Risk Assessments (ERAs). All RQs > 1 were calculated for algae or cyanobacteria in both surface water and effluents, and in one case for the bacterium *P. putida* in surface water. In several cases, values of RQ were greater than 10 (Ferrari et al. 2004; Gros et al. 2010; Guo et al. 2015). Eugenia Valdes et al. (2021) assessed the risk of antibiotics to the urban river water of Suquía River, Argentina. The results indicated that norfloxacin, ofloxacin, clindamycin and metronidazole had moderate risk (RQs 0.14–0.41), while ciprofloxacin, cephalexin and clarithromycin presented high risk (RQ > 1), either for resistance selection (ciprofloxacin) or the ecosystem (cephalexin and clarithromycin) (Eugenia Valdes et al. 2021). PNECs based on the bacteria resistance were also proposed by (Bengtsson-Palme and Larsson 2016). Ciprofloxacin was found to have the highest RQ (17.3) in Mijares River, Spain, where the concentrations of ciprofloxacin, azithromycin, norfloxacin, trimethoprim and clarithromycin were exceeding of the resistance PNECs (Fonseca et al. 2020).

The existing NSAIDs (aspirin, ibuprofen, paracetamol, ranitidine, diclofenac and carbamazepine) residues in the Yamuna pose an insignificant environmental risk (RQ < 1) to the various organisms exposed to the river Yamuna. The maximum RQ was for ibuprofen in the algal species (RQ: 0.324), while all other NSAIDs were having RQ < 0.1 (Mutiyar et al. 2018). Diclofenac, naproxen and ketoprofen showed low ecotoxicity risk to the aquatic organisms in northwestern France (Minguez et al. 2016). During the process of agricultural fields' reuse of sewage sludge, the organisms (e.g., algae, fish and plants) can also be impacted by the presence of limed sludge due to soil-to-water transfer by leaching, as shown by the RQ values calculated for diclofenac (RQ: 0.15 to 0.85) and carbamazepine (RQ: 0.18–0.39) of between 0.1 and 1 (Bastos et al. 2020). The ibuprofen affected the development of sea urchin (*L. variegatus*) and bivalve mussel (*P. perna*) in Santos Bay's sediment, with RQ values of 326.6 and 32.4, respectively (Pusceddu et al. 2018). On the northern Antarctic Peninsula region, ibuprofen, diclofenac and acetaminophen were identified as the substances posing the highest risk to the Antarctic ecosystem, with the RQ values far in excess of 10 at several sampling points (González-Alonso et al. 2017).

Many psychoactive drugs have been detected in the aquatic environment, both legal (Fick et al. 2017; Schlüsener et al. 2015) and illegal. Most psychoactive drugs are hydrophobic because to have their desired effects they need to cross the blood-brain barrier to act in the brain; thus, once discharged into the aquatic environment they will be readily taken up by aquatic organisms. Various psychoactive drugs have been found in wild fish (Arnnok et al. 2017). The RQ values of codeine, morphine, methamphetamine, cocaine and its metabolite benzoylecgonine in the Rhine and Meuse rivers in the Netherlands were between 0.0002 and 0.38, indicating that the environmental risk was relatively small (van der Aa et al. 2013). The RQs of morphine and tetrahydro-cannabinolic acid fluctuated within a range of 1 and 10 in the rivers passing by the densely populated Madrid, Spain, which may require further investigation (Mendoza et al. 2014). Cumulative toxicity was found in the surface water samples from four basins in Spain, most of which were associated with 2-ethylidene-1,5-dimethyl-3,3-diphenylpyrrolidine (RQs > 1), followed by compounds such as methamphetamine, and 3,4-methylenedioxymethamphetamine (Mastroianni et al. 2016). Fernandez-Rubio et al. (2019) found venlafaxine in 8 out of 22 coastal water samples and wastewater treatment plant influent samples in Spain, and 6 of them had venlafaxine RQs exceeding 10 (Fernandez-Rubio et al. 2019). Except for venlafaxine, the RQs of sertraline were also estimated to be greater than 1. In the Mediterranean rivers near Llobregat, Spain, the RQs of all psychoactive substances (benzoylecgonine, cocaine, morphine, heroin, ephedrine, amphetamine, methamphetamine, 3, 4-methylenedioxymethamphetamine, lysergic acid diethylamide, and tetrahydrocannabinol) were less than 0.015 (López-Serna et al. 2012). Alygizakis et al. (2016) evaluated the risks of the maximum concentrations of the studied substances (e.g., caffeine, tramadol, 2-ethylidene-1,5-dimethyl-3, 3-diphenylpyrrolidine, 3, 4-methylenedioxymethamphetamine, diazepam and ephedrine) at different depths in the Mediterranean Sea on fish, *Daphnia* magna and algae, and only caffeine may be toxic to algae (RQ = 5.2) (Alygizakis et al. 2016). In general, most ecological risk assessments introduced the risk evaluation of psychoactive substances to the aquatic environment. However, the risk assessments of sediments or other media were rarely mentioned, which is closely related to the lack of relevant literature and the lack of hazard data for sediment-dwelling organisms. The risk assessment of multiple pollutants is also of potential concern since the simple addition of RQ values may not reflect the true risk level (Yadav et al. 2017; Zuccato et al. 2008). Most illicit drugs (e.g., amphetamines and opiates) are chiral compounds and chirality of these compounds may be a major parameter determining their potency and possible toxicity (Kasprzyk-Hordern et al. 2010). Besides, the environmental risks of emerging pollutants during the treatment process of psychoactive substances should be a cause with concern (Jin et al. 2022).

There also some other pharmaceuticals in the environment, such as cardiovascular agents and antiviral drugs. In northwestern France, fenofibrate was estimated at medium ecotoxicity risk with RQ of 0.76, while gemfibrozil, bezafibrate, fenofibric acid, propranolol, metoprolol, sotalol, acebutolol, bisoprolol, atenolol and pravastatin showed low ecotoxicity risk with RQs lower than 0.01 (Minguez et al. 2016). Antiviral drugs are among the most common and important classes of pharmaceuticals to treat viral infections; however their continuous emission and persistence in the receiving environment has attracted increasing attention about their potential ecological risks. The occurrence of antiviral drugs in the environment has shown high bioactive and adverse effects on organisms (Jain et al. 2013; Nannou et al. 2020). A recent study indicated that zidovudine, ritonavir, lopinavir and telbivudine showed high risk (RQ > 1) to aquatic ecosystems downstream of wastewater treatment plants (Yao et al. 2021). More importantly, antiviral drugs also produce virus resistance, which is similar to the bacterial resistance induced by antibiotics (Gillman et al. 2015). Thus, the ecosystem alteration and evolution of antiviral resistance strains in animals and humans through the unconscious exposure to trace contaminated water have aroused much concern (Gillman et al. 2015; Nannou et al. 2020; Singer et al. 2007).

11.3.3 Control Measures for Pharmaceutical Risks

The advanced treatment of effluents has been investigated using (photochemical) oxidation processes, filtration, application of powdered charcoal and constructed wetlands. It has been found that each type of advanced effluent treatment has its specific limitations (Patel et al. 2019). Therefore, it is very necessary to comprehensively consider the risk of drugs and the cost of advanced treatment technology for pharmaceutical removal. Other strategies need to be carefully considered (Kümmerer 2010). In other words, environmental protection has to include the shareholders, stakeholders and people using the compounds (i.e., patients, doctors, nurses and pharmacists) when seeking workable solutions. Also, green chemistry seems to be the promising strategy in long sustainability development.

11.4 SUMMARY

1. Pharmaceuticals displayed different ecological effects on aquatic organisms, including algae, duckweed, crustaceans, fish and benthos. The acute effect values and chronic effect values can range from low ng/L to high mg/L.
2. Specific toxicological effects were dependent on pharmaceutical class and species. Hormones showed endocrine-disrupting effects to vertebrates, and antibiotics could accelerate the spread of antibiotic-resistant genes. Psychotropic drugs obviously displayed neurotoxicity to fish and other organisms.
3. Some of the pharmaceuticals were assessed as high risks to aquatic organisms due to their extensive existence in the aquatic environment. Control measures should be considered in order to reduce the risk of pharmaceuticals to aquatic organisms.

REFERENCES

Ahmed, KT, Amin, MR, Shah, P and Ali, DW, 2018. Motor neuron development in zebrafish is altered by brief (5-hr) exposures to THC (Δ 9-tetrahydrocannabinol) or CBD (cannabidiol) during gastrulation. *Scientific Reports* 8(1), 1–14.

Alygizakis, NA, Gago-Ferrero, P, Borova, VL, Pavlidou, A, Hatzianestis, I and Thomaidis, NS, 2016. Occurrence and spatial distribution of 158 pharmaceuticals, drugs of abuse and related metabolites in offshore seawater. *Science of the Total Environment* 541, 1097–1105.

Arnnok, P, Singh, RR, Burakham, R, Pérez-Fuentetaja, A and Aga, DS, 2017. Selective uptake and bioaccumulation of antidepressants in fish from effluent-impacted Niagara River. *Environmental Science & Technology* 51(18), 10652–10662.

Bastos, MC, Soubrand, M, Le Guet, T, Le Floch, E, Joussein, E, Baudu, M and Casellas, M, 2020. Occurrence, fate and environmental risk assessment of pharmaceutical compounds in soils amended with organic wastes. *Geoderma* 375.

Bengtsson-Palme, J and Larsson, DGJ, 2016. Concentrations of antibiotics predicted to select for resistant bacteria: Proposed limits for environmental regulation. *Environment International* 86, 140–149.

Bouchard, B, Gagné, F, Fortier, M and Fournier, M, 2009. An in-situ study of the impacts of urban wastewater on the immune and reproductive systems of the freshwater mussel Elliptio complanata. *Comparative Biochemistry and Physiology Part C: Toxicology & Pharmacology* 150(2), 132–140.

Caldwell, DJ, Mastrocco, F, Hutchinson, TH, Lange, R, Heijerick, D, Janssen, C, Anderson, PD and Sumpter, JP, 2008. Derivation of an aquatic predicted no-effect concentration for the synthetic hormone, 17 alpha-ethinyl estradiol. *Environmental Science & Technology* 42(19), 7046–7054.

Carty, DR, Thornton, C, Gledhill, JH and Willett, KL, 2018. Developmental effects of cannabidiol and Δ9-tetrahydrocannabinol in zebrafish. *Toxicological Sciences* 162(1), 137–145.

Chen, Y-F, Peng, J, Fang, M, Liu, Y, Nie, L-H, Mo, Z-X and Zhu, L-L, 2016. Effect of rhynchophylline on behaviors of methamphetamine-dependent zebrafish and the mechanism. *Nan Fang yi ke da xue xue bao* (*Journal of Southern Medical University*) 36(11), 1541–1545.

Cunha, V, Rodrigues, P, Santos, M, Moradas-Ferreira, P and Ferreira, M, 2016. *Danio rerio* embryos on Prozac: Effects on the detoxification mechanism and embryo development. *Aquatic Toxicology* 178, 182–189.

Duarte, IA, Reis-Santos, P, Novais, SC, Rato, LD, Lemos, MFL, Freitas, A, Pouca, ASV, Barbosa, J, Cabral, HN and Fonseca, VF, 2020. Depressed, hypertense and sore: Long-term effects of fluoxetine, propranolol and diclofenac exposure in a top predator fish. *Science of the Total Environment* 712, 136564.

EC, 2003. Technical Guidance Document in support of Commission Directive 93/67/EEC on risk assessment for new notified substances and Commission Regulation (EC) No 1488/94 on risk assessment for existing substances Part II, Office for official publications of the European communities, Italy.

EMEA, 2006. *Guideline on the environmental risk assessment of medicinal products for human use* (Ref EMEA/CRMP/SWP/4447/00). European Medicines Agency, London.

Eugenia Valdes, M, Santos, LHMLM, Carolina Rodriguez Castro, M, Giorgi, A, Barcelo, D, Rodriguez-Mozaz, S and Valeria Ame, M, 2021. Distribution of antibiotics in water, sediments and biofilm in an urban river (Cordoba, Argentina, LA). *Environmental Pollution* 269.

Félix, LM, Serafim, C, Martins, MJ, Valentim, AM, Antunes, LM, Matos, M and Coimbra, AM, 2017. Morphological and behavioral responses of zebrafish after 24 h of ketamine embryonic exposure. *Toxicology and Applied Pharmacology* 321, 27–36.

Fent, K, 2015. Progestins as endocrine disrupters in aquatic ecosystems: Concentrations, effects and risk assessment. *Environment International*, 84, 115–130.

Fernandez-Rubio, J, Luis Rodriguez-Gil, J, Postigo, C, Mastroianni, N, Miren, LDA, Barcelo, D and Valcarcel, Y, 2019. Psychoactive pharmaceuticals and illicit drugs in coastal waters of North-Western Spain: Environmental exposure and risk assessment. *Chemosphere* 224, 379–389.

Ferrari, B, Mons, R, Vollat, B, Fraysse, B, Paxeus, N, Lo Giudice, R, Pollio, A and Garric, J, 2004. Environmental risk assessment of six human pharmaceuticals: Are the current environmental risk assessment procedures sufficient for the protection of the aquatic environment? *Environmental Toxicology and Chemistry* 23(5), 1344–1354.

Fick, J, Brodin, T, Heynen, M, Klaminder, J, Jonsson, M, Grabicova, K, Randak, T, Grabic, R, Kodes, V, Slobodnik, J, Sweetman, A, Earnshaw, M, Barra Caracciolo, A, Lettieri, T and Loos, R, 2017. Screening of benzodiazepines in thirty European rivers. *Chemosphere* 176, 324–332.

Flippin, JL, Huggett, D and Foran, CM, 2007. Changes in the timing of reproduction following chronic exposure to ibuprofen in Japanese medaka, Oryzias latipes. *Aquatic Toxicology* 81(1), 73–78.

Fonseca, E, Hernández, F, Ibáñez, M, Rico, A, Pitarch, E and Bijlsma, L, 2020. Occurrence and ecological risks of pharmaceuticals in a Mediterranean river in Eastern Spain. *Environment International* 144, 106004.

Frankel, T, Yonkos, L and Frankel, J, 2017. Exposure effects of levonorgestrel on oogenesis in the fathead minnow (*Pimephales promelas*). *Environmental Toxicology and Chemistry* 36(12), 3299–3304.

Gagné, F, Bouchard, B, André, C, Farcy, E and Fournier, M, 2011. Evidence of feminization in wild Elliptio complanata mussels in the receiving waters downstream of a municipal effluent outfall. *Comparative Biochemistry and Physiology Part C: Toxicology & Pharmacology* 153(1), 99–106.

Gao, S and Yang, F, 2022. Behavioral changes and neurochemical responses in Chinese rare minnow exposed to four psychoactive substances. *Science of the Total Environment*, 808, 152100.

Gillman, A, Nykvist, M, Muradrasoli, S, Soderstrom, H, Wille, M, Daggfeldt, A, Brojer, C, Waldenstrom, J, Olsen, B and Jarhult, JD, 2015. Influenza A(H7N9) virus acquires resistance-related neuraminidase I222T substitution when infected mallards are exposed to low levels of oseltamivir in water. *Antimicrobial Agents and Chemotherapy* 59(9), 5196–5202.

Goeury, K, Duy, SV, Munoz, G, Prevost, M and Sauve, S, 2019. Data supporting the optimization of liquid chromatography tandem mass spectrometry conditions to analyze EPA-priority hormones and bisphenol A in water samples. *Data in Brief* 24.

González-Alonso, S, Merino, LM, Esteban, S, López de Alda, M, Barceló, D, Durán, JJ, López-Martínez, J, Aceña, J, Pérez, S, Mastroianni, N, Silva, A, Catalá, M and Valcárcel, Y, 2017. Occurrence of pharmaceutical, recreational and psychotropic drug residues in surface water on the northern Antarctic Peninsula region. *Environmental Pollution* 229, 241–254.

Gros, M, Petrovic, M, Ginebreda, A and Barcelo, D, 2010. Removal of pharmaceuticals during wastewater treatment and environmental risk assessment using hazard indexes. *Environment International* 36(1), 15–26.

Guo, J, Boxall, A and Selby, K, 2015. Do Pharmaceuticals pose a threat to primary producers? *Critical Reviews in Environmental Science & Technology* 45(23), 2565–2610.

Hagino, S, Kagoshima, M and Asida, S, 2001. Effects of ethinylestradiol, diethylstilbestrol, 4-t-pentylphenol, 17β-estradiol, methyltestosterone and flutamide on sex reversal in S-rR strain medaka (Oryzias latipes). *Environmental Science & Technology* 8(1), 75–87.

Hahlbeck, E, Griffiths, R and Bengtsson, B-E, 2004. The juvenile three-spined stickleback (Gasterosteus aculeatus L.) as a model organism for endocrine disruption: I. Sexual differentiation. *Aquatic Toxicology* 70(4), 287–310.

Han, S, Choi, K, Kim, J, Ji, K, Kim, S, Ahn, B, Yun, J, Choi, K, Khim, JS and Zhang, X, 2010. Endocrine disruption and consequences of chronic exposure to ibuprofen in Japanese medaka (*Oryzias latipes*) and freshwater cladocerans *Daphnia magna* and *Moina macrocopa*. *Aquatic Toxicology* 98(3), 256–264.

He, LY, Ying, GG, Liu, YS, Su, HC, Chen, J, Liu, SS and Zhao, JL, 2016. Discharge of swine wastes risks water quality and food safety: Antibiotics and antibiotic resistance genes from swine sources to the receiving environments. *Environment International* 92–93, 210–219.

Hellou, J, 2011. Behavioural ecotoxicology, an "early warning" signal to assess environmental quality. *Environmental Science and Pollution Research* 18(1), 1–11.

Henry, TB and Black, MC, 2008. Acute and chronic toxicity of fluoxetine (selective serotonin reuptake inhibitor) in western mosquitofish. *Archives of Environmental Contamination and Toxicology*, 54(2), 325–330.

Hua, J, Han, J, Guo, Y and Zhou, B, 2015. The progestin levonorgestrel affects sex differentiation in zebrafish at environmentally relevant concentrations. *Aquatic Toxicology* 166, 1–9.

Jain, S, Kumar, P, Vyas, RK, Pandit, P and Dalai, AK, 2013. Occurrence and removal of antiviral drugs in environment: A review. *Water Air and Soil Pollution* 224(2).

Jin, H, Yang, D, Wu, P and Zhao, M, 2022. Environmental occurrence and ecological risks of psychoactive substances. *Environment International* 158, 106970.

Kalichak, F, Idalencio, R, Rosa, JGS, de Oliveira, TA, Koakoski, G, Gusso, D, de Abreu, MS, Giacomini, ACV, Barcellos, HH and Fagundes, M, 2016. Waterborne psychoactive drugs impair the initial development of Zebrafish. *Environmental Toxicology and Pharmacology* 41, 89–94.

Kasprzyk-Hordern, B, Kondakal, V and Baker, DR, 2010. Enantiomeric analysis of drugs of abuse in wastewater by chiral liquid chromatography coupled with tandem mass spectrometry. *Journal of Chromatography A* 1217(27), 4575–4586.

Khor, BS, Amar Jamil, MF, Adenan, MI and Chong Shu-Chien, A, 2011. Mitragynine attenuates withdrawal syndrome in morphine-withdrawn zebrafish. *PLoS One*, 6(12), e28340.

Kim, Y, Choi, K, Jung, J, Park, S, Kim, P-G and Park, J, 2007. Aquatic toxicity of acetaminophen, carbamazepine, cimetidine, diltiazem and six major sulfonamides, and their potential ecological risks in Korea. *Environment International* 33(3), 370–375.

Kümmerer, K, 2010. Pharmaceuticals in the environment. *Annual Review of Environment and Resources* 35, 57–75.

Leet, JK, Gall, HE and Sepúlveda, MS, 2011. A review of studies on androgen and estrogen exposure in fish early life stages: Effects on gene and hormonal control of sexual differentiation. *Journal of Applied Toxicology* 31(5), 379–398.

Leung, HW, Jin, L, Wei, S, Tsui, MMP, Zhou, B, Jiao, L, Cheung, PC, Chun, YK, Murphy, MB and Lam, PKS, 2013. Pharmaceuticals in tap water: Human health risk assessment and proposed monitoring framework in China. *Environmental Health Perspectives* 121(7), 839–846.

Li, Q, Gao, J, Zhang, Q, Liang, L and Tao, H, 2017. Distribution and risk assessment of antibiotics in a typical river in North China Plain. *Bulletin of Environmental Contamination & Toxicology* 98(4), 478–483.

Liao, PH, Yang, WK, Yang, CH, Lin, CH, Hwang, CC and Chen, PJ, 2018. Illicit drug ketamine induces adverse effects from behavioral alterations and oxidative stress to p53-regulated apoptosis in medaka fish under environmentally relevant exposures. *Environmental Pollution* 237, 1062–1071.

López-Serna, R, Postigo, C, Blanco, J, Pérez, S, Ginebreda, A, de Alda, ML, Petrović, M, Munné, A and Barceló, D, 2012. Assessing the effects of tertiary treated wastewater reuse on the presence emerging contaminants in a Mediterranean river (Llobregat, NE Spain). *Environmental Science and Pollution Research* 19(4), 1000–1012.

Marti, E, Variatza, E and Balcazar, JL, 2014. The role of aquatic ecosystems as reservoirs of antibiotic resistance. *Trends in Microbiology* 22(1), 36–41.

Martin, JM, Bertram, MG, Saaristo, M, Fursdon, JB, Hannington, SL, Brooks, BW, Burket, SR, Mole, RA, Deal, ND and Wong, BB, 2019. Antidepressants in surface waters: Fluoxetine influences mosquitofish anxiety-related behavior at environmentally relevant levels. *Environmental Science & Technology* 53(10), 6035–6043.

Massart, S, Milla, S and Kestemont, P, 2014. Expression of gene, protein and immunohistochemical localization of the estrogen receptor isoform ERα1 in male rainbow trout lymphoid organs; indication of the role of estrogens in the regulation of immune mechanisms. *Comparative Biochemistry and Physiology Part B: Biochemistry and Molecular Biology* 174, 53–61.

Mastroianni, N, Bleda, MJ, López de Alda, M and Barceló, D, 2016. Occurrence of drugs of abuse in surface water from four Spanish river basins: Spatial and temporal variations and environmental risk assessment. *Journal of Hazardous Materials* 316, 134–142.

Mendoza, A, Rodríguez-Gil, JL, González-Alonso, S, Mastroianni, N, López de Alda, M, Barceló, D and Valcárcel, Y, 2014. Drugs of abuse and benzodiazepines in the Madrid Region (Central Spain): Seasonal variation in river waters, occurrence in tap water and potential environmental and human risk. *Environment International* 70, 76–87.

Minguez, L, Pedelucq, J, Farcy, E, Ballandonne, C, Budzinski, H and Halm-Lemeille, M-P, 2016. Toxicities of 48 pharmaceuticals and their freshwater and marine environmental assessment in northwestern France. *Environmental Science and Pollution Research* 23(6), 4992–5001.

Mutiyar, PK, Gupta, SK and Mittal, AK, 2018. Fate of pharmaceutical active compounds (PhACs) from River Yamuna, India: An ecotoxicological risk assessment approach. *Ecotoxicology and Environmental Safety* 150, 297–304.

Nannou, C, Ofrydopoulou, A, Evgenidou, E, Heath, D, Heath, E and Lambropoulou, D, 2020. Antiviral drugs in aquatic environment and wastewater treatment plants: A review on occurrence, fate, removal and ecotoxicity. *Science of the Total Environment* 699.

Näslund, J, Fick, J, Asker, N, Ekman, E, Larsson, DJ and Norrgren, L, 2017. Diclofenac affects kidney histology in the three-spined stickleback (*Gasterosteus aculeatus*) at low μg/L concentrations. *Aquatic Toxicology* 189, 87–96.

Nie, X-P, Liu, B-Y, Yu, H-J, Liu, W-Q and Yang, Y-F, 2013. Toxic effects of erythromycin, ciprofloxacin and sulfamethoxazole exposure to the antioxidant system in Pseudokirchneriella subcapitata. *Environmental Pollution* 172, 23–32.

Ostroumov, A and Dani, JA, 2018. Inhibitory plasticity of mesocorticolimbic circuits in addiction and mental illness. *Trends in Neurosciences* 41(12), 898–910.

Patel, M, Kumar, R, Kishor, K, Mlsna, T, Pittman, CU and Mohan, D, 2019. Pharmaceuticals of emerging concern in aquatic systems: Chemistry, occurrence, effects, and removal methods. *Chemical Reviews* 119(6), 3510–3673.

Pusceddu, FH, Choueri, RB, Pereira, CDS, Cortez, FS, Santos, DRA, Moreno, BB, Santos, AR, Rogero, JR and Cesar, A, 2018. Environmental risk assessment of triclosan and ibuprofen in marine sediments using individual and sub-individual endpoints. *Environmental Pollution* 232, 274–283.

Robinson, B, Gu, Q, Ali, SF, Dumas, M and Kanungo, J, 2019. Ketamine-induced attenuation of reactive oxygen species in zebrafish is prevented by acetyl l-carnitine in vivo. *Neuroscience Letters* 706, 36–42.

Runnalls, TJ, Beresford, N, Losty, E, Scott, AP and Sumpter, JP, 2013. Several synthetic progestins with different potencies adversely affect reproduction of fish. *Environmental Science & Technology* 47(4), 2077–2084.

Saaristo, M, McLennan, A, Johnstone, CP, Clarke, BO and Wong, BB, 2017. Impacts of the antidepressant fluoxetine on the anti-predator behaviours of wild guppies (Poecilia reticulata). *Aquatic Toxicology* 183, 38–45.

Schlüsener, MP, Hardenbicker, P, Nilson, E, Schulz, M, Viergutz, C and Ternes, TA, 2015. Occurrence of venlafaxine, other antidepressants and selected metabolites in the Rhine catchment in the face of climate change. *Environmental Pollution* 196, 247–256.

Schultz, MM, Painter, MM, Bartell, SE, Logue, A, Furlong, ET, Werner, SL and Schoenfuss, HL, 2011. Selective uptake and biological consequences of environmentally relevant antidepressant pharmaceutical exposures on male fathead minnows. *Aquatic Toxicology* 104(1–2), 38–47.

Self, DW and Nestler, EJ, 1998. Relapse to drug-seeking: Neural and molecular mechanisms. *Drug and Alcohol Dependence* 51, 49–60.

Shi, W-J, Jiang, Y-X, Huang, G-Y, Zhao, J-L, Zhang, J-N, Liu, Y-S, Xie, L-T and Ying, G-G, 2018. Dydrogesterone causes male bias and accelerates sperm maturation in zebrafish (*Danio rerio*). *Environmental Science & Technology* 52(15), 8903–8911.

Shi, W-J, Ma, D-D, Fang, G-Z, Zhang, J-G, Huang, G-Y, Xie, L, Chen, H-X, Hou, L-P and Ying, G-G, 2021. Levonorgestrel and dydrogesterone affect sex determination via different pathways in zebrafish. *Aquatic Toxicology* 240, 105972.

Singer, AC, Nunn, MA, Gould, EA and Johnson, AC, 2007. Potential risks associated with the proposed widespread use of Tamiflu. *Environmental Health Perspectives* 115(1), 102–106.

Sourbron, J, Schneider, H, Kecskés, AI, Liu, Y, Buening, EM, Lagae, L, Smolders, I and de Witte, P, 2016. Serotonergic modulation as effective treatment for Dravet syndrome in a zebrafish mutant model. *ACS Chemical Neuroscience* 7(5), 588–598.

Suter, GW, III, 2008. Ecological risk assessment in the United States Environmental Protection Agency: A historical overview. *Integrated Environmental Assessment and Management* 4(3), 285–289.

van der Aa, M, Bijlsma, L, Emke, E, Dijkman, E, van Nuijs, ALN, van de Ven, B, Hernández, F, Versteegh, A and de Voogt, P, 2013. Risk assessment for drugs of abuse in the Dutch watercycle. *Water Research* 47(5), 1848–1857.

Vane, JR and Botting, RM, 1998. Anti-inflammatory drugs and their mechanism of action. *Inflammation Research* 47(2), 78–87.

Wang, Z, Gao, S, Dai, Q, Zhao, M and Yang, F, 2020. Occurrence and risk assessment of psychoactive substances in tap water from China. *Environmental Pollution* 261, 114163.

Wollenberger, L, Halling-Sørensen, B and Kusk, KO, 2000. Acute and chronic toxicity of veterinary antibiotics to Daphnia magna. *Chemosphere* 40(7), 723–730.

Yadav, MK, Short, MD, Aryal, R, Gerber, C, van den Akker, B and Saint, CP, 2017. Occurrence of illicit drugs in water and wastewater and their removal during wastewater treatment. *Water Research* 124, 713–727.

Yang, HT, Li, X, Chen, HH, Mao, ZG, Gu, XH and Liang, XF, 2021. Research progress on toxic effects of typical SSRIs antidepressants on fish. *Asian Journal of Ecotoxicology* 16(3), 28–39.

Yao, L, Chen, Z-Y, Dou, W-Y, Yao, Z-K, Duan, X-C, Chen, Z-F, Zhang, L-J, Nong, Y-J, Zhao, J-L and Ying, G-G, 2021. Occurrence, removal and mass loads of antiviral drugs in seven wastewater treatment plants with various treatment processes. *Water Research* 207, 117803.

Zeilinger, J, Steger-Hartmann, T, Maser, E, Goller, S, Vonk, R and Länge, R, 2009. Effects of synthetic gestagens on fish reproduction. *Environmental Toxicology and Chemistry* 28(12), 2663–2670.

Zhang, X, Zhao, H, Du, J, Qu, Y, Shen, C, Tan, F, Chen, J and Quan, X, 2017. Occurrence, removal, and risk assessment of antibiotics in 12 wastewater treatment plants from Dalian, China. *Environmental Science and Pollution Research* 24(19), 16478–16487.

Zhao, Y, Castiglioni, S and Fent, K, 2015. Synthetic progestins medroxyprogesterone acetate and dydrogesterone and their binary mixtures adversely affect reproduction and lead to histological and transcriptional alterations in zebrafish (*Danio rerio*). *Environmental Science & Technology* 49(7), 4636–4645.

Zhong, R, Zou, H, Gao, J, Wang, T, Bu, Q, Wang, Z-L, Hu, M and Wang, Z, 2021. A critical review on the distribution and ecological risk assessment of steroid hormones in the environment in China. *Science of the Total Environment* 786.

Zounkova, R, Kliemesova, Z, Nepejchalova, L, Hilscherova, K and Blaha, L, 2011. Complex evaluation of ecotoxicity and genotoxicity of antimicrobials oxytetracycline and flumequine used in aquaculture. *Environmental Toxicology and Chemistry* 30(5), 1184–1189.

Zuccato, E, Castiglioni, S, Bagnati, R, Chiabrando, C, Grassi, P and Fanelli, R, 2008. Illicit drugs, a novel group of environmental contaminants. *Water Research* 42(4), 961–968.

Zucchi, S, Mirbahai, L, Castiglioni, S and Fent, K, 2014. Transcriptional and physiological responses induced by binary mixtures of drospirenone and progesterone in zebrafish (*Danio rerio*). *Environmental Science & Technology* 48(6), 3523–3531.

12 Alternative Approaches for Safety and Toxicity Assessment of Personal Care Products

Carmine Merola, Monia Perugini, and Giulia Caioni

CONTENTS

12.1 INTRODUCTION

Alternatives to animal testing have been implemented to override the drawbacks associated with the classical animal models and to promote the 3Rs strategy which is based on three fundamental concepts: Replacement, Refinement and Reduction. In 1959, the 3Rs principle was first discussed in the book *The Principles of Humane Experimental Technique*, which was published by two English researchers, namely Russell (1925–2006) and Burch (1926–1996). The basic recommendation of this strategy is to reduce the number of animals used (reduction), minimize the stress to such "low" number of experimental animals in medical and biological research (refinement) and to replace, as much as possible, the use of experimental animals by encouraging the use of *in vitro* models and *in silico* computational models, and when the use of *in vivo* model is strictly required, utilizing animals which have a lower potential for pain perception.

The industry of pharmaceuticals and personal care products (PPCPs) is a science-driven sector, characterized by continuous research aimed at selecting appropriate ingredients, additives and preservatives for the realization of high-quality goods. However, safety assessment represents a critical component of this mission, contributing to create the awareness of hazards. These kinds of evaluations require a comprehensive study, which begins with the identification of the risks related to the exposure to every ingredient used. Safety standards and the identification of safety levels may vary in different countries, although a common ground in the development of new products is represented by a screening methodology that involves toxicity tests. Toxicological examination is fundamental to identify which substances are hazardous—for the environment and humans—and how they can be harmful, providing detailed information. The performance of an exhaustive safety assessment should rely on alternative and innovative methods, in order to replace the traditional approaches

DOI: 10.1201/9781003361091-12

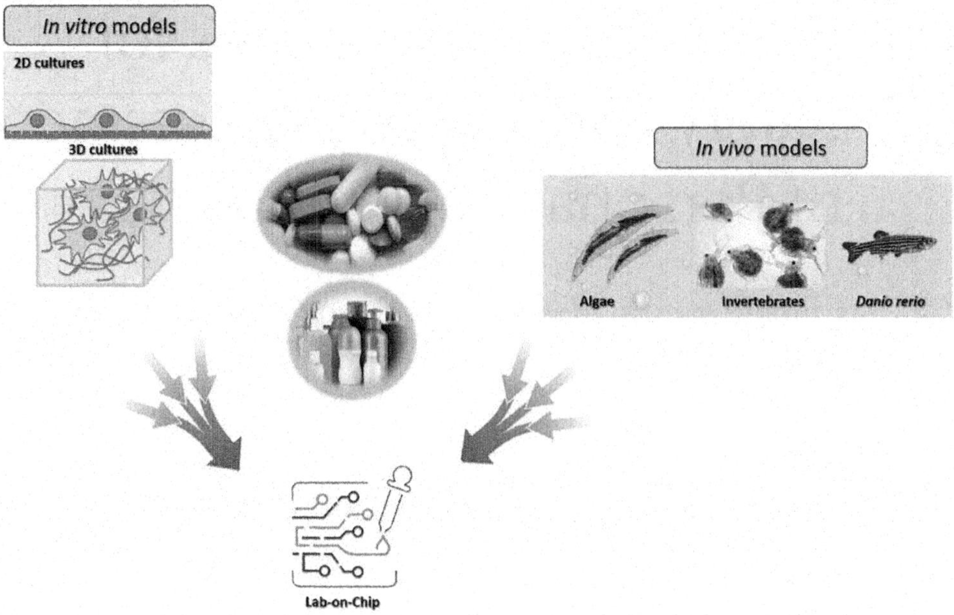

FIGURE 12.1 Schematic representation of available "alternative" tools used to investigate PPCPs toxicity.

(Figure 12.1). The ecotoxicological risk assessment comprehends an integrated vision deriving from the separate information given by cell toxicology, molecular biology, ecology, chemistry, physics, and virtual modeling.

12.2 *IN VITRO* MODELS

12.2.1 A Preliminary Approach: The Use of Two-Dimensional Models in Ecotoxicology

Pharmacologists and toxicologists have always used two-dimensional (2D) models for the comprehension of the mechanisms of action of a substance. The advantages include the possibility to control the culture conditions and perform genetic manipulation. Table 12.1 summarizes the main advantages, limitations and biological endpoints of the classical *in vitro* models. However, from an ecotoxicological point of view, the information given by 2D or other *in vitro* models could appear useless and little applicable in practice. Actually, they could be a useful tool, especially as a starting point in the comprehension of toxicological effects of chemicals, using a mechanistic approach that considers the investigation of the general and specific mechanisms of chemicals-promoted toxicity in target and non-target organisms. Literature offers different examples of mammalian and non-mammalian specialized cells used in toxicology (Stammati et al. 1981). Starting from the 1970–1980s, toxicology availed itself of specialized and non-specialized cells. General toxicology investigations were performed using fibroblastic and epithelioid cells, such as monkey kidney fibroblasts, human embryo lung diploid cells, hamster lung cells, normal mouse fibroblasts, human epithelial carcinoma of cervix (Stammati et al. 1981). However, the investigation of the effects related to a particular drug or chemical on the target tissue or organ employed specialized cells, such as chick embryo ganglia, brain cells and hepatocytes, mouse cerebellum cells, lymphocytes and erythrocytes, rat alveolar macrophages and hepatocytes, and also human lymphocytes (Bourdeau 1990). Among the non-mammalian cell lines, a particular attention was given to piscine cell lines in environmental

TABLE 12.1
Overview of 2D Cell Models in Ecotoxicology

Advantages	Limitations	Main endpoints investigated
Controllable conditions	Loss of tissue-specific architecture	Viability
Scalability	Alterations in biochemical signaling	Mutagenesis and carcinogenicity
No ethical issues	Reduction in cell-to-cell interactions	Teratogenicity
Easy genetic manipulation of the cells	Low heterogeneity (homogenous configuration)	DNA methylation
High-throughput screenings	Absence of complexity	Protein markers
Reproducibility	Absence of vascularization	Interactions receptor-ligand
Affordability	Overestimation of the toxicological effects	Calcium signaling
	Genetic drift and other mutations	
	Short culture period (especially for primary cell lines)	

toxicology. In fact, many ecotoxicants influence the quality of aquatic environment and have an impact on fish populations (Bols et al. 2005). These cell lines were useful to get information about the main biological activities or potency of chemicals, and to perform different bioassays for cytotoxicants, AhR-active compounds, genotoxicants, environmental estrogens and oxidants chemicals. In other words, water samples can be tested for the presence of the different classes of ecotoxicants. In comparison with analytical chemistry methods, these bioassays are less expensive and quicker, and sometimes more sensitive. The use of piscine cell lines also contributed to the development and improvement of biomarkers, which specific alterations used for the assessment of toxicity. Some of the main biomarkers for fish are stress proteins (Whyte et al. 2000), metallothioneins (Viarengo et al. 1999) and vitellogenin (van der Oost et al. 2003).

For example, in the field of aquatic toxicology, the evaluation of bioaccumulation, which is the process of uptake of the chemicals from the environmental medium (water, for example) in aquatic organisms, is particularly important. The OECD Test Guideline No. 305 presents the standard method to measure the bioconcentration factor in fish, which involves the use of living specimens (OECD 1996). The alternative of using living fish was also represented by the *in vitro* assays (Weisbrod et al. 2009), which were based on the use of cell suspensions (Fay et al. 2015) and s9 (Johanning et al. 2012) fractions from fish liver.

However, the scientific enthusiasm for the possibility to quickly and easily gain an overview of toxicants had to deal with the impossibility to overlap the results with *in vivo* conditions, and that means not all results are realistic. Cell culture procedures may lead to artifice, since cells grow in an artificial context, with an excess of nutrients and growth factors. These conditions are enough to select subpopulations (Pamies and Hartung 2017). Regarding the external factors, the experimental reproducibility could be affected by the presence of contaminations. As far as all the external factors can be controlled, cellular physiology may be deeply impaired by *Mycoplasma* spp., gram-negative pyrogens (endotoxins), viruses, prions or other chemicals in media. In addition, the *in vitro* metabolism of chemicals and the expression of the enzymes required may have qualitative and quantitative insufficiencies in comparison to *in vivo* models. Before performing any type of screening, the assessment of the metabolic competence should be carried out, since tissue-specific functions could not be maintained *in vitro* (Coecke et al. 2006). 2D models are a simplified version of what happens in physiological conditions, so many experimental set-ups could not contribute to give real results. A critical issue in the use of bidimensional models is related to the distortion of dose-response relation. In many cases, the study conducted led to an overestimation of toxicological effects. At this point, we could ask the reasons behind the use of *in vitro* models, especially in light of all these limitations. Culture-based methods are useful tools to investigate the main basal and specific toxicity endpoints, which represent the first approach in the study of environmental toxicant effects. Among the basal endpoints, the identification of morphological indicators has an important role. Formation of vacuoles, apoptotic particles, blebs or nuclear fragments belong to that series of observations that are made using a microscope. Other methods allow focus on the biochemistry and metabolic alterations of the cells: proliferation rate, mitochondrial activity and the generation of reactive oxygen species, caspase activity, integrity of the plasma membrane and nuclear acids, release of lactate dehydrogenase (LDH), measurement of DNA repair mechanisms, sister chromatid exchange analysis—only a few examples of the parameters from which it is possible to start analysis. Specific endpoints may include study on mutagenesis and carcinogenicity, DNA methylation, calcium-mediated signaling pathways and specific protein analysis. Regarding proteomics, the proteins of interest are related to oxidative or reticulum stress, enzymes involved in detoxification, scavengers, receptors or immunologically activated markers (Mahto et al. 2010).

12.2.2 A Step Closer to Reality: Three-Dimensional Models

The advantages, disadvantages and limitations deriving from using traditional cell-systems and animal models are reported previously. This introduction was necessary to define the scientific background from which the need for alternative methods and approaches arises. In fact, until recently, the debate involved only the choice of the best method between *in vitro* and *in vivo*. Actually, the research and the development of new technologies allowed the introduction of the three-dimensional (3D) models, which may represent the possibility to overcome the limitations of 2D models and fill the gap with animals used. In fact, cell monolayers could not reproduce all communications which cells establish in their physiological environment. The network of extracellular elements, interactions with cells from other tissues are lost in these models.

The history of 3D cultures begins with the fabrication of synthetic scaffolds, made with different kinds of polymers (Jafari et al. 2017). Initially, the cells were seeded in these pre-formed structures with the traditional "top-down" approach. In this case, cells were seeded on biodegradable polymers, waiting for the settlement on the structures (Lu et al. 2013) and the creation of extracellular matrix mimics (ECM) and the right architecture. This technique was optimized and the difficulties were overcome through the "bottom-up" approach, which differs from the first method for the generations of tissue blocks, that can be assembled to form a single complex construct (Lu et al. 2013). However, the use of specific polymers implies several limitations, which are related to the compatibility of the chosen support with the cells and the influence of the experimental outcome with the microenvironment provided (Freshney 2005). The 3D cell culture strategies provide an environment where cells are allowed to grow and interact in three dimensions. The reasons to prefer them are related to the possibility to better mimic tissue-like structures, co-culture different cell types, simulating the microenvironmental conditions and physiological interactions. This complex architecture can be achieved with or without the use of supporting scaffolds. Here the main examples of 3D models are reported.

Organotypic models: in the organotypic culture, cells are collected from the tissue and then reassembled *in vitro* (Freshney 2005). The advantages of this model are related to the main characteristics possessed: 3D structure, with tissue-like polarization and architecture, presence of inter- and extracellular interactions and physiological metabolic capacity (Salas et al. 2020). One of the most diffused models used in pharmaco-toxicological studies is the "skin equivalent", deriving from the reassembly of cells isolated from intact tissue. It consists of collagen and fibroblast, which form the dermal equivalent, and keratinocytes, cultured on the surface of the dermal equivalent. In this way, the system has an exposed surface accessible for the application of test substances (Bell et al. 1991). Other variants are characterized by the addition of different cells (such as immune cells) to make the system more complex (Oh et al. 2013). The skin equivalents can be used to assess the dermatotoxicity of some irritants, for example (Gay et al. 1992). Skin irritation was determined *in vivo* using Draize rabbit skin irritation test, following the OECD guideline No. 404 (OECD 2015). Since the systemic response does not have a key role in skin irritation, it would only take an *in vitro* model to mimic the effects of chemicals. Moreover, reconstructed cornea-like epithelium, such as EpiOcular™ (MatTek, USA) can be used for the assessment of eye damage/irritation, representing an alternative to the use of *in vivo* models (rabbit). Human corneal epithelial (HCE™) model is a reconstructed epithelial tissue from SkinEthic (Nice, France) and it has been included in OECD guideline No. 492 (OECD 2019). It involves the immortalized human corneal epithelial cells, and allows the development of a 3D tissue resembling normal human corneal epithelium (Leblanc et al. 2019). Eye irritation tests can be performed using specific protocol for solid and liquid substances. Also, EpiOcular (MatTek, USA) is an example of reconstructed corneal epithelium, accepted by OECD guideline No. 492, as an example of *in vitro* tests (Jung et al. 2011). These considerations are really applicable in ecotoxicology. In fact, skin models represent a useful tool for the evaluation of how chemicals interact with the external part of the body. For example, recently attention was given to the ecotoxicological assessment of metal nanoparticles, which contaminate aquatic ecosystems through the release from silver-coated textiles and fabrics (Thanigaivel et al. 2021).

Organoids: these structures are miniatures of organs, prepared from adult tissue-resident stem cells (ASCs) or pluripotent stem cells (PSCs) (Bartfeld and Clevers 2017), through a process of self-organization, and represent a useful tool for predictive toxicology research (Matsui and Shinozawa 2021). The advantages of using ASCs are related to the reduced time of organoid generation (compared to PSCs), and the presence of only organ-specific epithelial and stem cells in the structure. Human and mouse organoids have been used to investigate the effects of different agents. For example, mouse intestinal crypt organoids were used to evaluate the effects of physical agents (X-rays and UV) and chemicals (cisplatin and 5-fluorouracil), also investigating the genes involved in death process. Surprisingly, the results of these analyses were closer to the ones obtained in primary cell cultures and *in vivo* experimental models, rather than in immortalized cell lines (Grabinger et al. 2014). Human mammary organoids were used to determine the effects of cadmium exposure on patient-derived breast stem cells, and the study showed the ability of cadmium to inhibit human mammary stem cell proliferation and differentiation (Rocco et al. 2018). Mammalian organoids are widely used to perform toxicity assessment, although they showed some limitations related to the long developmental time and lack of comparison with other species. Actually, organoids can be generated from different species, representing a useful tool to examine the impact of chemicals on ecosystems. There are studies about the organoid formation from farm animals and companion animals (Augustyniak et al. 2019), which demonstrate the possibility to apply the traditional technology in cases of other species. Moreover, organoids were obtained also from teleosts, for example from medaka and zebrafish primary embryonic pluripotent cells (Zilova et al. 2021).

Spheroids: they are cell aggregates which are formed in particular non-adhesive conditions. Based on the cell type used they can be homotypic (one cell type) or heterotypic (different cell types from the same tissue). The development of 3D spheroids allowed for better investigation of the function of cancer immunotherapy agents (Herter et al. 2017), for example, in tissue engineering for organ reconstruction (Lin et al. 2008). In addition, they showed a great potential in toxicology. For example, the determination of drug toxicity was performed using spheroids deriving from an immortal human hepatocyte cell line (Fey and Wrzesinski 2012). Regarding liver toxicity in particular, HepG2 spheroids were considered to provide better results than 2D cultures (Ramaiahgari et al. 2017). Spheroids could represent a useful tool to investigate in detail the effect related to xenobiotic exposure, as in the case of human primary urine-derived stem cells used to evaluate the drug-induced mitochondrial toxicity (Ding et al. 2022). As reported earlier the technologies for the mammalian toxicity are particularly developed and well-established, so that the studies on the effects of compounds and xenobiotics detected in the environment are able to gather exhaustive information. However, if it were necessary to provide a more comprehensive vision of the environmental impact of chemicals, other species should be considered. Aquatic toxicology definitely needs a new tool for testing substances. In this context, the study by Baron and colleagues (2012) was really important as one of the first approaches to understanding new methods for spheroids deriving from fish. In particular, they reported a detailed protocol for the formation of rainbow trout primary hepatocyte spheroids (Baron et al. 2012). The advantages include the rapid formation of 3D structures (they reach maturity in 6–8 days) and their long survival (up to 40 days). These characteristics enable the performance of acute toxicity tests and investigations on bioaccumulation. The authors proposed an exhaustive alternative to the use of living fish for toxicity tests, posing the attention on the possibility to obtain a large number of spheroids from a single fish, in line with the reduction principle. The fabrication of 3D scaffold-free models can be performed also using fish liver cell line PLHC-1, which originally derived from a hepatocellular carcinoma induced in the topminnow *Poeciliopsis lucida* (Rodd et al. 2017). The authors confirmed the long survival of spheroids, and the possibility to perform extended exposures. These model and primary fish hepatocytes appeared to have a similar physiology, and both had increased levels of *cyp1a*, the function of which is critical in the assessment of toxicity (Thibaut et al. 2009). A recent application of hepatocyte spheroids (from rainbow trout) came from the study by Hultman and colleagues (Hultman et al. 2019). They used the 3D models in order to evaluate the biotransformation of pyrene, demonstrating the reproducibility of data. Spheroids had competence to metabolize pyrene, displaying the formation of metabolites, such as OH-PYR-Glu.

12.3 ALTERNATIVE *IN VIVO* MODELS

Ethical concerns have posed many limitations over the experimental use of vertebrates as *in vivo* models for testing the ecotoxicity of PPCPs; therefore, several alternative methodological approaches have been proposed. Among these, the use of standardized guidelines using algae, invertebrates, and lower vertebrates is preferable to the execution of standard acute toxicity tests which involved adult fish.

The Organisation for Economic Co-Operation and Development (OECD) is the focal point for the harmonization of laboratory tests applied to the risk assessment of chemicals. OECD Test Guidelines are covered by the principle of the Mutual Acceptance of Data (M.A.D.). This consideration made possible the acceptance of toxicological data arising from chemical tests developed for regulatory purposes in another country if these data were produced following the OECD Test Guidelines. This means new data for notifications or registrations of a chemical only have to be generated once, harmonizing the toxicological data, optimizing the use of experimental animals, and reducing laboratory costs. Moreover, since the adoption in 1981 of the first set of Test Guidelines, many of the short- and long-term toxicity tests, as well as the genetic toxicity tests, have been developed or revised to introduce aspects of the 3R principles. In the field of PPCPs toxicity, the execution of OECD guidelines is strongly recommended and their application to alternative *in vivo* models such as algae, invertebrates and lower vertebrates could be the perfect compromise for ensuring animal welfare and obtaining validated toxicological data.

12.3.1 ALGAE

Algae can be found in several aquatic environments, providing an efficient and realistic model which is directly in contact with the released PPCPs. Algae play an essential role in aquatic ecosystems, through the production of the greatest amount of oxygen for living organisms, capturing carbon dioxide during the process of photosynthesis, and also representing a vital source of food and energy for other organisms. For these reasons, if PPCPs cause toxicological effects on algae, the life of higher trophic organisms would be indirectly affected through the oxygen/carbon dioxide imbalance and the disruption of the food web. Although algae play a significant role in aquatic ecotoxicology, there are fewer PPCP toxicity studies on algae compared to higher trophic organisms. Among OECD guidelines that use algae as a model to study the toxicological properties of chemicals, it is worth mentioning the OECD guideline No. 201 (OECD 2011). This test aims to determine the effects of a substance on the growth of freshwater microalgae and/or cyanobacteria. Growing test organisms are exposed to at least five concentrations of the tested substance in batch cultures over a period of normally 72 hours. Growth and growth inhibition are quantified from measurements of the algal biomass as a function of time.

12.3.2 Invertebrates

Invertebrates also represent suitable model organisms for ecotoxicological studies. They are the most widely distributed living organisms on Earth and they offer the opportunity to explore several ecological niches. Moreover, they have a relatively short life span, reproduce quickly at higher rates and are very sensitive to environmental chemicals. Invertebrates, as also stated for algae, largely enter the food chain at intermediate levels. As predators of bacteria, plants, algae and other invertebrates they become the preferential prey of higher vertebrate organisms, including fish and birds, which, in turn, represent a great deal of human diets. For their essential role in aquatic food webs and for the potential for sediments to serve as a repository for anthropogenic contaminants, the use of invertebrates in the evaluation of PPCPs toxicity is highly relevant. Several OECD guidelines have been developed using invertebrates. Among copepods, the harpacticoid and calanoid groups are covered, and development and reproduction tests with *Amphiascus tenuiremis*, *Nitocra spinipes*, *Tisbe battagliai* and *Acartia tonsa* have been implemented. The Mysid shrimp *Americamysis bahia* is also used by OECD as an animal model to evaluate the effects of emerging contaminants on the survival and reproduction rate of this invertebrate species. The water flea *Daphnia magna* is a small planktonic cladoceran and is used as model species in several fields of biological research, including ecology, ecotoxicology, evolution and reproductive biology due to their important position in the aquatic food chain, a high degree of phenotypic plasticity and higher sensitivity to environmental stimuli (Guilhermino et al. 2000). In ecotoxicology, *D. magna* has been considered a good model organism and has been used for reproduction tests, acute toxicity studies and chronic toxicity tests in response to various chemicals by OECD guidelines. Among these, TG 211 is an OECD validated test that assesses the effect of chemicals on the reproductive output of *Daphnia magna* (OECD 2012). To this end, the tested chemical is added to water at a range of concentrations, and young females of *Daphnia magna* are treated using a semi-static exposure technique. The test duration is 21 days. The total number of living offspring produced per parent animal and the number of living offspring produced per surviving parent animal at the end of the test are reported. Optionally other effects can be reported, including the sex ratio of the offspring. Indeed, the OECD guideline No. 202 is based on the use of young daphnids, aged less than 24 hours at the start of the test, which are exposed to the test substance at a range of concentrations for a period of 48 hours (OECD 2004). Immobilization is recorded at 24 hours and 48 hours and compared with control values.

12.3.3 The Zebrafish Model

The zebrafish (*Danio rerio*) is a lower vertebrate organism widely used in several fields of biomedical research, including molecular biology, developmental biology, genetics, oncology, neurobiology and toxicology. George Streisinger started to use zebrafish in 1981 at the University of Oregon, with his first research on genetic mutations affecting nervous system development. The zebrafish offers several advantages for biological studies, including short generation time, relatively large amounts of embryos for performing genetics and cell biology experiments and overall low maintenance cost (Choi et al. 2021). Moreover, about 75% of known human disease genes have a recognizable match in the zebrafish genome. Hence, this model can be used to mimic several human diseases and conditions, including cancer, cardiovascular diseases, neurodegenerative diseases and inflammation (Sarasamma et al. 2017). Worldwide, more than 1000 laboratories use zebrafish as a research model. In Europe, the tightly interconnected zebrafish scientific community consists of more than 350 laboratories that use zebrafish as a model in different research fields. Zebrafish, especially zebrafish's early-life stages, represent an alternative strategy to classical animal experimentation, complying with the principles of Russell and Burch. Indeed, a key issue within the European Directive that regulates animal experimentation, is the time point when fish larvae can be regarded as independently feeding and free-living, thus falling under the scope of the legislation. The development of poikilothermic animals is temperature dependent, and for zebrafish, this critical time point is 120 hours post fertilization (hpf) at 28.5°C (Aleström et al. 2020). For this reason, research projects with zebrafish embryos and zebrafish larvae before the crucial time of 120 hpf are exempted from registration under European animal experimentation legislation. This legal aspect is particularly suitable for use of zebrafish in the field of developmental toxicology and ecotoxicology, including the characterization of PPCPs toxicity. An important tool in the field of zebrafish toxicology and ecotoxicology is represented by the execution of standardized acute toxicity tests, namely Fish Embryo Acute Toxicity (FET) Tests. In May 2013, the FET was approved by the Working Group of the National Coordinators (WNT) of the OECD Test Guideline Program and published as OECD test guideline (TG) No. 236 on July 26, 2013 (OECD 2013). Officially, the test guideline intends to determine the acute or lethal toxicity of chemicals in the embryonic stages of fish *Danio rerio*; however, originally, the FET was designed as an alternative tool to the fish acute test which used adult fish (e.g. OECD TG 203) in the context of the Registration, Evaluation, Authorisation and Restriction of Chemicals (REACH) regulation (Sobanska et al. 2018). According to the procedure specified by OECD TG 236 (OECD 2013), newly fertilized zebrafish embryos are exposed to the test chemical for a total of 96 h. Every 24 h, up to four apical observations are recorded as indicators of lethality (Braunbeck et al. 2005): (i) coagulation of fertilized eggs, (ii) lack of somite formation, (iii) lack of detachment of the tail bud from the yolk sac and (iv) lack of heartbeat. In order not to miss the possible very early adverse effects on zebrafish development, particular care has to be taken to initiate exposure as early as possible (at latest 1.5 h after fertilization). At the end of the exposure period, acute toxicity is determined based on a positive outcome in any of the four apical observations recorded, and the toxicological endpoints are calculated (Braunbeck et al. 2005). However, the versatility of FET tests has thus prompted a massive expansion of the scope of these tests leading to the integration of numerous further endpoints into the original FET protocol and resulting in a rapidly growing list of not only morphological observations, but also physiological, biochemical and molecular endpoints (von Hellfeld et al. 2020). Indeed, sublethal alterations are valuable endpoints that could integrate the Adverse Outcome Pathway (AOP) of the tested chemicals, correlating the exposure to the potential mechanisms of toxicity. However, some of these sublethal endpoints are unspecific and are related to a systemic response of the organism to the highest concentration tested. A recent study aimed to extend the use of FET tests to the identification of different classes of toxicants via a "fingerprint" of more specific morphological observations (von Hellfeld et al. 2022). Among sublethal alterations recorded during FET tests, it is worth mentioning reduced heartbeat, lordosis, reduced yolk resorption, reduced blood circulation, kyphosis, reduced pigmentation, lack of somite formation, blood congestion, scoliosis, notochord malformation, yolk oedemata, impaired fin development, behavioral changes, pericardial oedemata, reduced eye development, tremors, lack of otolith formation and craniofacial deformations.

Although there is a robust correlation between FET test results and toxicological endpoints derived from conventional acute fish toxicity tests, there are several factors that restrict the domain of application for the FET. In particular: (i) toxic substances with a very high molecular weight will not be able to pass the chorion of the embryo before hatch; (ii) there is some minor evidence of limitations in the biotransformation capacity of early embryonic stages of zebrafish development; (iii) there is increasing evidence that certain neurotoxic substances are less toxic in the embryo than in adult fish. However, these "limitations" have been recently discussed and overcome through different techniques. The use of zebrafish early-life stages in the field of PCPPs toxicology is rising sharply and a comprehensive list of published studies using this animal model is presented in Table 12.2, which lists the chemical tested, the endpoints used for the evaluation of toxicity and, where applicable, the value(s) of the toxicological endpoint.

TABLE 12.2
Experimental Studies Using Zebrafish Organisms as Animal Models in the Field of PPCPs Toxicology

PPCPs	Toxicological endpoints	Value(s) of toxicological endpoints	References
Gemfibrozil	Nothocord alterations, changes in swim behavior	LC50 (11.9 mg/L)	(Henriques et al. 2016)
Clofibrate	Lethality	LC50 (11.01 mg/L)	(Raldúa et al. 2008)
Diclofenac	Lethality	LC50 (10.13 μM)	(Chen et al. 2014)
Diclofenac	Changes in swim behavior	EC50 (500 μg/L)	(Xia et al. 2017)
Diclofenac	Oxidative stress	-	(Bartoskova et al. 2013)
Ibuprofen	Hatching	EC50 (500 μg/L)	(Xia et al. 2017)
Acetaminophen	Lethality/developmental alterations	0.707 nmol/mL	(Weigt et al. 2010)
17-β-estradiol	Sexual development	-	(Brion et al. 2004)
17-α-ethinylestradiol	Reproduction/sexual development	5 ng/L	(Nash et al. 2004)
Triclosan	Energy metabolism	-	(Fu et al. 2020)
Triclosan	Cardiotoxicity	≥40 μg/L	(Saley et al. 2016)
Triclosan	Lethality	LC50 embryo (0.42 mg/L) LC50 adult (0.34 mg/L)	(Oliveira et al. 2009)
Triclosan	Lethality/developmental alterations	LC10 (168 μg/L) LC20 (197.2 μg/L) LC50 (267.8 μg/L) LOEC (200 μg/L) NOEC (300 μg/L)	(Iannetta et al. 2022)
Triclocarban	Lethality/developmental alterations	LC10 (75.2 μg/L) LC20 (94.9 μg/L) LC50 (148 μg/L) LOEC (100 μg/L) NOEC (50 μg/L)	(Caioni et al. 2021)
Triclocarban	Lethality/developmental alterations	NOEC (100 μg/L)	(Torres et al. 2016)
Methylparaben	Lethality/developmental alterations	LC50 (428 μM)	(Dambal et al. 2017)
Methylparaben	Lethality/developmental alterations	LC50 (72.67 mg/L)	(Merola et al. 2020b)
Ethylparaben	Lethality/developmental alterations	LC50 (20.86 mg/L)	(Merola et al. 2020a)

(Continued)

TABLE 12.2 (Continued)

PPCPs	Toxicological endpoints	Value(s) of toxicological endpoints	References
Butylparaben	Lethality/developmental alterations	LC50 (2.34 mg/L)	(Merola et al. 2020a)
Propylparaben	Lethality/developmental alterations	LC50 (3.98 mg/L)	(Perugini et al. 2020)
Propylparaben	Lethality/developmental alterations	NOEC (1000 μg/L)	(Torres et al. 2016)
Oxybenzone	Lethality	LC50 (4.74 mg/L)	(Zhang et al. 2021)
Benzophenone	Lethality	LC50 (9.54 mg/L)	(Zhang et al. 2021)
Benzophenone 2	Developmental alterations	>9.85 mg/L	(Fong et al. 2016)
4-Methylbenzylidene Camphor	Lethality/developmental alterations	NOEC (50 μg/L)	(Torres et al. 2016)
Methyl-triclosan	Energy metabolism	400 μg/L	(Fu et al. 2020)
Musk ketone	Lethality	LOEC (10 μg/L)	(Carlsson and Norrgren 2004)
Musk xylene	Cardiotoxicity/developmental toxicity	LOEC (heart rate) 33 μg/L; LOEC (survival) 33 μg/L;	(Carlsson and Norrgren 2004)

12.4 THE REVOLUTION OF MICROFLUIDICS AND LAB-ON-CHIP

2D and 3D models have been intensively used for toxicological investigations and these methods can be also applied to the risk assessment of environmental pollutants and contaminants. The estimation of toxic effects deriving from a certain substance is influenced by several factors, such as concentrations, susceptibility of target organisms, cell types, route of administration, interaction with other chemicals, which leads to an enhancement or attenuation of toxicity. Thus, the evaluation of toxicological effects requires a major complexity in models, since the consequences of the exposure cannot be intuitive or predictable. Microfluidics respond to the need to reproduce in an exhaustive manner the interactions between the organisms and the substances. In order to maximize the performances, it is necessary to create screening platforms, which enable testing in parallel hundreds or thousands of substances. All these reasons explain the importance of miniaturization and the reduction in the working volumes. In this context, microtoxicology has a peculiar relevance. This term was used to define the field of toxicology focused on the investigations deriving from small doses of toxicants. Actually, it also refers to the miniaturized devices used to measure toxic effects. Miniaturization is the process to scale down in size the tools used for biological assays. The advantages are different and include the possibility to use small amounts of biological samples and substances to test, with significant economic savings, especially in terms of reagents and plastic supplies. Professor George Whitesides had a key role in the field of microfluidics and nanotechnology. He said that microfluidics had "four parents: molecular analysis, biodefence, molecular biology and microelectronics" (Whitesides 2006). The contribution coming from molecular analysis was given by the development of microanalytical methods, such as gas-phase chromatography, high-pressure liquid chromatography and capillary electrophoresis. In fact, these techniques allow the achievement of higher sensitivity and resolution. Biodefense dealt with the possibility to have devices able to detect biological and chemical threats. The development of microfluidics technologies made possible the high-throughput screenings of several PPCP toxicological properties; however, the choice of the appropriate materials and the fabrication methods represented the major challenge. The success of this technology is linked to the presence of a flow, which supplies cells and reproduces mechanical forces present *in vivo*, and this is a favorable characteristic in toxicological investigations. In particular, much attention was given to the droplet-based microfluidics. This technique relies on the use of immiscible phases, a higher number of droplets, with diameter in order of nano- or micrometer. Thus, there is a unique flow, but a series of discrete droplets, each of which can be independently controlled. The formation of the droplet requires different methodologies, as well as their sorting or fusion. The advantages of using them instead of a unique flow are related to the possibility to make different tests in a unique experimental run. These analytical microsystems are known as Lab-on-Chip (LOC) technologies, which offer several advantages, related to the possibility to manipulate biological specimens, and obtain a major degree of resolution in comparison with traditional approaches. LOC is an integrated technology which arises from the combination between microscale engineering, physics and biology. Moreover, miniaturization contributes to the rapidity to perform assays and allows high-throughput screening, which is particularly important in routine procedures. The small amounts of fluids pass through miniaturized channels. They can be fabricated using 3D printing technologies, which enables the production of 3D structures in an automated manner, and uses computer-assisted deposition and assembly with geometric control. In Table 12.3 the main advantages and disadvantages of LOC are reported. If we wanted to summarize the applications of lab-on-chip (microfluidic devices) in aquatic toxicology, we should consider the extreme facilitation in testing molecules on cells, organs, organisms firstly. The LOC technology using bacteria and protozoa is the simplest example and enables the comprehension of the potentialities. In particular, bioluminescent-based microfluidics (bacteria or yeast) can be used for the detection of specific compounds and specific toxic mechanisms. The natural bioluminescent properties of *Vibrio fischeri* have been exploited to realize the standard ISO 11384, for the monitoring of toxicants (heavy metals and phenols) in drinking water (Zhao and Dong 2013). *Saccharomyces cerevisiae* and *Escherichia coli* were genetically modified to produce green fluorescence protein in presence of particular contaminants (metals, antibiotics, pharmaceuticals) and they were able to detect DNA damage (Buffi et al. 2011). Starting from a simple design, it is possible to modulate the original scheme to create devices capable of handling and analyzing complex responses in Chordata. In fact, they are not only suitable to test effects on cells, but also on living embryos, realizing the well-known lab-on-chip living embryo array, which allows the automatization of fish embryo toxicity test procedures (Akagi et al. 2012). Microfluidic technology could offer a dynamic environment for zebrafish, which requires excretion-free culture conditions and fresh water, and adapts to the age of the specimens. Moreover, the devices can be integrated with other technologies in order to realize miniaturized *in situ* analyzers capable of multiparameter data acquisition. The implantable sensing devices allow the performance of *q*PCR (Ahrberg et al. 2016), PCR, immunochemical assays such as ELISA (sandwich enzyme-linked immunoassay) (Dong and Ueda 2017), flow cytometry (Piyasena and Graves 2014), blotting and acid nucleic analysis. The possibility to integrate all the methodologies in a single platform reduces time and costs compared to traditional processes. Miniaturized systems were proposed as alternatives to study behavioral responses of small aquatic organisms in presence of toxicants. The behavioral ecotoxicity studies are an important tool to understand how organisms could adapt in the presence of chemical stressors, since the alteration in individuals could influence the ecological organization at different levels. In fact, behavior influences population dynamics, feeding habits, interactions between species; however, these data are considered to have a low relevance (Ford et al. 2021). Moreover, the acquisition of information is particularly laborious, and requires time-costing procedures. Microfluidics represents a great opportunity to realize platforms for real-time behavioral studies, and the reasons are related to the physical properties of the fluids, and the possibility to set up laminar flow conditions under low Reynolds number. For example, LOC systems were used to test negative chemotaxis in small invertebrates, as reported by Huang et al. (2015), demonstrating the possibility to minimize the operator intervention through the fabrication of a customized device, where a camera has been installed (Huang et al. 2015). However, the application of LOC is not limited to the evaluation of toxicological effects, but also to the different aspect of risk assessment, starting from the identification of the substances and going to the implantation of sensors in moving animals.

TABLE 12.3
The Advantages and Disadvantages of LOC for Ecotoxicological Studies

Advantages	Disadvantages
Small volumes of reagents required	Not suitable for larger volumes
Reduction in diffusion time of reagents	Surface adsorption (losses of tested xenobiotics)
Low operational costs	Interaction between polymers used in fabrication with the chemicals
Automatized procedures in fabrication and analysis	Lack of standardization
High-throughput screenings	Not suitable for complex specimens or sediment tests
Faster experimental run	Accumulation of air bubbles in microchannels
Customed (integration with sensors or devices for *in situ* analysis)	Trained personnel required

The water quality monitoring, for example, could employ sensors for the detection of chemicals released in the environment and even deriving from accidental events. There are many situations in which the sampling is difficult, with an operator unable to reach the interested zone, or the substances are intermittently present. For all these reasons, the use of miniaturized systems may facilitate the sampling of water and at the same time accelerate the biological or chemical analysis. Microfluidics principles and LOC are put at the service of biological environmental monitoring (BEM), the requirements of which include: automatized processes, ability to perform analysis on highly diluted samples and integration of several methods (Delattre et al. 2012).

The monitoring could be also performed through the use of implantable devices (Meng and Sheybani 2014), for remote monitoring of free-moving animals. As seen earlier, ecotoxicology employs the collection of samples exposed to a particular substance or the simulation of the exposure to the chemical. However, the use of sensors could facilitate the acquisition of data in real time and directly from species of interest. This kind of technology is still in its infancy, but it would transform the ecotoxicology studies, creating a network of continuous information, without sacrificing animals and, overall, without artifice, since the studies would be conducted in the natural environment.

12.5 SUMMARY AND PERSPECTIVES

The safety characterization of PPCPs is a challenging and complex task due to the large amounts of chemicals belonging to this category, their different combinations and their unpredictable toxicological effects. Several efforts have been made in the field of toxicology and ecotoxicology to overcome the indiscriminate use of animals as *in vivo* models and to obtain experimental data from alternative *in vitro* and *in vivo* models. Among these, the use of three-dimensional models, microfluidics and lower vertebrates such as zebrafish is strongly recommended. However, in addition to these *in vitro* and *in vivo* strategies, computational methods can also be used as alternatives to animal testing. In particular, mathematical approaches such as Quantitative Structure-Activity Relationship (QSAR) modeling and Physiologically Based Kinetic and Dynamic (PBK/D) modeling can be applied to replace and reduce the use of animals in safety and efficacy testing. The integration between *in vitro*, *in vivo* and *in silico* tools could be considered a reasonable approach to the traditional evaluation of chemical toxicity. The implementation of such a tiered strategy could maximize the robustness and reliability of toxicological data in the field of PPCPs toxicity evaluation and could preserve animal welfare in the context of animal experimentation.

REFERENCES

Ahrberg, C.D., A. Manz, B.G. Chung. 2016. "Polymerase chain reaction in microfluidic devices". *Lab-on-a-Chip* 16: 3866–3884. https://doi.org/10.1039/C6LC00984K

Akagi, J., K. Khoshmanesh, B. Evans, C.J. Hall, K.E. Crosier, J.M. Cooper, P.S. Crosier, D. Wlodkowic. 2012. "Miniaturized embryo array for automated trapping, immobilization and microperfusion of zebrafish embryos". *PLoS One* 7: e36630. https://doi.org/10.1371/journal.pone.0036630

Aleström, P., L. D'Angelo, P.J. Midtlyng, D.F. Schorderet, S. Schulte-Merker, F. Sohm, S. Warner. 2020. "Zebrafish: Housing and husbandry recommendations". *Laboratory Animals* 54: 213–224. https://doi.org/10.1177/0023677219869037

Augustyniak, J., A. Bertero, T. Coccini, D. Baderna, L. Buzanska, F. Caloni. 2019. "Organoids are promising tools for species-specific in vitro toxicological studies". *Journal of Applied Toxicology* 39: 1610–1622. https://doi.org/10.1002/jat.3815

Baron, M.G., W.M. Purcell, S.K. Jackson, S.F. Owen, A.N. Jha. 2012. "Towards a more representative in vitro method for fish ecotoxicology: Morphological and biochemical characterisation of three-dimensional spheroidal hepatocytes". *Ecotoxicology* 21: 2419–2429. https://doi.org/10.1007/s10646-012-0965-5

Bartfeld, S., H. Clevers. 2017. "Stem cell-derived organoids and their application for medical research and patient treatment". *Journal of Molecular Medicine* 95: 729–738. https://doi.org/10.1007/s00109-017-1531-7

Bartoskova, M., R. Dobsikova, V. Stancova, D. Zivna, J. Blahova, P. Marsalek, L. Zelníckova, M. Bartos, F.C. di Tocco, C. Faggio. 2013. "Evaluation of ibuprofen toxicity for zebrafish (Danio rerio) targeting on selected biomarkers of oxidative stress". *Neuroendocrinology Letters* 34(Suppl 2): 102–108.

Bell, E., N. Parenteau, R. Gay, C. Nolte, P. Kemp, P. Bilbo, B. Ekstein, E. Johnson. 1991. "The living skin equivalent: Its manufacture, its organotypic properties and its responses to irritants". *Toxicology in Vitro* 5: 591–596. https://doi.org/10.1016/0887-2333(91)90099-Y

Bols, N.C., V.R. Dayeh, L.E.J. Lee, K. Schirmer. 2005. "Use of fish cell lines in the toxicology and ecotoxicology of fish. Piscine cell lines in environmental toxicology", in: Mommsen, T.P., T.W. Moon (Eds.), *Biochemistry and Molecular Biology of Fishes, Environmental Toxicology*. Elsevier. https://doi.org/10.1016/S1873-0140(05)80005-0

Bourdeau, P. 1990. *Short-Term Toxicity Tests for Non-Genotoxic Effects*. Wiley.

Braunbeck, T., M. Boettcher, H. Hollert, T. Kosmehl, E. Lammer, E. Leist, M. Rudolf, N. Seitz. 2005. "Towards an alternative for the acute fish LC(50) test in chemical assessment: The fish embryo toxicity test goes multi-species—an update". *ALTEX* 22: 87–102.

Brion, F., C.R. Tyler, X. Palazzi, B. Laillet, J.M. Porcher, J. Garric, P. Flammarion. 2004. "Impacts of 17beta-estradiol, including environmentally relevant concentrations, on reproduction after exposure during embryo-larval-, juvenile- and adult-life stages in zebrafish (Danio rerio)". *Aquatic Toxicology* 68: 193–217. https://doi.org/10.1016/j.aquatox.2004.01.022

Buffi, N., D. Merulla, J. Beutier, F. Barbaud, S. Beggah, H. van Lintel, P. Renaud, J.R. van der Meer. 2011. "Miniaturized bacterial biosensor system for arsenic detection holds great promise for making integrated measurement device". *Bioengineering Bugs* 2: 296–298. https://doi.org/10.4161/bbug.2.5.17236

Caioni, G., M. d'Angelo, G. Panella, C. Merola, A. Cimini, M. Amorena, E. Benedetti, M. Perugini. 2021. "Environmentally relevant concentrations of triclocarban affect morphological traits and melanogenesis in zebrafish larvae". *Aquatic Toxicology* 236: 105842. https://doi.org/10.1016/j.aquatox.2021.105842

Carlsson, G., L. Norrgren. 2004. "Synthetic musk toxicity to early life stages of zebrafish (Danio rerio)". *Archives of Environmental Contamination Toxicology* 46: 102–105. https://doi.org/10.1007/s00244-003-2288-2

Chen, J.-B., H.W. Gao, Y.L. Zhang, Y. Zhang, X.F. Zhou, C.Q. Li, H.P. Gao. 2014. "Developmental toxicity of diclofenac and elucidation of gene regulation in zebrafish (Danio rerio)". *Scientific Reports* 4: 4841. https://doi.org/10.1038/srep04841

Choi, T.-Y., T.I. Choi, Y.R. Lee, S.K. Choe, C.H. Kim. 2021. "Zebrafish as an animal model for biomedical research". *Experimental Molecular Medicine* 53: 310–317. https://doi.org/10.1038/s12276-021-00571-5

Coecke, S., H. Ahr, B.J. Blaauboer, S. Bremer, S. Casati, J. Castell, R. Combes, R. Corvi, C.L. Crespi, M.L. Cunningham, G. Elaut, B. Eletti, A. Freidig, A. Gennari, J.F. Ghersi-Egea, A. Guillouzo, T. Hartung, P. Hoet, M. Ingelman-Sundberg, S. Munn, W. Janssens, B. Ladstetter, D. Leahy, A. Long, A. Meneguz, M. Monshouwer, S. Morath, F. Nagelkerke, O. Pelkonen, J. Ponti, P. Prieto, L. Richert, E. Sabbioni, B. Schaack, W. Steiling, E. Testai, E., J.A. Vericat, A. Worth. 2006. "Metabolism: A bottleneck in in vitro toxicological test development. The report and recommendations of ECVAM workshop 54". *Alternative Laboratory Animals* 34: 49–84. https://doi.org/10.1177/026119290603400113

Dambal, V.Y., K.P. Selvan, C. Lite, S. Barathi, W. Santosh. 2017. "Developmental toxicity and induction of vitellogenin in embryo-larval stages of zebrafish (Danio rerio) exposed to methyl Paraben". *Ecotoxicology and Environmental Safety* 141: 113–118. https://doi.org/10.1016/j.ecoenv.2017.02.048

Delattre, C., C.P. Allier, Y. Fouillet, D. Jary, F. Bottausci, D. Bouvier, G. Delapierre, M. Quinaud, A. Rival, L. Davoust, C. Peponnet. 2012. "Macro to microfluidics system for biological environmental monitoring". *Biosensors & Bioelectronics* 36: 230–235. https://doi.org/10.1016/j.bios.2012.04.024

Ding, H., K. Jambunathan, G. Jiang, D.M. Margolis, I. Leng, M. Ihnat, J.X. Ma, J. Mirsalis, Y. Zhang. 2022. "3D spheroids of human primary urine-derived stem cells in the assessment of drug-induced mitochondrial toxicity". *Pharmaceutics* 14: 1042. https://doi.org/10.3390/pharmaceutics14051042

Dong, J., H. Ueda. 2017. "ELISA-type assays of trace biomarkers using microfluidic methods". *Wiley Interdisciplinary Reviews: Nanomedicine and Nanobiotechnology* 9. https://doi.org/10.1002/wnan.1457

Fay, K.A., D.L. Nabb, R.T. Mingoia, I. Bischof, J.W. Nichols, H. Segner, K. Johanning, X. Han. 2015. "Determination of metabolic stability using cryopreserved hepatocytes from rainbow trout (Oncorhynchus mykiss)". *Current Protocolos in Toxicology* 65: 4.42.1–4.42.29. https://doi.org/10.1002/0471140856.tx0442s65

Fey, S.J., K. Wrzesinski. 2012. "Determination of drug toxicity using 3D spheroids constructed from an immortal human hepatocyte cell line". *Toxicology Sciences* 127: 403–411. https://doi.org/10.1093/toxsci/kfs122

Fong, H.C.H., J.C.H. Ho, A.H.Y. Cheung, K.P. Lai, W.K.F. Tse. 2016. "Developmental toxicity of the common UV filter, benophenone-2, in zebrafish embryos". *Chemosphere* 164: 413–420. https://doi.org/10.1016/j.chemosphere.2016.08.073

Ford, A.T., M. Ågerstrand, B.W. Brooks, J. Allen, M.G. Bertram, T. Brodin, Z. Dang, S. Duquesne, R. Sahm, F. Hoffmann, H. Hollert, S. Jacob, N. Klüver, J.M. Lazorchak, M. Ledesma, S.D. Melvin, S. Mohr, S. Padilla, G.G. Pyle, S. Scholz, M. Saaristo, E. Smit, J.A. Steevens, S. van den Berg, W. Kloas, B.B.M. Wong, M. Ziegler, G. Maack. 2021. "The role of behavioral ecotoxicology in environmental protection". *Environmental Science and Technology* 55: 5620–5628. https://doi.org/10.1021/acs.est.0c06493

Freshney, R.I. 2005. "Organotypic culture", in: *Culture of Animal Cells*. John Wiley & Sons. https://doi.org/10.1002/0471747599.cac025

Fu, J., Y.X.R. Tan, Z. Gong, S. Bae. 2020. "The toxic effect of triclosan and methyl-triclosan on biological pathways revealed by metabolomics and gene expression in zebrafish embryos". *Ecotoxicology and Environmental Safety* 189: 110039. https://doi.org/10.1016/j.ecoenv.2019.110039

Gay, R., M. Swiderek, D. Nelson, A. Ernesti. 1992. "The living skin equivalent as a model in vitro for ranking the toxic potential of dermal irritants". *Toxicology in Vitro* 6: 303–315. https://doi.org/10.1016/0887-2333(92)90020-r

Grabinger, T., L. Luks, F. Kostadinova, C. Zimberlin, J.P. Medema, M. Leist, T. Brunner. 2014. "Ex vivo culture of intestinal crypt organoids as a model system for assessing cell death induction in intestinal epithelial cells and enteropathy". *Cell Death Disease* 5: e1228–e1228. https://doi.org/10.1038/cddis.2014.183

Guilhermino, L., T. Diamantino, M. Carolina Silva, A.M.V.M. Soares. 2000. "Acute toxicity test with Daphnia magna: An alternative to mammals in the prescreening of chemical toxicity?". *Ecotoxicology and Environmental Safety* 46: 357–362. https://doi.org/10.1006/eesa.2000.1916

Henriques, J.F., A.R. Almeida, T. Andrade, O. Koba, O. Golovko, A.M.V.M. Soares, M. Oliveira, I. Domingues. 2016. "Effects of the lipid regulator drug gemfibrozil: A toxicological and behavioral perspective". *Aquatic Toxicology* 170: 355–364. https://doi.org/10.1016/j.aquatox.2015.09.017

Herter, S., L. Morra, R. Schlenker, J. Sulcova, L. Fahrni, I. Waldhauer, S. Lehmann, T. Reisländer, I. Agarkova, J.M. Kelm, C. Klein, P. Umana, M. Bacac. 2017. "A novel three-dimensional heterotypic spheroid model for the assessment of the activity of cancer immunotherapy agents". *Cancer, Immunology and Immunotherapy* 66: 129–140. https://doi.org/10.1007/s00262-016-1927-1

Huang, Y., G. Persoone, D. Nugegoda, D. Wlodkowic. 2015. "Enabling sub-lethal behavioral ecotoxicity biotests using microfluidic Lab-on-a-Chip technology". *Sensors and Actuators B: Chemical* 226. https://doi.org/10.1016/j.snb.2015.11.128

Hultman, M.T., K.B. Løken, M. Grung, M.J. Reid, A. Lillicrap. 2019. "Performance of three-dimensional rainbow trout (Oncorhynchus mykiss) hepatocyte spheroids for evaluating biotransformation of pyrene". *Environmental Toxicology and Chemistry* 38: 1738–1747. https://doi.org/10.1002/etc.4476

Iannetta, A., G. Caioni, V. Di Vito, E. Benedetti, M. Perugini, C. Merola. 2022. "Developmental toxicity induced by triclosan exposure in zebrafish embryos". *Birth Defects Research* 114: 175–183. https://doi.org/10.1002/bdr2.1982

Jafari, M., Z. Paknejad, M.R. Rad, S.R. Motamedian, M.J. Eghbal, N. Nadjmi, A. Khojasteh. 2017. "Polymeric scaffolds in tissue engineering: A literature review". *Journal of Biomedical Material Research Part B* 105: 431–459. https://doi.org/10.1002/jbm.b.33547

Johanning, K., G. Hancock, B. Escher, A. Adekola, M.J. Bernhard, C. Cowan-Ellsberry, J. Domoradzki, S. Dyer, C. Eickhoff, M. Embry, S. Erhardt, P. Fitzsimmons, M. Halder, J. Hill, D. Holden, R. Johnson, S. Rutishauser, H. Segner, I. Schultz, J. Nichols. 2012. "Assessment of metabolic stability using the rainbow trout (Oncorhynchus mykiss) liver S9 fraction". *Current Protocols in Toxicology* 14(10): 1–28. https://doi.org/10.1002/0471140856.tx1410s53

Jung, K.-M., S.H. Lee, Y.H. Ryu, W.H. Jang, H.S. Jung, J.K. Han, J.H. Seok, J.H. Park, Y. Son, Y.H. Park, K.M. Lim. 2011. "A new 3D reconstituted human corneal epithelium model as an alternative method for the eye irritation test". *Toxicology in Vitro* 25: 403–410. https://doi.org/10.1016/j.tiv.2010.10.019

Leblanc, V., M. Yokota, M.H. Grandidier, D. Yoshida, E. Adriaens, J. Cotovio, D. Kyoutani, N. Alépée. 2019. "SkinEthic™ HCE eye irritation test: Similar performance demonstrated after long distance shipment and extended storage conditions". *Toxicology in Vitro* 54: 202–214. https://doi.org/10.1016/j.tiv.2018.10.003

Lin, R.-Z., H.Y. Chang. 2008. "Recent advances in three-dimensional multicellular spheroid culture for biomedical research". *Biotechnology Journal* 3: 1172–1184. https://doi.org/10.1002/biot.200700228

Lu, T., Y. Li, T. Chen. 2013. "Techniques for fabrication and construction of three-dimensional scaffolds for tissue engineering". *International Journal of Nanomedicine* 8: 337–350. https://doi.org/10.2147/IJN.S38635

Mahto, S.K., P. Chandra, S.W. Rhee. 2010. "In vitro models, endpoints and assessment methods for the measurement of cytotoxicity". *Toxicology and Environmental Health Sciences* 2: 87–93. https://doi.org/10.1007/BF03216487

Matsui, T., T. Shinozawa. 2021. "Human organoids for predictive toxicology research and drug development". *Frontiers in Genetics* 12.

Meng, E., R. Sheybani. 2014. "Insight: Implantable medical devices". *Lab-on-a-Chip* 14: 3233–3240. https://doi.org/10.1039/c4lc00127c

Merola, C., O. Lai, A. Conte, C. Crescenzo, T. Torelli, M. Alloro, M. Perugini. 2020a. "Toxicological assessment and developmental abnormalities induced by butylparaben and ethylparaben exposure in zebrafish early-life stages". *Environmental Toxicology and Pharmacology* 80: 103504. https://doi.org/10.1016/j.etap.2020.103504

Merola, C., M. Perugini, A. Conte, G. Angelozzi, M. Bozzelli, M. Amorena. 2020b. "Embryotoxicity of methylparaben to zebrafish (Danio rerio) early-life stages". *Comparative Biochemistry and Physiology Part C: Toxicology and Pharmacology* 236: 108792. https://doi.org/10.1016/j.cbpc.2020.108792

Nash, J.P., D.E. Kime, L.T.M. Van der Ven, P.W. Wester, F. Brion, G. Maack, P. Stahlschmidt-Allner, C.R. Tyler. 2004. "Long-term exposure to environmental concentrations of the pharmaceutical ethynylestradiol causes reproductive failure in fish". *Environmental Health Perspectives* 112: 1725–1733. https://doi.org/10.1289/ehp.7209

OECD. 1996. *Test No. 305: "Bioconcentration: Flow-through Fish Test"*. Organisation for Economic Co-operation and Development.

OECD. 2004. *Test No. 202: "Daphnia sp. Acute Immobilisation Test"*. Organisation for Economic Co-operation and Development.

OECD. 2011. *Test No. 201: "Freshwater Alga and Cyanobacteria, Growth Inhibition Test"*. Organisation for Economic Co-operation and Development.

OECD. 2012. *Test No. 211: "Daphnia Magna Reproduction Test"*. Organisation for Economic Co-operation and Development.

OECD. 2013. *Test No. 236: "Fish Embryo Acute Toxicity (FET) Test"*. Organisation for Economic Co-operation and Development.

OECD. 2015. *Test No. 404: "Acute Dermal Irritation/Corrosion"*. Organisation for Economic Co-operation and Development.

OECD. 2019. *Test No. 492: "Reconstructed Human Cornea-like Epithelium (RhCE) Test Method for Identifying Chemicals Not Requiring Classification and Labelling for Eye Irritation or Serious Eye Damage"*. Organisation for Economic Co-operation and Development.

Oh, J.W., T.C. Hsi, C.F. Guerrero-Juarez, R. Ramos, M.V. Plikus. 2013. "Organotypic Skin Culture". *Journal of Investigative Dermatology* 133: e14. https://doi.org/10.1038/jid.2013.387

Oliveira, R., I. Domingues, C. Koppe Grisolia, A.M.V.M. Soares. 2009. "Effects of triclosan on zebrafish early-life stages and adults". *Environmental Science and Pollution Research International* 16: 679–688. https://doi.org/10.1007/s11356-009-0119-3

Pamies, D., T. Hartung. 2017. "21st century cell culture for 21st century toxicology". *Chemical Research in Toxicology* 30: 43–52. https://doi.org/10.1021/acs.chemrestox.6b00269

Perugini, M., C. Merola, M. Amorena, M. D'Angelo, A. Cimini, E. Benedetti. 2020. "Sublethal exposure to propylparaben leads to lipid metabolism impairment in zebrafish early-life stages". *Journal of Applied Toxicology* 40: 493–503. https://doi.org/10.1002/jat.3921

Piyasena, M.E., S.W. Graves. 2014. "The intersection of flow cytometry with microfluidics and microfabrication". *Lab-on-a-Chip* 14: 1044–1059. https://doi.org/10.1039/C3LC51152A

Raldúa, D., M. André, P.J. Babin. 2008. "Clofibrate and gemfibrozil induce an embryonic malabsorption syndrome in zebrafish". *Toxicology and Applied Pharmacology* 228: 301–314. https://doi.org/10.1016/j.taap.2007.11.016

Ramaiahgari, S.C., S. Waidyanatha, D. Dixon, M.J. DeVito, R.S. Paules, S.S. Ferguson. 2017. "From the cover: Three-dimensional (3D) HepaRG spheroid model with physiologically relevant xenobiotic metabolism competence and hepatocyte functionality for liver toxicity screening". *Toxicological Sciences* 159: 124–136. https://doi.org/10.1093/toxsci/kfx122

Rocco, S.A., L. Koneva, L.Y.M. Middleton, T. Thong, S. Solanki, S. Karram, K. Nambunmee, C. Harris, L.S. Rozek, M.A. Sartor, Y.M. Shah, J.A. Colacino. 2018. "Cadmium exposure inhibits branching morphogenesis and causes alterations consistent with HIF-1α inhibition in human primary breast organoids". *Toxicological Sciences* 164: 592–602. https://doi.org/10.1093/toxsci/kfy112

Rodd, A.L., N.J. Messier, C.A. Vaslet, A.B. Kane. 2017. "A 3D fish liver model for aquatic toxicology: Morphological changes and Cyp1a induction in PLHC-1 microtissues after repeated benzo(a)pyrene exposures". *Aquatic Toxicology* 186: 134–144. https://doi.org/10.1016/j.aquatox.2017.02.018

Salas, A., J. López, R. Reyes, C. Évora, F.M. de Oca, D. Báez, A. Delgado, T.A. Almeida. 2020. "Organotypic culture as a research and preclinical model to study uterine leiomyomas". *Scientific Reports* 10: 5212. https://doi.org/10.1038/s41598-020-62158-w

Saley, A., M. Hess, K. Miller, D. Howard, T.C. King-Heiden. 2016. "Cardiac toxicity of triclosan in developing zebrafish". *Zebrafish* 13: 399–404. https://doi.org/10.1089/zeb.2016.1257

Sarasamma, S., M.M. Varikkodan, S.T. Liang, Y.C. Lin, W.P. Wang, C.D. Hsiao. 2017. "Zebrafish: A premier vertebrate model for biomedical research in Indian scenario". *Zebrafish* 14: 589–605. https://doi.org/10.1089/zeb.2017.1447

Sobanska, M., S. Scholz, A.M. Nyman, R. Cesnaitis, S. Gutierrez Alonso, N. Klüver, R. Kühne, H. Tyle, J. de Knecht, Z. Dang, I. Lundbergh, C. Carlon, W. De Coen. 2018. "Applicability of the fish embryo acute toxicity (FET) test (OECD 236) in the regulatory context of Registration, Evaluation, Authorisation, and Restriction of Chemicals (REACH)". *Environmental Toxicology and Chemistry* 37: 657–670. https://doi.org/10.1002/etc.4055

Stammati, A.P., V. Silano, F. Zucco. 1981. "Toxicology investigations with cell culture systems". *Toxicology* 20: 91–153. https://doi.org/10.1016/0300-483x(81)90046-9

Thanigaivel, S., A.S. Vickram, K. Anbarasu, G. Gulothungan, R. Nanmaran, D. Vignesh, K. Rohini, V. Ravichandran. 2021. "Ecotoxicological assessment and dermal layer interactions of nanoparticle and its routes of penetrations". *Saudi Journal of Biological Sciences* 28: 5168–5174. https://doi.org/10.1016/j.sjbs.2021.05.048

Thibaut, R., S. Schnell, C. Porte. 2009. "Assessment of metabolic capabilities of PLHC-1 and RTL-W1 fish liver cell lines". *Cell Biology and Toxicology* 25: 611–622. https://doi.org/10.1007/s10565-008-9116-4

Torres, T., I. Cunha, R. Martins, M.M. Santos. 2016. "Screening the toxicity of selected personal care products using embryo bioassays: 4-MBC, propylparaben and triclocarban". *International Journal of Molecular Sciences* 17: 1762. https://doi.org/10.3390/ijms17101762

van der Oost, R., J. Beyer, N.P.E. Vermeulen. 2003. "Fish bioaccumulation and biomarkers in environmental risk assessment: A review". *Environmental Toxicology and Pharmacology* 13: 57–149. https://doi.org/10.1016/s1382-6689(02)00126-6

Viarengo, A., F. Dondero, A. Marro, R. Fabbri. 1999. "Metallothionein as a tool in biomonitoring programmes". *Biomarkers* 4: 455–466. https://doi.org/10.1080/135475099230615

von Hellfeld, R., K. Brotzmann, L. Baumann, R. Strecker, T. Braunbeck. 2020. "Adverse effects in the fish embryo acute toxicity (FET) test: A catalogue of unspecific morphological changes versus more specific effects in zebrafish (Danio rerio) embryos". *Environmental Sciences Europe* 32: 122. https://doi.org/10.1186/s12302-020-00398-3

von Hellfeld, R., P. Pannetier, T. Braunbeck. 2022. "Specificity of time- and dose-dependent morphological endpoints in the fish embryo acute toxicity (FET) test for substances with diverse modes of action: The search for a "fingerprint". *Environmental Science Pollution Research* 29: 16176–16192. https://doi.org/10.1007/s11356-021-16354-4

Weigt, S., N. Huebler, T. Braunbeck, F. von Landenberg, T.H. Broschard. 2010. "Zebrafish teratogenicity test with metabolic activation (mDarT): Effects of phase I activation of acetaminophen on zebrafish Danio rerio embryos". *Toxicology* 275: 36–49. https://doi.org/10.1016/j.tox.2010.05.012

Weisbrod, A.V., J. Sahi, H. Segner, M.O. James, J. Nichols, I. Schultz, S. Erhardt, C. Cowan-Ellsberry, M. Bonnell, B. Hoeger. 2009. "The state of in vitro science for use in bioaccumulation assessments for fish". *Environmental Toxicology and Chemistry* 28: 86–96. https://doi.org/10.1897/08-015.1

Whitesides, G.M. 2006. "The origins and the future of microfluidics". *Nature* 442: 368–373. https://doi.org/10.1038/nature05058

Whyte, J.J., R.E. Jung, C.J. Schmitt, D.E. Tillitt. 2000. "Ethoxyresorufin-O-deethylase (EROD) activity in fish as a biomarker of chemical exposure". *Critical Reviews in Toxicology* 30: 347–570. https://doi.org/10.1080/10408440091159239

Xia, L., L. Zheng, J.L. Zhou. 2017. "Effects of ibuprofen, diclofenac and paracetamol on hatch and motor behavior in developing zebrafish (Danio rerio)". *Chemosphere* 182: 416–425. https://doi.org/10.1016/j.chemosphere.2017.05.054

Zhang, Y., P. Shah, F. Wu, P. Liu, J. You, G. Goss. 2021. "Potentiation of lethal and sub-lethal effects of benzophenone and oxybenzone by UV light in zebrafish embryos". *Aquatic Toxicology* 235: 105835. https://doi.org/10.1016/j.aquatox.2021.105835

Zhao, X., T. Dong. 2013. "A microfluidic device for continuous sensing of systemic acute toxicants in drinking water". *International Journal of Environmental Research and Public Health* 10: 6748–6763. https://doi.org/10.3390/ijerph10126748

Zilova, L., V. Weinhardt, T. Tavhelidse, C. Schlagheck, T. Thumberger, J. Wittbrodt. 2021. "Fish primary embryonic pluripotent cells assemble into retinal tissue mirroring in vivo early eye development". *eLife* 10: e66998. https://doi.org/10.7554/eLife.66998.

Index

U

V

W

Z